IJPHM

International Journal of Prognostics and Health Management

I0033353

The International Journal of Prognostics and Health Management (IJPHM) is the premier online journal related to multidisciplinary research on Prognostics, Diagnostics, and System Health Management. IJPHM is the archival journal of the Prognostics and Health Management (PHM) Society. It exists to serve the following objectives:

- To provide a focal point for dissemination of peer-reviewed PHM knowledge.
- To promote multidisciplinary collaboration in PHM education and research.
- To encourage and assure establishment of professional standards for the practice of PHM.
- To improve the professional and academic standing of all those engaged in the practice of PHM.
- To encourage governmental and industrial support for research and educational programs that will improve the PHM process and practice.

The Journal supports these goals by providing a venue for archival publication of peer-reviewed results from research and development in the area of PHM. We define PHM as a system engineering discipline focused on assessing the current status and well as predicting the future condition of a component and/or system of components. PHM is broader than any single field of engineering: it draws from electrical, electronics, mechanical, civil, and chemical engineering, computer and materials science, reliability, test and measurement, artificial intelligence, physics, and economics. IJPHM seeks to publish multidisciplinary articles from industry, academia, and government in diverse application areas such as energy, aerospace, transportation, automotive, and industrial automation. IJPHM is dedicated to all aspects of PHM: technical, management, economic, and social.

International Journal of Prognostics and Health Management

2013 Vol. 4 Special Issue

Table of Contents

Full Papers

Technical Briefs

Author Index

http://phmsociety.org
Free and open access to full text papers worldwide.

Editorial

Special Issue on Wind Turbine Prognostics and Health Management

David He, Eric Bechhoefer, and Abhinav Saxena

WIND POWER generating capacity was 239 GW at the end of 2011, with a further 46 GW of installed capacity to be operational by the end of 2012. While only providing 2.8% of the energy produced in the United States, it is anticipated that by 2030, almost 20% of the total electrical energy will come from wind. This widespread deployment of industrial wind projects will require a more proactive maintenance strategy in order to be more cost competitive with traditional energy systems, such as natural gas or coal. This will be particularly true for offshore wind projects, where availability of the site for maintenance can be restricted for extended periods of time due to weather conditions. Prognostics and Health Management (PHM) of these assets can improve operational availability while reducing the cost of unscheduled maintenance.

This Special Issue on Wind Turbine Prognostics and Health Management contains 14 excellent papers that highlight a wide range of current research and application topics related to wind turbine PHM. Fault diagnostics is an important aspect of wind turbine PHM. Eight papers included in this special issue deal with fault diagnostics of different parts of a wind turbine. Each of these papers presents different fault diagnostic techniques and sensing technologies. The paper by Waters, Beaujean, & Vendittis presents a vibrations-based method to detect, localize, and identify a faulty bearing in an ocean turbine electric motor. In their paper, Li & Frogley develop a vibration based method using adaptive filtering technique for fault detection in wind turbine gear transmission systems. The two papers by Godwin & Matthews and Zhao, Siegel, Lee, & Su deal with fault diagnosis using SCADA data. In particular, Godwin & Matthews present a data driven classification system for the diagnosis of wind turbine pitch faults, while the paper by Zhao, Siegel, Lee, & Su discusses the CMS and SCADA data and assesses drivetrain degradation over its lifecycle. The paper by Tamilselvan, Wang, Sheng, & Twomey presents a vibration-based, two-stage fault detection framework for failure diagnosis of rotating components in wind turbines. Chase, Danai, Lackner, & Manwell, in their paper, introduce a damage estimation method for blade and tower damage detection in operating wind turbines. In addition to the fault diagnostic methods based on vibration signals, two papers present methods developed using acoustic emission (AE) signals. The paper by Niknam, Thomas, Hines, & Sawhney presents an AE based method for detecting a faulty bearing subject to unbalance. In the paper by Qu, Bechhoefer, He, & Zhu, a new acoustic emission sensor based gear fault detection approach is presented. This approach combines a heterodyne based frequency reduction technique with time synchronous average and spectral kurtosis to process AE sensor signals and extract features as condition indictors for gearbox fault detection.

Prognostics is another important aspect of wind turbine PHM. Two papers included in this special issue specifically deal with prognostics of different components in a wind turbine. In their paper, Hussain & Gabbar present vibration analysis and time series prediction based methods for wind turbine gearbox prognostics. In the paper by Zhu, Yoon, He, Qu, & Bechhoefer, an online lubrication oil condition monitoring and remaining useful life prediction method using particle filtering technique and commercially available online sensors is presented. One important issue in developing wind turbine PHM systems is to develop an operational-condition-independent condition monitoring technique. The paper by Yang, Sheng, & Court addresses this issue by developing three operational-condition-independent criteria. One special characteristic of wind turbine PHM is that wind turbines are more likely to be subjected to considerable stresses due to unpredictable environmental conditions resulting from rapidly changing local dynamics than other types of machinery. The paper by Frost, Goebel & Obrecht explores the integration of condition monitoring of wind turbine blades with contingency control to balance the trade-offs between maintaining system health and energy capture. Risk assessment represents an important step toward successful implementation of wind turbine PHM. In their paper, Dinmohammadi & Shafiee present a fuzzy-FMEA approach for risk and failure mode analysis in offshore wind turbine systems. Life cycle cost has to be considered in developing an effective wind turbine PHM, and the paper by Lesmerises & Crowley presents a reliability-based statistical analysis to determine which PHM strategy will yield the lowest life cycle cost for wind turbine gearboxes.

We, the editors, are confident that this special issue containing research papers on wind turbine prognostics and health management with both an academic and industrial focus will push further the wind turbine research PHM and help to bring more advanced PHM technologies into the industrial applications.

We would also like to thank the authors for their contributions and express our sincere appreciation to the reviewers for their time and expertise in providing valuable feedback.

DAVID HE, *Guest Editor*
Dept. of Mechanical and Industrial Engineering
University of Illinois- Chicago
Chicago, IL 60607 USA

ERIC BECHHOEFER, *Guest Editor*
Green Power Monitoring Systems, LLC
Burlington, VT 05401 USA

ABHINAV SAXENA, *Editor*
Intelligent System Division
NASA Ames Research Center
Moffett Field, CA 94035 USA

David He received his B.S. degree in metallurgical engineering from Shanghai University of Technology, China, MBA degree from The University of Northern Iowa, and Ph.D. degree in industrial engineering from The University of Iowa in 1994. Dr. He is a Professor and Director of the Intelligent Systems Modeling & Development Laboratory in the Department of Mechanical and Industrial Engineering at The University of Illinois- Chicago. Dr. He's research areas include: machinery health monitoring, diagnosis and prognosis, complex systems failure analysis, quality and reliability engineering, and manufacturing systems design, modeling, scheduling and planning.

Eric Bechhoefer received his B.S. in Biology from the University of Michigan, his M.S. in Operations Research from the Naval Postgraduate School, and a Ph.D. in General Engineering from Kennedy Western University. His is a former Naval Aviator who has worked extensively on condition based maintenance, rotor track and balance, vibration analysis of rotating machinery and fault detection in electronic systems. Dr. Bechhoefer is a board member of the Prognostics Health Management Society, and a member of the IEEE Reliability Society.

Abhinav Saxena is a Research Scientist with SGT Inc. at the Prognostics Center of Excellence NASA Ames Research Center, Moffett Field CA. His research focus lies in developing and evaluating prognostic algorithms for engineering systems. He is a PhD in Electrical and Computer Engineering from Georgia Institute of Technology, Atlanta. He earned his B.Tech in 2001 from Indian Institute of Technology (IIT) Delhi.

Dr. Saxena is a Technical Fellow for Prognostics at SGT Inc. and serving as Editor-in-Chief of International Journal of Prognostics and Health Management since 2011.

Operational-Condition-Independent Criteria Dedicated to Monitoring Wind Turbine Generators

Wenxian Yang[1], Shuangwen Sheng[2], and Richard Court[3]

[1]*School of Marine Science and Technology, Newcastle University, Newcastle upon Tyne, NE1 7RU, UK*
wxwyyang@gmail.com

[2]*National Renewable Energy Laboratory, Golden, CO, 80401, USA*
shuangwen.sheng@nrel.gov

[3]*National Renewable Energy Center, Blyth, Northumberland, NE24 1LZ, UK*
richard.court@narec.co.uk

ABSTRACT

Condition monitoring is beneficial to the wind industry for both onshore and offshore plants. However, due to the variations in operational conditions, its potential has not been fully explored. There is a need to develop an operational-condition-independent condition monitoring technique, which has motivated the research presented here. In this paper, three operational-condition-independent criteria are developed. The criteria accomplish the condition monitoring by analyzing the wind turbine electrical signals in the time domain. Therefore, they are simple to calculate and ideal for online use. All proposed criteria were tested through both simulated and practical experiments. The experiments have shown that these criteria not only provide a solution for detecting both mechanical and electrical faults that occur in wind turbine generators, but provide a potential tool for diagnosing generator winding faults.

1. INTRODUCTION

The wind industry continues to grow worldwide. In particular, offshore wind is attracting increasing interest because of its high and stable wind speed, and lack of significant visual impact and noise issues. Take the offshore wind resource in the United Kingdom (UK) as an example. With 18 sites and a total electric-generating capacity of 1.5 gigawatts (GW), the UK government launched the first round of offshore wind farm development plans in 2000. To date, 11 Round 1 offshore wind farms are fully operational, with a combined electric-generating capacity of 876 megawatts (MW). The 90-MW Teesside wind farm has been recently consented. Following Round 1, the government launched a plan for a second round of larger sites in 2003. In Round 2, 16 sites were awarded, totaling a combined capacity of up to 7.2 GW. Today, two Round 2 offshore wind farms (Gunfleet Sands II 64 MW and Thanet 300 MW) are fully operational, five are under construction, three were consented, and the other six are in planning. Subsequently, the UK offshore wind Round 3 scheme was announced in 2008, offering nine development zones and a total electric-generating capacity of 32 GW—equaling one-fourth of the UK's electricity needs. From these data, it can be inferred that large megawatt-scaled wind turbines are being increasingly deployed offshore. Moreover, the locations where the wind farms are deployed tend to increase in size and distance from shore. Hence, wind turbine operators are very concerned about the reliability and availability issues of these offshore giants. The early experience of UK Round 1 offshore wind farms (Feng, Tavner and Long, 2010) shows that although these farms achieve satisfactory capacity factors (about 29.5%), their annual average availability is only 80.2%. This is fairly low in comparison to the average availability of UK onshore wind farms (97%).

Apparently, the low availability has become a barrier to reducing the cost of energy from offshore wind. For example, the Barrow site, a standard UK Round 1 offshore wind farm, includes:

- A rated power of 90 MW
- An annual average capacity factor of 29.5%
- A tariff rate of $117 (USD) per megawatt-hour (MWh).

The annual revenue of this wind farm can be estimated by 90MW × 365 days/year × 24 hours/day × 29.5% ×

117USD/MWh ≈ 27.2 million USD. Then it can be readily inferred that this wind farm will yield $272,000 USD more annual revenue if its production can be improved by 1%. For the future 32-GW UK Round 3 offshore wind development scheme, a 1% improvement in production could yield more than 9.7 billion USD per year. Undoubtedly, this is a significant economic profit to operators. In practice, there are many methods that can be used to improve wind turbine availability. Specifically, condition monitoring (CM) has been proven as one of the most efficient methods (Yang, Tavner, Crabtree, Feng and Qiu, 2012, McMillan and Ault, 2007). The additional value of a proper wind turbine condition-monitoring system (CMS) is observable from detecting the incipient faults occurring in wind turbines and their subassemblies, protecting the defective parts from second-damage, and preventing fatal, catastrophic wind turbine accidents. The additional value can also be justified by the following actual wind turbine operation and maintenance (O&M) data. In the commercial market today, the component replacement costs for a 5-MW turbine include the following [McMillan and Ault, 2007, BVG Associates, 2011]:

- For a rotor: 1.9–2.3 million USD
- For a blade: 391–547 thousand USD
- For a blade bearing: 62.5–78.2 thousand USD
- For a gearbox: 628 thousand USD
- For a generator: 314 thousand USD
- For electronic modules: 16 thousand USD.

By contrast, the current average market price of a CMS is only around $16,000 USD per unit. As a result, it may be worthwhile to equip offshore wind turbines with a CMS to ensure the turbine's anticipated availability and economic return.

However, wind practice shows that the wind turbine CMSs that are currently available in the market have not been as successful as expected at improving wind turbine availability. The reasons are complicated (Yang, Tavner, Crabtree, Feng and Qiu, 2012), but one of the major reasons is that the available CMSs are using the general-purpose CM techniques, which are unable to provide reliable CM results for wind turbines. Because wind turbines operate at variable rotational speeds and are constantly subjected to varying loads, these influences and the effects of potential faults are combined in wind turbine CM signals, which makes it difficult to interpret the signals and reduces the reliability of the CM results. Although there have been some recent efforts (Yang, Court, Tavner and Crabtree, 2011, Yang, Tavner, Crabtree and Wilkinson, 2010), frequent false alarms generated by the wind turbine CMS still overwhelm the operators and make them hesitant to use CMS extensively on their machines. Therefore, it is necessary to remove the negative influences of varying loads and rotational speeds on wind turbine CM results in order to use CMS to improve wind turbine availability. Two straightforward approaches could be taken for a solution: 1) introduce operational-condition decoupling procedures into the current wind turbine CM techniques, and 2) develop new operational-condition-independent wind turbine CM techniques. Considering the complexities and difficulties in realizing the first approach, this paper demonstrates the second approach for simplicity. Thereby, the work presented in the paper will focus on monitoring wind turbine induction generators under continuously varying operational conditions.

2. OPERATIONAL-CONDITION-INDEPENDENT CONDITION MONITORING CRITERIA

To achieve a reliable CM technique that is dedicated for wind turbine induction generators that experience constantly varying operational conditions, the following criteria were researched.

2.1. Criterion δ

For a wind turbine induction generator, neither varying load nor potential faults are able to influence the primary frequency of its stator current signal. But the sub- or higher-order of fault-related harmonics will distort the time waveform of the stator current signal and thus disturb its zero-crossing behavior. Whereas, the zero-crossing behavior of stator current signal is less dependent on load because varying load mainly affects the amplitude of the signal. Inspired by this knowledge, the first CM criterion, δ, was developed to describe the fault effect on the zero-crossing behavior of the stator current signal (see Figure 1). The figure shows the relative phase angles of the generator stator current signal with respect to the corresponding voltage signal in various scenarios. I_1 represents the line current signal measured at phase 1 of generator stator, V_1 represents the corresponding phase voltage signal, and Δt represents the leading/lagging time of I_1, with respect to V_1.

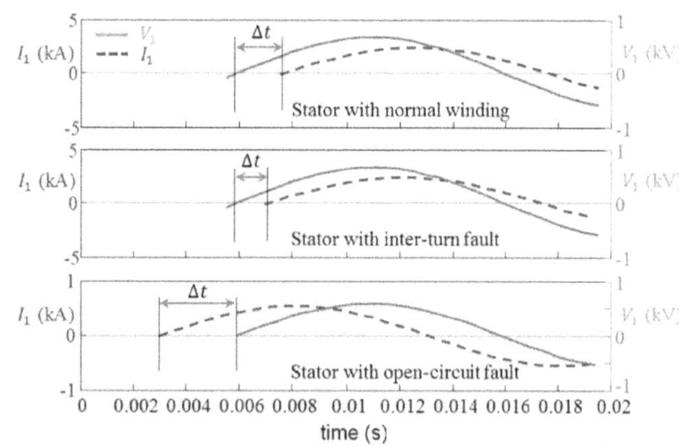

Figure 1. Zero-crossing behavior of the stator current signal in different scenarios

As Figure 1 shows, the potential generator winding faults will either increase or decrease the value of Δt , thus disturbing the zero-crossing behavior of its stator current signal. So potentially Δt can be an indicator of the health of a wind turbine induction generator. However, Δt is not accurate enough to be a CM criterion. It only reflects the shift of the current signal, however, it does not exactly show how much the phase shift is with respect to the corresponding voltage signal. In other words, the same value of Δt might indicate a different level of phase shift of the current signal when the grid frequency is different [either 50 hertz (Hz) or 60 Hz, depending on the areas]. Therefore, an improvement of Δt occurs by normalizing Δt using the frequency of mains electricity. Hence, a new CM criterion, δ, is designed as:

$$\delta = \frac{\Delta t}{T} \times 360^o \tag{1}$$

where

$$T = \frac{1}{f_{grid}} \tag{2}$$

f_{grid} refers to the frequency of mains electricity.

From a physics point of view, δ is less dependent on the external load applied to the generator because it is only related to the impedance status and the magnetizing behavior of the generator. However, the potential faults occurring in an induction generator will disturb its circuit impedance or ability to magnetize. In other words:

- Either an inter-turn or open-circuit winding fault occurring in the generator rotor or stator will change the impedance of the generator circuit, thus leading to the change of criterion δ
- The generator mechanical faults, usually related to rotor shaft, bearings, and air-gap eccentricity, will alter the rotor-to-stator air gap statically or dynamically. Then, the modified air gap will result in the change of the magnetizing current of the generator. As a consequence, criterion δ will change correspondingly.

Therefore, criterion δ should be able to send an alert of both electrical and mechanical faults occurring in the generator. However, it should be awared that Double-Fed Induction Generators (DFIG) have capability to correct power factor on the grid therefore the value of δ, particularly when the grid has a large inductive load. This would more or less affect the capability of the proposed method in detecting incipient faults occurring in the generators. So further research could be needed to improve the efficiency of the proposed technique in DFIG cases.

2.2. Criterion *r*

As mentioned above, both varying load and generator abnormalities will influence the power (and power quality) output from a wind turbine induction generator. The difference is that varying load can alter the power but not the physical performance of the generator, however, the generator abnormalities can alter both. For example, a generator winding fault can change the circuit impedance of an induction generator, whereas an external load cannot. So the circuit impedance of an induction generator is load-independent. However, it can be a challenge to accurately measure generator circuit impedance online. In addition, there is a lack of control over uncertainties related to slip rings or carbon brushes. The metering accuracy and sensitivity of the data acquisition instruments are also concerns. In response to these issues, an alternative criterion, r, was developed for the purpose of indicating the status of the circuit impedance of an induction generator. It is based on a statistical calculation of the generator stator line current and the corresponding voltage signals. The criterion $r(t)$, expressed in terms of line current $I_1(t)$ and the corresponding phase voltage $V_1(t)$, can be written as:

$$r(t) = \sqrt{\frac{\sum_{\delta t=-\tau/2}^{\tau/2}[V_1(t+\delta t)]^2}{\sum_{\delta t=-\tau/2}^{\tau/2}[I_1(t+\delta t)]^2}} \tag{3}$$

where, a sliding window with the width of τ is used in the calculation. Subscript '1' indicates the phase 1 of the generator stator. The criterion can also be assessed by using the electrical signals measured from phase 2 or phase 3 of the generator stator.

Inter-turn or open-circuit faults that occur in the winding of the generator rotor or stator will change the circuit impedance of the generator. Therefore, researchers believe that this change should be observable from criterion r. The generator shaft and bearing-related faults will statically or dynamically alter the rotor-to-stator eccentricity, which will affect the magnetizing current of the generator. As a result, criterion r will change as well.

In principle, criterion r is load-independent and valid for detecting both electrical and mechanical faults that occur in wind turbine induction generators.

2.3. Criterion *e*

Park's vector (Cardoso and Saraiva, 1993) successfully describes the three-phase phenomenon in a two-dimensional coordinate system that allows the pattern-based CM of three-phase induction motors (Nejjari and Benbouzid, 2000). This technique can be applied to the CM of three-phase induction generators. The current Park's vector components (i_d, i_q) can be calculated by:

$$\begin{cases} i_d = \sqrt{\frac{2}{3}} I_1 - \frac{1}{\sqrt{6}} I_2 - \frac{1}{\sqrt{6}} I_3 \\ i_q = \frac{1}{\sqrt{2}} I_2 - \frac{1}{\sqrt{2}} I_3 \end{cases} \tag{4}$$

where I_1, I_2, and I_3 represent the line currents measured from the stator of a three-phase generator/motor.

When the health condition of the generator is without any defect, (i_d, i_q) is characterized by a circle pattern that is centered at the origin of the coordinate system. The size of the circle is related to the power generated by the generator. For example, the larger the power, the bigger the circle. Apparently, the pattern depicted by (i_d, i_q) is load-dependent. So if i_d and i_q are applied directly to CM application (as done by Nejjari and Benbouzid (2000)), it is hard to get a creditable CM result under a varying loading condition. To help alleviate this issue, a further improvement is provided below.

Because the external load will change the values of I_1, I_2, and I_3, it will affect the values of i_d and i_q and the size of the pattern depicted by (i_d, i_q). However, external load is unable to change the eccentricity and roughness of the pattern curve, which are only dependent on the balance status of the generator load (i.e., the balance of three phases) and the harmonics (either sub- or higher order) that are induced by the faults. Therefore, the pattern eccentricity and the roughness of its curve are load-independent and have the potential to be used for generator CM. To provide a comprehensive description of the pattern eccentricity and curve roughness, a new CM criterion, e, is developed:

$$e(t) = \frac{R_{max}(t) - R_{min}(t)}{\bar{R}(t)} \times 100\% \tag{5}$$

where

$$\bar{R}(t) = \frac{1}{T}\int_{t-T/2}^{t+T/2} R(\tau)d\tau \tag{6}$$

$$R(\tau) = |i_d(\tau) + j \times i_q(\tau)| \tag{7}$$

$$R_{max}(t) = max(R(\tau)), \tau \in [t - T/2, t + T/2] \tag{8}$$

$$R_{min}(t) = min(R(\tau)), \tau \in [t - T/2, t + T/2] \tag{9}$$

T stands for the frequency period of grid.

3. FAULT SIMULATIONS

To verify the CM criteria, both inter-turn and open-circuit faults are simulated on the rotor and stator windings of an induction generator by using Simulink. The winding fault simulation model is shown in Figure 2, and the parameters of the generator being simulated are listed in Table 1. In the experiment, the winding faults are simulated by introducing electrical asymmetry into the generator rotor and stator circuit, and the CM signals (i.e., line currents, phase voltages, and total power) are measured from the stator of the generator.

In Figure 2, L refers to the inductance and R_1 refers to the resistance of the transmission line. R_2 =1e3Ω represents the phase resistance to the ground. In the model, resistances R_3 and R_4 are specially designed to introduce electrical asymmetries into the stator and rotor of the generator. The initial values are R_3 =1.16e-3Ω and R_4 =3.92e-2Ω in the absence of electrical asymmetry.

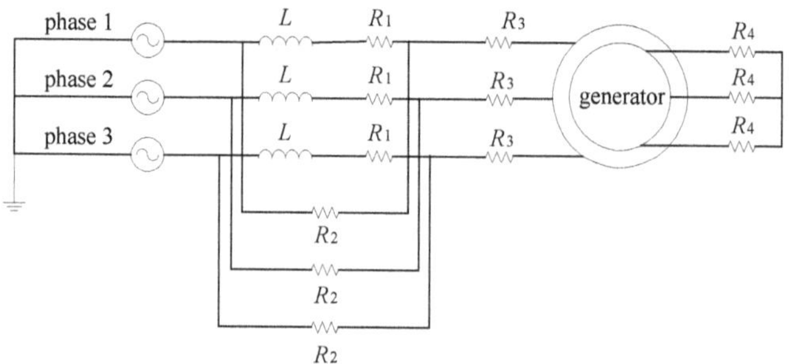

Figure 2. Model for simulating generator winding faults

Parameter	Value	Parameter	Value
Rated power, S_b	2 MW	Rotor leakage inductance, L_r^*	1.493e-1 p.u.
Line-line voltage, V_n	690 volts (V)	Magnetizing inductance, L_m^*	3.9527 p.u.
Rated frequency, f_b	50 Hz	Winding connection (stator/rotor)	Y-Y
Stator resistance, R_s^*	4.88e-3 p.u.	Turns ratio (N_s/N_r)	0.45
Stator leakage inductance, L_s^*	1.386e-1 p.u.	Inertia	32 Kg·m^2
Rotor resistance, R_r^*	5.49e-3 p.u.	Friction	2.985e-4 N·m·s

Table 1. The parameters used for simulating an induction generator (Xiang, Ran, Tavner and Yang, 2006).

Based on the parameters listed in Table 1, the transmission line impedance can be estimated as:

$$R_1 + jL = 0.01Z_{sb} + j \times 0.1L_{sb} \tag{10}$$

where

$$Z_{sb} = \frac{V_{sb}}{I_{sb}} \tag{11}$$

$$V_{sb} = \sqrt{\frac{2}{3}}V_n \tag{12}$$

$$I_{sb} = \sqrt{\frac{2}{3}} \times \frac{S_b}{V_n} \tag{13}$$

$$L_{sb} = \frac{Z_{sb}}{2\pi f_b} \tag{14}$$

By submitting (11)-(14) into (10), obtain R_1 =2.3805e-3Ω and L =7.6e-5H.

3.1. Winding faults in rotor

The inter-turn and open-circuit faults are simulated first in the generator rotor by setting the resistance in the second phase to R_4=3.92e-3Ω and R_4=3.92e-1Ω. The resulting electrical signals and the corresponding CM results are shown in Figure 3.

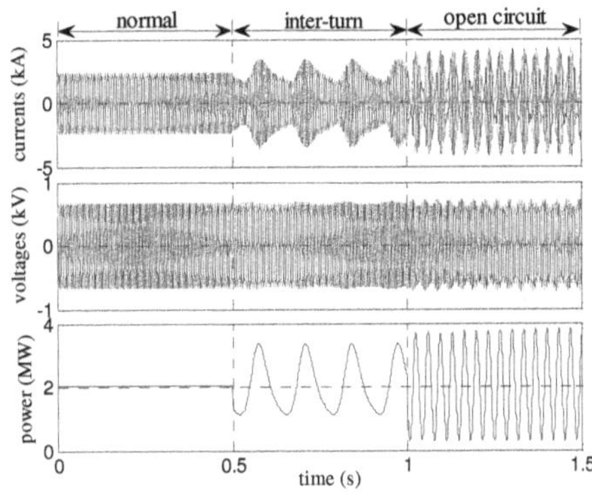

(a) Electrical signals

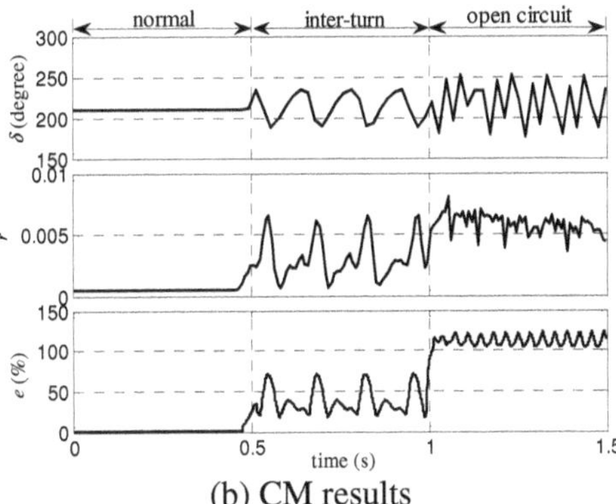

(b) CM results

Figure 3. Simulation and CM of generator rotor winding faults

Figure 3(a) shows that both the inter-turn and open-circuit faults that occur in the rotor significantly distort the time-waveforms of stator current signals, but have less of an effect on the time-waveforms of stator voltage signals. This is because the generator voltage is mainly dominated by the grid's voltage, whereas the generator stator current is a different matter. The generator stator current is not only dependent on the generator's external load and operational condition, but on its actual health status. Researchers also noticed that the total power signal appears in direct current (DC) form when the generator is healthy and without any winding defect, and appears in alternating current (AC) form time-waveforms in the presence of either inter-turn or open-circuit winding faults. In addition, researchers found that the power signals that were collected in two fault scenarios show distinctly different pattern features, for example, the power fluctuates slowly along time in the presence of an inter-turn fault, but fluctuates quickly in the presence of an open-circuit fault. The similar phenomena can also be observed from the time-waveforms of the current signals. The two types of rotor winding faults have different characteristic frequencies in generator electrical signals. This could be important information, not only to the generator's CM but also to its fault diagnosis.

Figure 3(b) shows that all three proposed CM criteria are aware of the presence of both types of rotor winding faults, although the faults display different variation characteristics in different scenarios. The CM results shown in Figure 3b prove that the winding faults that occur in the generator rotor will:

- Disturb the zero-crossing behaviour of generator stator current signal (characterized by criterion δ)
- Change the circuit impedance of the generator (characterized by criterion r)
- Cause load imbalance to the generator (characterized by criterion e).

3.2. Winding faults in stator

The inter-turn and open-circuit faults are simulated in the stator of the generator by setting the resistance in its second phase to R3=1.16e-4Ω and R3=1.16e-2Ω. The electrical signals and corresponding calculation results of CM criteria are shown in Figure 4.

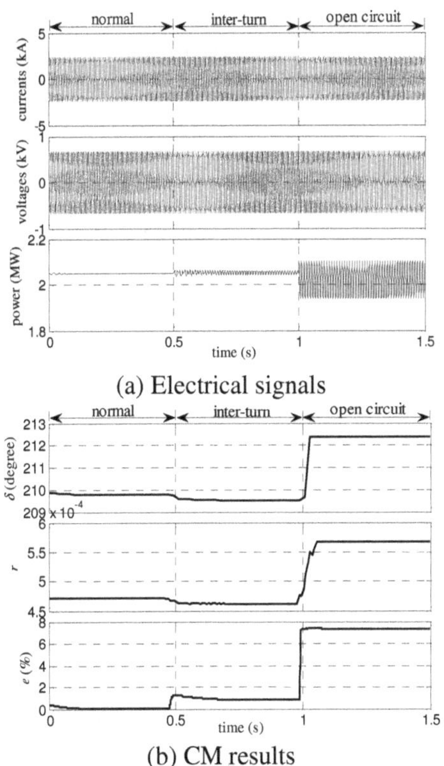

(a) Electrical signals

(b) CM results

Figure 4. Simulation and CM of generator stator winding faults

Figure 4(a) shows that, compared to the winding faults that occur in the generator rotor, both inter-turn and open-circuit faults that occur in the generator stator do not have a significant effect on the time-waveforms of stator current and voltage signals. However, like a generator rotor winding fault, they create a load imbalance in the generator, which is characterized by the AC form of power signal.

Figure 4(b) shows that all three proposed CM criteria are equally sensitive to both types of stator winding faults. For example:

- An inter-turn circuit fault decreases the value of criterion δ, while an open-circuit fault increases it. This finding further supports the phase shift phenomena of the current signal shown in Figure 1.
- An inter-turn circuit fault decreases the circuit impedance of the generator, while an open-circuit fault increases it.
- Both inter-turn and open-circuit faults change the pattern depicted by Park's vector (i_d, i_q), thus changing criterion e.

4. VERIFICATION EXPERIMENTS

The verification experiments were conducted on a 30-kilowatt (kW), three-phase, four-pole induction generator, which was driven by a 54-kW, DC-variable speed motor through a two-stage gearbox with a gear ratio of 1:5. To emulate faults, the rotor circuit of the generator was coupled to an externally connected, three-phase resistive load bank (see Figure 5). As a result, electrical asymmetry could be readily applied to the generator rotor by adjusting the phase resistances of the load bank. The test rig was instrumented and controlled using LabVIEW, allowing researchers to apply a variety of wind speed inputs to the test rig via the DC motor. The rotational speed of the DC motor was controlled by a mathematical model that incorporates the properties of natural wind at different wind speeds and the mechanical behavior of a 2-MW wind turbine operating under closed-loop control conditions. More details about this test rig are provided in (Yang, Tavner, Court, 2013). In the experiment, both fixed and variable rotational speeds were applied to the DC motor and generator to verify the capabilities of the three proposed CM criteria under various loading conditions.

Researchers verified and applied the CM criteria when monitoring the generator at a fixed rotational speed of 1,750 revolutions per minute (rpm). In addition, researchers periodically applied a winding fault to the generator's rotor by repeatedly adjusting the phase resistances of the load bank. The electrical signals were collected from the generator terminals by using a sampling frequency of 5 kilohertz (kHz). The time-waveforms of the electrical signals and the corresponding CM results are shown in Figure 6. In the figure, normal represents electrical symmetry and abnormal represents electrical asymmetry.

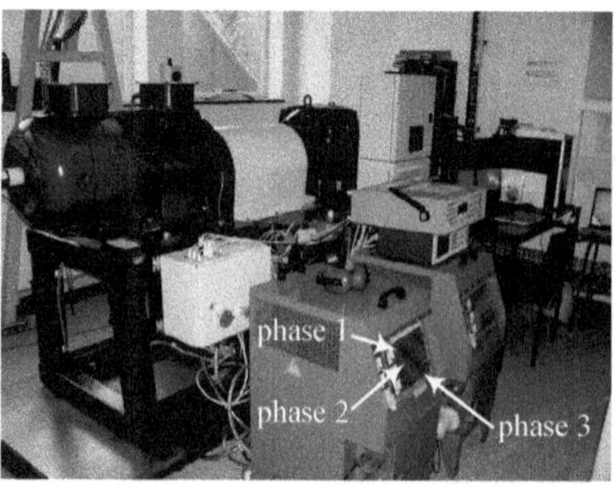

Figure 5. Test rig for conducting verification experiments

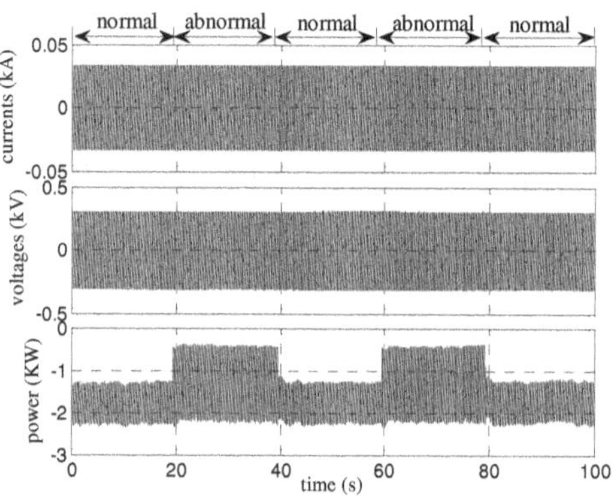

(a) Electrical signals

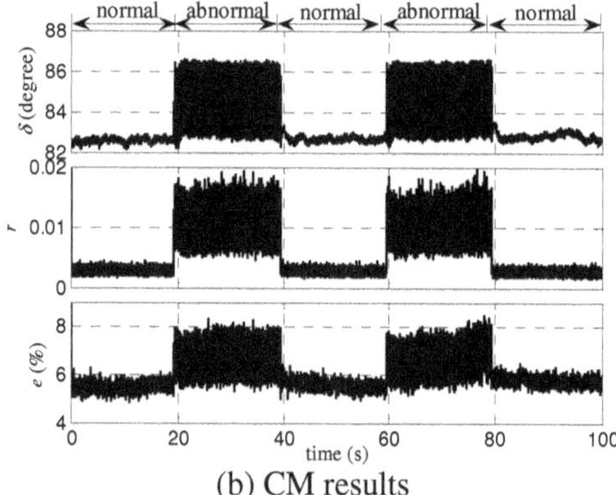

(b) CM results

Figure 6. CM when generator rotates at constant rotational speed

Compared to Figure 3(a), the emulated rotor winding fault shown in Figure 6(a) does not disturb the time-waveform of the stator current signal as anticipated. This is because the asymmetry that is introduced by the load bank is too small to be observable from the measured current signal. To confirm this, the current signal shown in Figure 6(a) was amplified 40 times. The amplified time-waveform of the current signal is shown in Figure 7, from which the waveform distortion that was caused by the asymmetry has already been clearly visible. So it is demonstrated, once again, that the winding fault that occurs in the induction generator's rotor distorts the time-waveform of its stator current signal. Compared to the effect of the winding fault on generator stator current signal, the effect of the distortion on generator total power is much more significant. In Figure 6(a), a big distortion of generator power signal happens as soon as the fault occurs, and the power returns back to its normal waveform immediately when the fault goes away. This suggests that, in comparison to the stator current signal, the generator total power signal is more alert to the fault. This is why the generator power signal, rather than the stator current signal, was adopted in (Yang, Tavner, Crabtree and Wilkinson, 2010) for wind turbine condition monitoring, even though the majority of existing motor/generator CM techniques are based on the analysis of stator current signals (Tavner, 2008).

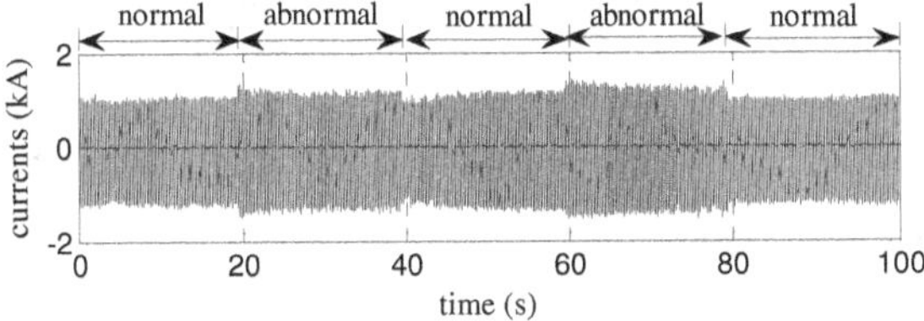

Figure 7. Amplified time-waveform of the stator current signal

In Figure 6(b), the winding fault has been successfully detected by all three proposed CM criteria, although these criteria are developed based on the analysis of the stator current and voltage signals.

To further verify the CM capability of the proposed criteria under varying loading conditions, as observed from real wind turbines, researchers repeated the experiment when the DC motor was controlled to operate at constantly varying speeds. As part of this process, the electrical signals (i.e., stator current, voltage, and total power) were collected from the generator terminals by using the same sampling frequency of 5 kHz. The time-waveforms of these CM signals and the corresponding results are shown in Figure 8.

Figure 8(a) shows that, not only is the winding fault undetected by the time-waveforms of the stator current and voltage signals, but it is undetected by the generator power signal, due to the negative influence of varying loads. Nonetheless, the fault is detected successfully by using the three proposed CM criteria [see Figure 8(b)]. However, the result of criterion δ is not completely load-independent, therefore a de-trending process is necessary.

5. CONCLUSIONS

In this paper, researchers first discussed the influences of various types of winding faults on an induction generator. Three operational-condition-independent condition monitoring (CM) criteria were researched to achieve a reliable CM of a wind turbine generator that is usually subjected to constantly varying loads. Based on this work, researchers came to the following conclusions:

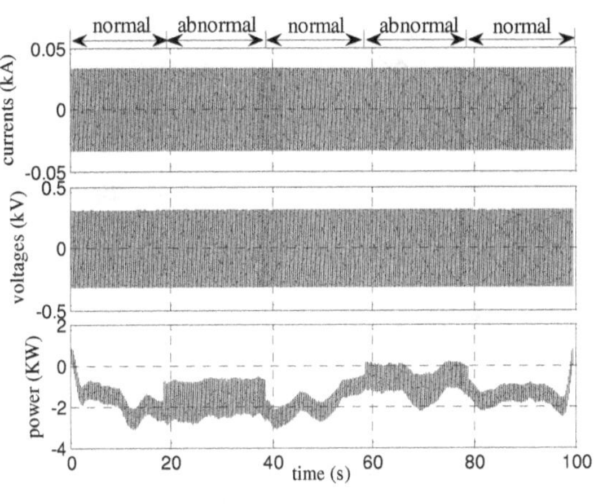

(a) Electrical signals

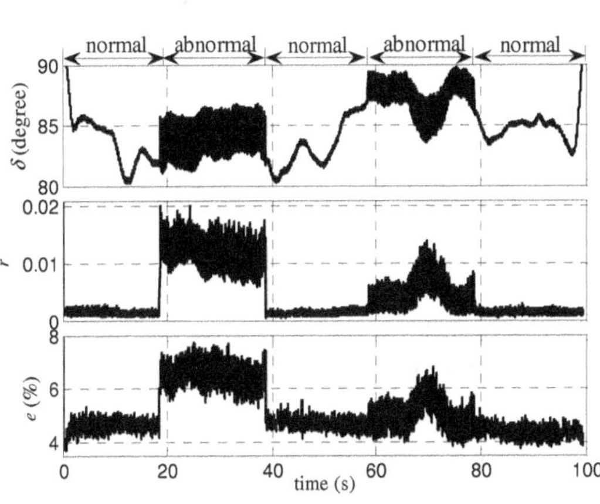

(b) CM results

Figure 8. CM when the generator rotates at a varying rotational speed

- Winding faults that occur in a generator rotor will distort the time-waveform of the generator stator current signal, but the faults have less influence on the stator voltage signal.
- Compared to the stator current and voltage signals, the generator total power signal is more alert to the presence of generator faults.
- All three proposed CM criteria are able to successfully detect the generator winding faults, despite the loading conditions. However, experimental results indicate that criterion δ is not completely load-independent. Under varying loading conditions, an additional de-trending of the criterion could be necessary to further improve its fault detection capability.
- Experimental results suggest that different types of rotor winding faults show different characteristics in the stator current signal. This is potentially important information to both CM and fault diagnosis of the induction generator.

Following this work, the future effort will continue validate the proposed techniques in the cases of (1) when a DFIG is conncted to the grid with large inductive load; (2) when various types of generator faults happen simultaneously; and (3) when the generator faults deteriorate with time.

ACKNOWLEDGEMENT

The work presented in this paper was funded by the European FP7 project OPTIMUS 322430 and China Natural Science Foundation, with the project reference number of 51075331.

REFERENCES

BVG Associates (2011), *A guide to an offshore wind farm.*

Cardoso, A.J.M., & Saraiva, E.S. (1993), Computer-aided detection of airgap eccentricity in operating three-phase induction motors by Park's vector approach. *IEEE Transactions on Industry Applications*, vol.29, no.5, pp.897–901.

Feng, Y., Tavner, P., & Long, H. (2010), Early experiences with UK Round 1 offshore wind farms. *Proceedings of the Institution of Civil Engineers: Energy*, vol.163, no.4, pp.167–181.

McMillan, D., & Ault, G.W. (2007), Quantification of condition monitoring benefit for offshore wind turbines. *Wind Engineering*, vol.31, no.4, pp.267–285.

Nejjari, H., & Benbouzid, M.E.H. (2000), Monitoring and diagnosis of induction motors electrical faults using a current Park's vector pattern learning approach. *IEEE Transactions on Industry Applications*, vol.36, no.3, pp.730–735.

Tavner, P. (2008), Review of condition monitoring of rotating electrical machines. *IET Electric Power Applications*, vol.2, no.4, pp.215–247.

Xiang, D., Ran, L., Tavner, P., & Yang, S. (2006), Control of doubly fed induction generator in a wind turbine during grid fault ride-through. *IEEE Transactions on Energy Conversion*, vol.21, no.3, pp.652–662.

Yang, Wenxian, Tavner, P., Crabtree, C., & Wilkinson, M.R. (2010), Cost-effective condition monitoring for wind turbines. *IEEE Transactions on Industrial Electronics*, vol.57, no.1, pp.263–271.

Yang, Wenxian, Court, R., Tavner, P., & Crabtree, C. (2011), Bivariate empirical mode decomposition and its contribution to wind turbine condition monitoring. *Journal of Sound and Vibration*, vol.330, no.15, pp.3766–3782.

Yang, Wenxian, Tavner, P., Crabtree, C., Feng, Y., & Qiu, Y. (2012), Wind turbine condition monitoring: Technical and commercial challenges. *Wind Energy*, In Press.

Yang, Wenxian, Tavner, P., & Court, R. (2013), An online technique for condition monitoring the induction generators used in wind and marine turbines. *Mechanical Systems and Signal Processing*, vol.38, no.1, pp.103-112.

BIOGRAPHIES

Dr. Wenxian Yang received the Ph.D. degree in mechanical engineering from Xi'an Jiaotong University, Xi'an, China, in 1999. He is currently a lecturer in offshore renewable energy at Newcastle University. Before joining Newcastle University, he worked for UK National Renewable Energy Centre (narec) as a technical specialist. Dr Yang worked in the areas of new and renewable energy, marine hydrodynamics, offshore structures and installation, signal processing, machine condition monitoring and fault diagnosis, non-destructive testing and evaluation, and artificial intelligence in both industry and academia.

Dr. Shuangwen (Shawn) Sheng is a senior engineer at National Renewable Energy Laboratory (NREL). He has B.S. and M.S. degrees both in electrical engineering and a Ph.D. in mechanical engineering. Shawn is currently leading wind turbine condition monitoring, failure database and wind plant operation & maintenance research at NREL. Shawn also has experience in mechanical and electrical system modeling and analysis, soft computing techniques, and automatic control. He has published his work in various journals, conference proceedings, and book chapters.

Dr. Richard Court received the Ph.D. degree in material engineering from Cambridge University, in 2001. He is currently the principal wind specialist at UK National Renewable Energy Centre (narec). Dr Court's main ecpertise is on testing of wind turbine blades, with secondary experience of other technologies within wind sector.

Integrating Structural Health Management with Contingency Control for Wind Turbines

Susan A. Frost [1], Kai Goebel[2] and Léo Obrecht[3]

[1,2]*NASA Ames Research Center, Moffett Field, CA, 94035, USA*
susan.frost@nasa.gov
kai.goebel@nasa.gov

[3]*FEMTO-ST Institute/ENSMM, Besançon, France*
leo.obrecht@ens2m.org

ABSTRACT

Maximizing turbine up-time and reducing maintenance costs are key technology drivers for wind turbine operators. Components within wind turbines are subject to considerable stresses due to unpredictable environmental conditions resulting from rapidly changing local dynamics. In that context, systems health management has the aim to assess the state-of-health of components within a wind turbine, to estimate remaining life, and to aid in autonomous decision-making to minimize damage to the turbine. Advanced contingency control is one way to enable autonomous decision-making by providing the mechanism to enable safe and efficient turbine operation. The work reported herein explores the integration of condition monitoring of wind turbine blades with contingency control to balance the trade-offs between maintaining system health and energy capture. Results are demonstrated using a high fidelity simulator of a utility-scale wind turbine.

1. INTRODUCTION

System health monitoring provides useful information on the current state of a system that can be used to improve many operational objectives of a wind turbine (Doebling et al., 1996). Growing demand for improving the reliability and survivability of safety-critical systems (such as aerospace systems) has led to the accelerated development of prognostics and health management (PHM) and fault-tolerant control (FTC) systems. Active FTC techniques that are capable of retaining acceptable performance in the presence of faults are being developed for both inhabited and uninhabited air vehicles (Shore & Bodson, 2005 – Litt et al., 2003 – Zhang & Jiang, 2003) and researchers are exploring new paradigms and approaches for integrating PHM with controls (Balaban et al., 2013 – Farrar & Lieven, 2007 – Tang et al., 2008). Typically, a decision-making component reasons over the system health and the objectives and constraints of the system and then determines (and sometimes enacts) the optimal course of action. For instance, a component could be identified as having a fault that would eventually lead to component failure and system shutdown. Decision making using prognostic information on the estimated remaining useful life (RUL) of the component along with operational objectives and constraints may result in enacting changes to the operational mode of the system or to the system's controller.

Wind turbines operate in highly turbulent environments sometimes resulting in large aerodynamic loads, potentially causing component fatigue and failure. Two key technology drivers for turbine manufacturers are increasing turbine up-time and reducing maintenance costs. What is desirable from an operator's and original equipment manufacturers (OEMs) perspective is a turbine controller that is capable of adapting to damage and remaining useful life predictions provided by condition or health monitoring systems. The objective would be for the turbine to continue operating and producing power without exceeding some damage threshold resulting in unscheduled downtime. Operating limits would be prescribed by the operator to the health monitoring system. This paper will propose an integrated framework that uses structural health information to inform a contingency controller to enable a damaged turbine to operate in a reduced capacity under some operating conditions to mitigate further damage.

Recent advances in structural health monitoring allow for more accurate assessment of the structural health of a wind turbine, including the blades, tower, and gearbox (Butterfield et al., 2009). In this paper, we develop an approach to integrate system health monitoring with contingency control to enable safe operation of a utility-scale wind turbine with blade damage. The approach is being demonstrated on a high fidelity simulation of a horizontal axis utility-scale wind turbine. A blade fault is modeled in the wind turbine simulation. Characteristics of the fault are identified to provide an elementary fault classifier. An observer is developed that predicts potentially damaging operating conditions. Using this information, the contingency controller is able to de-rate the turbine, that is, it reduces the generator operating set-point which results in lower loads on the turbine blades (Frost et al., 2012).

The paper is organized as follows: section 2 provides motivation for the problem, section 3 describes the wind turbine simulation, section 4 describes the blade damage model and classifier, section 5 describes the generator de-rating scheme and its affect on the wind turbine, and section 6 describes the contingency operation and gives simulation results.

2. MOTIVATION

A wind farm is an interconnected group of wind turbines that collectively act as a power plant, supplying electrical power to the transmission grid. The wind farm operator manages the complex problem of safely and efficiently operating the turbines and the power supplied to the grid, in addition to determining maintenance schedules and coordinating unplanned repairs of the turbines (Manwell et al., 2009).

Original equipment manufacturers are building turbines with dramatically longer blades than just a few years ago because wind power is proportional to the swept area of the rotor and the cube of wind speed. This increases the turbines' power generation capacity. Since wind speed increases with distance from the ground, large rotors on tall towers means more energy is captured. Even though modern turbines are much larger than previous generation turbines, advances in materials, design, and manufacturing techniques has enabled the dimensions to increase without a corresponding increase in material. Consequently, the cost for new turbines is not scaling with the cube of their size, which is helping to drive down the cost of wind energy generation. A decrease in the cost of energy is one important factor that will sustain the demand for new turbine installations.

However, the rapid pace of OEM development of ever larger turbines and the entry of new stakeholders into the market could result in unforeseen reliability issues, possibly impacting the cost of energy. For example, larger turbines are inherently more flexible than smaller ones, resulting in lower frequency resonant modes that are more easily excited and more destructive to the turbine components. Most utility-scale turbines are variable speed with a gearbox connecting the low speed shaft to the high-speed shaft, in essence connecting the rotor hub to the generator. The drive train and gearbox are especially vulnerable to fatigue and failure loads. Most modern turbine blades are made from composite materials. Blades can be subject to destructive aerodynamic loads, cyclic loads, icing, insect and debris buildup (resulting in a roughness increase), and coupling of resonant modes. Any of these conditions can damage or contribute to damage progression of the composite material (Rumsey & Paquette, 2008). The power electronics of wind turbines are also vulnerable to several types of faults, especially if there is over-speeding of the generator. There are many other failure modes for turbines and their components not mentioned here (Lu et al., 2009). Even if turbines have a manufacturer's warranty, the timeliness of problem resolution can have a significant effect on operators' profits. In some cases, OEMs are providing contracts with turbine up-time guarantees, giving the operator more control over expected expenses and income. In any case, there is motivation for the manufacturer and the operator to monitor the health of the turbine and provide condition based maintenance.

Wind turbine operation is divided into several different *regions*, see Fig. (1). Region 1 represents the wind speeds below which the turbine does not operate. The wind speed at the start of region 2 is called the cut-in wind speed ($w_{cut\text{-}in}$). Rated wind speed (w_{rated}) is the velocity at which maximum power output, or rated power, of a wind turbine is achieved. Region 3 starts at the rated wind speed and extends to the cut-out wind speed ($w_{cut\text{-}out}$), above which the turbine is not allowed to operate. Turbines operating in region 2 use generator torque to maximize energy capture. In region 3, the turbine rotational speed is maintained constant at the rated speed by pitching the turbine blades. If a wind turbine were allowed to operate in an uncontrolled manner in region 3, the power output would increase in proportion to the cube of the wind speed, resulting in overheating of the generator and overstress of the power electronics system. An additional goal of operation in region 3 is to reduce the loads on the turbine due to aerodynamic forces. Multi-megawatt turbines typically have a control strategy for the transition region between region 2 and 3, also called region 2.5. In region 4, the turbine blades are locked down and the turbine is yawed out of the wind to prevent damage to the turbine and for safety.

The expected power output from a wind farm is a function of the installed name plate capacity of the turbines, e.g., a 2.5 MW turbine, and the expected capacity factor for the wind farm. The integration of system health monitoring of wind turbines with controls has the potential for significant payoff when applied to individual wind turbines and even more so when applied to large wind farms or wind parks. Contractual obligations to deliver power and the long lead time to replace a damaged turbine, requires wind farm operators to have contingency plans to manage the risk that one or more turbines will suffer damage between scheduled maintenance intervals. If a turbine suffers damage such as a blade delamination, it is crucial to be able to quickly detect the presence of damage. Next, the proper response needs to be determined. The easiest solution would be to shut a damaged turbine down, but that leads to lost output and, potentially, additional costs for the operator due to unscheduled maintenance, amongst others. Unscheduled maintenance is a considerable cost driver for wind turbines since wind turbines are often times in remote locations and using a crane or other means to access the blades tends to be expensive. Alternatively, if the degree of damage is known and if the damage mechanisms are understood, the turbine could potentially continue to operate safely (until an orderly maintenance can be performed), albeit at a reduced capacity for some period of time without the danger of catastrophic failure.

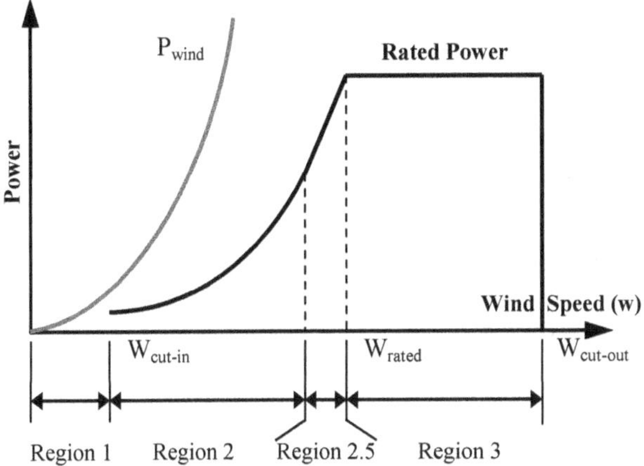

Figure 1. Regions of operation for a wind turbine.

that incorporates wind forecasts, historical data, contractual power output requirements, and maintenance schedules could be integrated with the health monitoring and controls. Such a system could allow the damaged turbine using contingency control to mitigate the blade stress to generate power under favorable wind conditions when the wind farm power requirements are the highest. The turbine health would be monitored to assess the damage and remaining useful life of the component, to ensure that, if the damage accelerated in an unexpected manner, the operating conditions would be further restricted. This includes a decision point where the turbine could no longer be operated safely and it would be shut down.

There is a continuum of solution techniques that include contingency controls, controller reconfiguration, planning and scheduling, and logistics optimization. This paper describes a contingency controller approach which acts on health information derived from blade strain gauge sensor data. It should be noted that in principle, the range of sensors is not limited to strain gauges. Accelerometers, thermo couples, or active piezoelectric sensors – to name a few – can all be part of a more comprehensive condition monitoring scheme that will allow to provide even more refined health information.

3. WIND TURBINE SIMULATION

3.1. Simulation overview

This study uses a nonlinear high-fidelity simulation of the 2-bladed Controls Advanced Research Turbine (CART2), an upwind, active-yaw, variable-speed horizontal axis wind turbine (HAWT) located at the National Renewable Energy Laboratory's (NREL) National Wind Technology Center (NWTC) in Golden, Colorado (Fingersh & Johnson, 2002 – Stol, 2004). CART2 is used as a test bed to study control algorithms for medium-scale turbines. The pitch system on the CART2 uses electromechanical servos that can pitch the blades up to ±18 deg/s. In Region 3, the CART2 uses a conventional variable-speed approach to maintain rated electrical power, which is 600 kW at a low-speed shaft [LSS] speed of 41.7 rpm and a high-speed shaft [HSS] speed of 1800 rpm. Power electronics are used to command constant torque from the generator and full-span blade pitch controls the turbine rotational speed. The maximum rotor-speed for the CART2 is 43 rpm (on the low-speed side) or 1856.1 rpm on the generator side. Whenever the rotor-speed reaches this value the turbine shuts down due to an over-speed condition.

The CART2 has been modeled using the Fatigue, Aerodynamics, Structures, and Turbulence Codes (FAST), a well-accepted simulation environment for HAWTs (Jonkman & Buhl, 2005). The FAST code is a comprehensive aeroelastic simulator capable of predicting both the extreme loads and the fatigue loads of two- and three-bladed horizontal axis wind turbines. Wind turbines can be modeled with FAST as a combination of rigid and flexible bodies connected by several degrees of freedom (DOFs) that can be individually enabled or disabled for analysis purposes. Kane's method is used by FAST to set up equations of motion that are solved by numerical integration. FAST computes the nonlinear aerodynamic forces and moments along the turbine blade using the AeroDyn subroutine package (Laino & Hansen, 2001). The FAST code with AeroDyn incorporated in the simulator was evaluated in 2005 by Germanischer Lloyd WindEnergie and found suitable for 'the calculation of onshore wind turbine loads for design and certification' (Manjock, 2005).

The parametric information for the FAST simulator as configured herein is available from NWTC Design Code (2013). An example of some FAST configuration parameters can be found in Table 1. The control objective is to regulate generator speed at 1800 rpm and to reject wind disturbances using collective blade pitch. The inputs to the FAST plant are generator torque, blade pitch angle, and nacelle yaw. The FAST simulator can be configured to output many different states or measurements of the plant, such as generator speed and low speed shaft velocity. In this study, the yaw is assumed fixed, and the mean wind inflow is normal to the rotor. A baseline torque controller operates to command the generator torque setting and a baseline pitch controller operates to command the blade pitch (Wright et al., 2006). These controllers will be described next.

Value	Variable	Description
True	FlapDOF1	First flapwise blade mode DOF
True	FlapDOF2	Second flapwise blade mode DOF
True	EdgeDOF	First edgewise blade mode DOF
False	TeetDOF	Rotor-teeter DOF
True	DrTrDOF	Drivetrain rotational-flexibility DOF
True	GenDOF	Generator DOF
False	YawDOF	Yaw DOF
True	TwFADOF1	First fore-aft tower bending-mode DOF
True	TwFADOF2	Second fore-aft tower bending-mode DOF
True	TwSSDOF1	First side-to-side tower bending-mode DOF
True	TwSSDOF2	Second side-to-side tower bending-mode DOF
True	CompAero	Compute aerodynamic forces
False	CompNoise	Compute aerodynamic noise

Table 1. FAST configuration parameters for CART2 simulations.

3.2. Torque controller design

The wind turbine simulation has independent generator torque control and collective pitch control. The purpose of the torque controller is to regulate the generator speed and torque to maximize the power capture below the rated generator speed. Once the turbine has reached rated speed, the torque controller commands constant generator torque and the pitch controller operates to maintain a constant rotational speed. If the turbine speed goes beyond a certain threshold, a shutdown procedure is commenced to protect the turbine.

To maximize the power capture, the relation between torque and speed is divided in four parts, called "Regions", that can be observed in Fig. (5). In region 1, the wind turbine does not operate. The start of region 2 is called the cut-in speed of the wind turbine. The generator torque versus wind speed curve of region 2 has a quadratic shape that is designed to track an optimal tip-speed ratio for maximum power capture. The relation governing this region is given in Eq. (1).

$$T_{R2} = k_2 V_G{}^2 \tag{1}$$

where T_{R2} is the generator torque (Nm) in region 2, k_2 is a coefficient and V_G is the generator speed in rpm.

Region 2.5 is a linear transition allowing the generator to reach the rated generator torque quickly. Its relation is given in Eq (2).

$$T_{R2.5} = \frac{V_G - BSS}{k_{S2.5}} + k_{T2.5} \tag{2}$$

where $T_{R2.5}$ is the generator torque in region 2.5, BSS is the Beginning Slope Speed in rpm (i.e., the speed at which the generator switch from region 2 to region 2.5), $k_{S2.5}$ is the coefficient of the slope in region 2.5 and $k_{T2.5}$ is the torque value (N.m) at the BSS speed.

When the generator speed reaches the set-point, the generator torque is set to a constant value. The Controls Advanced Research Turbine (CART2) has a constant torque set by the manufacturer to $T_m = 3524.36$ N.m.

The following values are used in the CART2 controller for all the following simulations:

$k_2 = 0.0008992$
$k_{S2.5} = 10.557$
$k_{T2.5} = 2574.2265$ N.m
$BSS = 1690.98\ rpm$

3.3. Pitch controller design

During region 3 operation, the pitch controller collectively varies blade pitch (i.e., both turbine blades receive the same blade pitch command) to regulate rotor speed to the rated speed and to mitigate aerodynamic loads due to gusts. The pitch controller designed for the wind turbine simulation is a proportional integral (PI) feedback controller with a low pass filter on the plant output signal. It is assumed that the plant is well modeled by the linear time invariant (LTI) system:

$$\begin{cases} \dot{x}_p = A_p x_p + B_p u_p + \Gamma_p u_D \\ y_p = C_p x_p;\ x_p(0) = x_0 \end{cases} \tag{3}$$

where the plant state x_p is an N_p-dimensional vector, the control input vector u_p is M-dimensional, and the sensor output vector y_p is P-dimensional. The disturbance input vector u_D is M_D-dimensional. The system given by Eq. (3) is sometimes referred to as the triple (A, B, C). The control objective will be to cause the plant output y_p to asymptotically track zero. Define the output error vector as:

$$e_y \equiv y_p - 0 \quad (4)$$

Hence, to achieve the desired control objective, the error must asymptotically approach zero:

$$e_y \xrightarrow[t\to\infty]{} 0 \quad (5)$$

Consider the plant given by Eq. (3). The control objective for this system is accomplished by a PI control law of the form:

$$u_p = G_P e_y + G_I \int e_y \quad (6)$$

where G_P and G_I are constant gains. It is well known that a system given by Eqs. (3)-(6) can achieve asymptotic stability if the plant given by Eq. (3) is controllable and observable and strict positive real (SPR). A system (A, B, C) is SPR when the matrix product CB is positive definite and the open-loop transfer function $P(s) = C(sI - A)^{-1}B$ is minimum phase, i.e., all of its zeros are in the left half of the complex plane.

In some cases the plant in Eq. (3) does not satisfy the controller's requirement of SPR. Instead, there may be a modal subsystem that inhibits this property during feedback control (Frost et al., June 2011). A low pass filter can be applied to the plant output to remove the effects of the modal subsystem.

The baseline pitch controller for the CART2 simulation was designed by researchers at NREL to regulate generator speed to the rated value of 1800 rpm and to alleviate aerodynamic loads on the turbine (Wright et al., 2006). The controller input is the generator operating speed and the controller output is the collective blade pitch. The baseline PI controller was designed for an operating point of 18 m/s wind inflow and has a proportional gain (G_P) of 0.38 and an integral gain (G_I) of 0.136.

The transfer function for the low pass filter used in this study is given by:

$$T(s) = \frac{10}{s + 10} \quad (7)$$

The low pass filter is placed in the feedback loop from the plant to the controller input in the Simulink™ turbine simulation. This has the effect of removing the modes from the plant output that inhibit the SPR property (Frost et al., 2011).

The baseline PI controller operating in turbulent winds is shown in Fig. (2). Simulation transients are removed from all results by omitting the first 60 seconds of the simulation. The generator speed set point is approximately 1800 rpm. Figure 2 shows the time series of average 18 m/s turbulent IEC 61400-1 class "C" wind in the direction normal to the wind turbine rotor plane. This wind file or one generated with the same characteristics is used for all turbulent wind simulations reported in this paper.

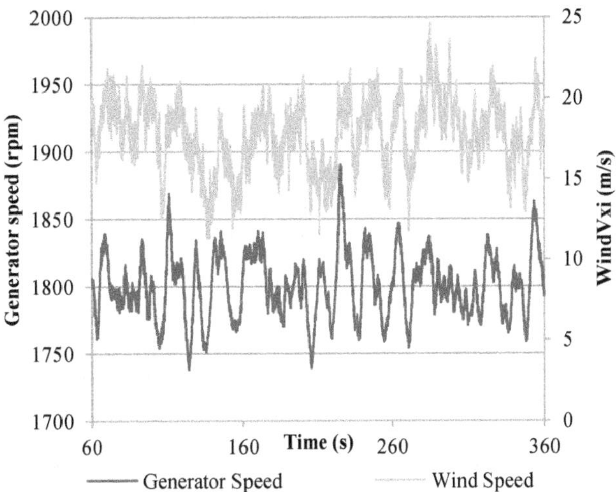

Figure 2. Time series of an 18m/s average turbulent wind (IEC 61400-1 class "C") and the generator speed.

Node (-)	BlFract (-)	BMassDen kg.m^{-1}	FlpStff kN.m^{-2}	EdgStff kN.m^{-2}
1	0.000	282.92	165000	283000
2	0.022	290.24	161000	318000
3	0.053	261.88	142000	328000
4	0.114	201.28	98700	307000
5	0.175	186.52	78400	340000
6	0.236	169.10	59200	342000
7	0.299	149.28	45400	278000
8	0.364	133.19	34100	237000
9	0.427	111.74	25000	169000
10	0.491	96.86	17900	138000
11	0.555	78.57	12300	94000
12	0.618	65.03	8190	72500
13	0.682	49.68	5140	46700
14	0.745	37.59	3020	32600
15	0.809	25.01	1620	189000
16	0.873	16.01	868	13000
17	0.936	10.73	468	8850
18	1.000	6.02	209	6800

Table 2. Distributed nodes and properties for CART2 blades.

4. TURBINE BLADE DAMAGE

An objective of this study is to explore the response of a utility scale wind turbine to blade damage. The FAST simulation of the CART2 allows configuration of blade properties at distributed stations along the blade span, thereby enabling the properties to be modified to simulate blade damage. The total beam length of the CART2 blade is 19.995 meters. Table 2 gives relevant information on the 21 blade nodes used in the CART2 FAST simulation, including the node number, the fraction of the total beam length which determines the node position measured from the root (BlFract), the local blade mass density (BMassDen), and the local flapwise stiffness (FlpStff) and edgewise stiffness (EdgStff).

Additional distributed blade information used by the simulation includes aerodynamic center, chord length, and structural twist. The first and second blade flapwise bending modes and structural damping in percent of critical and the first blade edgewise bending mode and its structural damping in percent of critical are specified in configuration files. Blade mode shapes are represented by sixth-order polynomials that are a function of spanwise position with boundary conditions at the blade root of zero displacement and zero first derivative.

This study assumes that blade damage can be represented by a decrease in the spanwise and edgewise stiffness at a blade station and be reflected in corresponding changes in strain

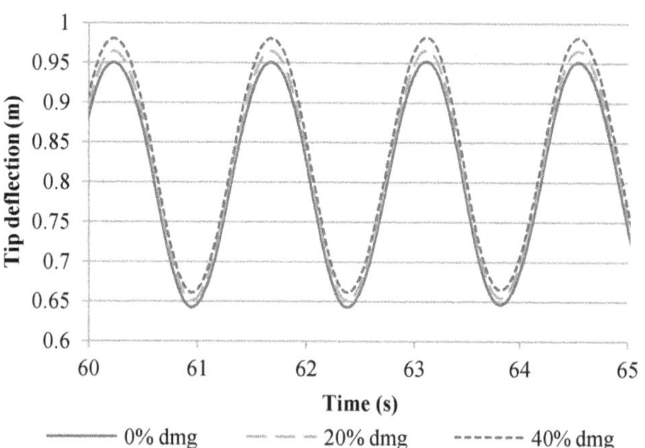

Figure 3. Tip deflection for simulation with blade having no damage, 20% and 40% reduction in stiffnesses at node 5, with constant 16 m/s wind inflow.

gauge data. Blade damage represented by local changes in stiffness includes cracks and delaminations (Nelson et al., 2011; Wahl et al., 2001; Yang et al., 1990). Simulations of the CART2 run with decreased blade stiffness and constant wind inflow show small changes in blade tip deflections. Figure (3) shows a plot of tip deflection for an undamaged turbine and a turbine with damage at node 5 consisting of a 20% and 40% reductions in stiffness in the edgewise and flapwise directions. The wind inflow for the results in Fig. (3) was 16 m/s constant wind with 3 m/s vertical shear. Simulations with 14-24 m/s wind inflow with 3 m/s vertical shear were performed giving similar results to those seen in Fig. (3). Negligible changes in the power spectral density (PSD) of the blade bending moments and rotor thrust were observed for the various damage cases. This means that damage detection using PSD or tip deflection measurements will not be trivial.

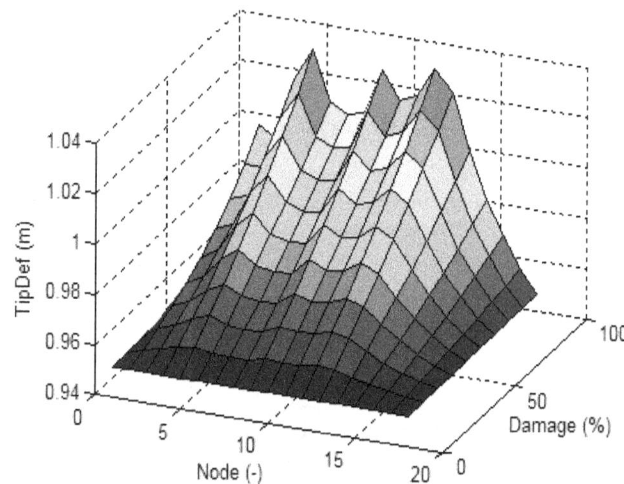

Figure 4. Tip Deflection with one damaged node, according to the node number and its damage level with a 16m/s steady wind.

Studies of the CART2 simulator are performed to investigate effects of changes in stiffness in the blades and to identify a possible feature for damage detection. A full factorial study with two parameters is performed to determine the spanwise blade point that is sensitive to changes in stiffness in the CART2 simulator. A Matlab program is developed to run multiple simulations, allowing one parameter to be modified at a time. For each simulation, a specific node of the blade is selected and its spanwise and edgewise stiffness are reduced by a damage coefficient expressed in percentage of the initial stiffness value.

The spanwise and edgewise tip deflections are recorded for each simulation. The norm of the two vectors is calculated to get the resultant tip deflection. The maximum value of the norm of each simulation is plotted according to the node and damage level. Figure 4 shows results of this study for wind inflow of 16 m/s with 3 m/s vertical shear.

This study shows that the most sensitive nodes to stiffness reduction, e.g., the nodes where stiffness reduction increases the tip deflection, are nodes 5, 12 and 9 (ordered by sensitivity). The same study was performed with different wind speeds and showed that these nodes were also the most sensitive.

In addition, it was observed that the minimum and maximum deflection of the damaged blade is a function of wind speed and blade pitch. Deflection is also a function of the stiffness, which in turn is a measure for the blade's degree of damage. Changes in deflection due to pitch or wind speed dominate changes due to stiffness reduction. What this means is that a damage detection tool must factor out the impact of pitch and wind speed to make an inference on the presence of damage in a blade (Frost et al., 2012).

A structural health management system should be designed to monitor relevant subsystems to detect blade faults. Using appropriate fault mode models, damage propagation models, and continued monitoring of relevant measurements, inference on blade damage progression can be performed with the intent of determining remaining useful life of the blade.

For the purposes of the example described in this paper, a simple tip displacement threshold logic is employed as a damage classifier.

5. GENERATOR DE-RATING

In an earlier study (Frost et al., Mar. 2011), it was hypothesized that aerodynamic loads on the turbine blades could be lowered by *de-rating* the generator, e.g., by lowering the generator torque and speed set-point used by the controller to regulate turbine rotational speed (Fig. 5). A controller was designed allowing the user to choose the rating of the generator.

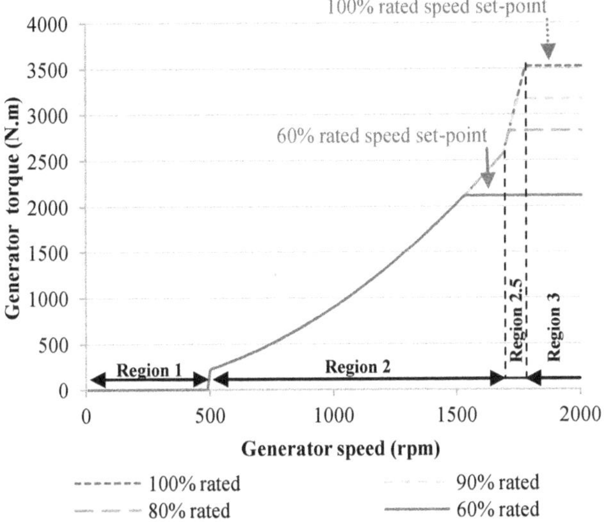

Figure 5. Correlation between generator torque and speed.

De-rating the generator means that the rated value of the generator speed set by the wind turbine manufacturer is reduced. When de-rating the generator, the value becomes the result of Eq. (8).

$$T_{R3} = T_m \times R \tag{8}$$

where T_{R3} is the generator torque (N.m) in region 3, T_m is the generator torque (N.m) set by the manufacturer and R is the rating coefficient between 0 (0% rated) and 1 (100% rated).

To keep the laws governing region 2 and 2.5 operation consistent when de-rating, the speed set-point is also changed. The new set-point is calculated through Eq. (9).

$$V_{Sp} = \frac{T_{R3} - k_{T2.5}}{k_{S2.5}} + BSS + \Delta_{S3} \tag{9}$$

where V_{Sp} is the generator speed set-point in rpm and Δ_{S3} is a constant speed value (rpm) shifting the set-point to strive to keep the speed in region 3.

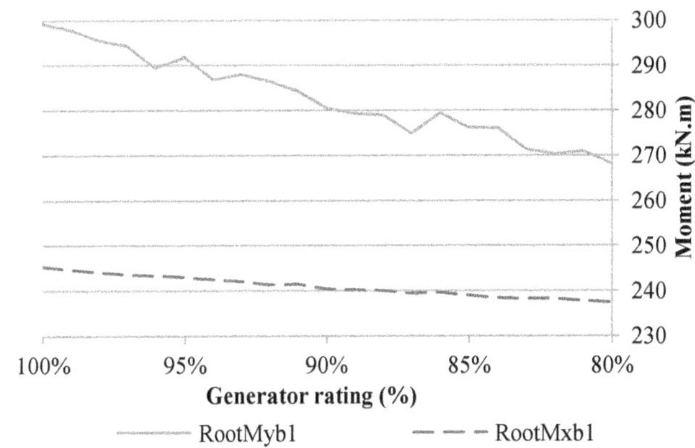

Figure 6. Effect of de-rating generator on damage equivalent loads at the root of the blade 1.

Simulations are run with de-rated generator values to see the effects on the blade root moments. Figure (6) shows the damage-equivalent loads of the root moments caused by flapwise and edgewise forces (RootMyb1 and RootMxb1, respectively), under an 18m/s average speed turbulent wind.

The results of this study show that de-rating the generator has a stronger impact on the damage-equivalent loads of the flapwise root moments. This can be explained by the fact that flapwise forces are mainly due to aerodynamic loads whereas edgewise forces are mainly due to gravity.

Simulations were run with the 18 m/s turbulent wind inflow resulting in primarily region 3 operation described in Section 3.3. The generator rating was constant during each simulation. Twenty simulations were run with rating between 100% and 80% with a 1% step. Each simulation ran for six minutes and the first minute was removed from results reported to avoid the transient period. The mean value of the root moment for each simulation and its single highest (extreme) value is reported.

To reduce the specific effects of a wind profile on the wind turbine, ten turbulent wind profiles were generated using TurbSim and the same study was repeated with each profile. TurbSim is a turbulent wind file generator for FAST available from NWTC Design Code (2012).

All the turbulent winds have a mean speed of 18 m/s and are following the third edition of the IEC 61400-1 standard. The turbulence characteristic is rated "C" in this standard.

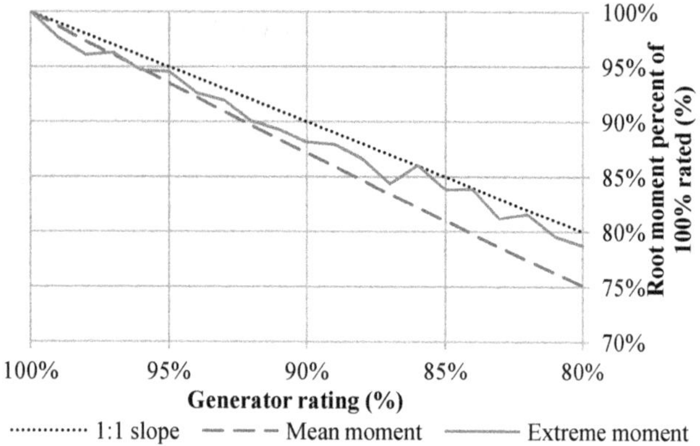

Figure 7. Effect of de-rating generator on blade 1 root moment

The mean values of the 10 studies were used to create Fig. (7). It represents the percentage of reduction of the root flapwise moment, i.e., the moment caused by flapwise forces, when de-rating the controller. The 1:1 slope shows the threshold where the de-rating percentage is equal to percentage of reduction of the root moment. Values under the threshold show that root moments are reducing faster than controller rating.

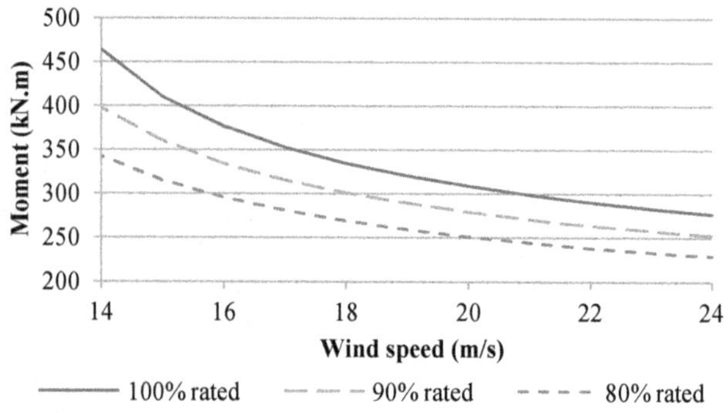

Figure 8. Extreme loads from flapwise root bending moment (RootMyb1) for various wind speeds and generator de-rating.

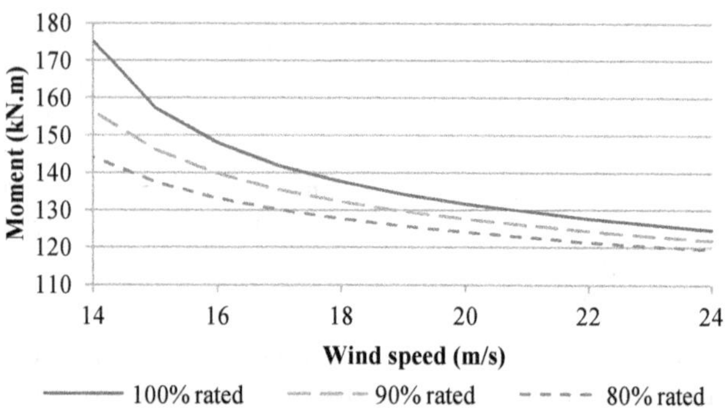

Figure 9. Extreme loads from edgewise root bending moment (RootMxb1) for various wind speeds and generator de-rating.

In Fig. (7) it can be seen that de-rating the controller has a positive impact on the reduction of the mean and the extreme moments in the root of the blade. Since the values are under the threshold, the root moment is reducing faster than the controller rating. The extreme moments do not monotonically decrease with generator de-rating, this is due to the periodicity of the wind turbine blades and the time-varying nature of the wind. For instance, a gust hitting a turbine blade at a given point of time in the wind time series would hit the blade at a different time if the rotational speed of the blade differed, as it does when the turbine is de-rated.

Studies were run with constant wind having 3 m/s vertical shear. Data were analyzed after the wind turbine had reached a trim state and the start-up transients had died off. Figures (8) and (9) show the extreme loads of flapwise and edgewise root bending moments for an undamaged blade for 3 levels of generator settings: rated generator speed, generator set-point de-rated by 10%, and generator set-point de-rated by 20%. As can be seen in the figures, the damage equivalent loads of the blades decrease when the generator is de-rated. The de-rating of the turbine results in the blades being collectively pitched at a lower wind speed. This has the effect of reducing the flapwise aerodynamic loads on the blade and hence the flapwise blade root bending moments.

6. CONTINGENCY OPERATION

This section describes the contingency operation of a wind turbine to reduce loads on damaged blades under certain operating conditions. At times, it may be desirable to limit loads on a damaged turbine to ensure continued safe

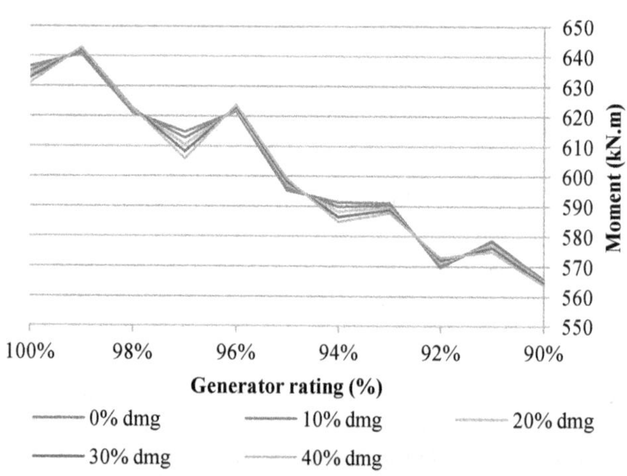

Figure 10. Extreme load for flapwise bending moment for blade with damage at node 5 for various damage levels and generator de-rating.

operation at a reduced level of energy capture from the turbine's nameplate generating capacity. In particular, controller operation might be modified when environmental or fault conditions exist that could result in high cyclic loads or extreme loads to the turbine blades resulting in a fault that causes the turbine to become disabled. As mentioned earlier, the success of wind energy depends on turbines being able to produce power whenever there is sufficient wind. Additionally, unscheduled turbine maintenance can be extremely expensive and disruptive, especially for blades, which require large, expensive cranes for access.

Figure 10 shows the resulting extreme loads in the flapwise direction for a blade with damage at node 5 under various generator de-rating values. It can be seen from the figure that de-rating the generator for a turbine with blade damage does lead to lower extreme loads on the damaged blade as was observed for undamaged blades. The mean loads were not included here since they were virtually unchanged across damage levels for the same generator rating.

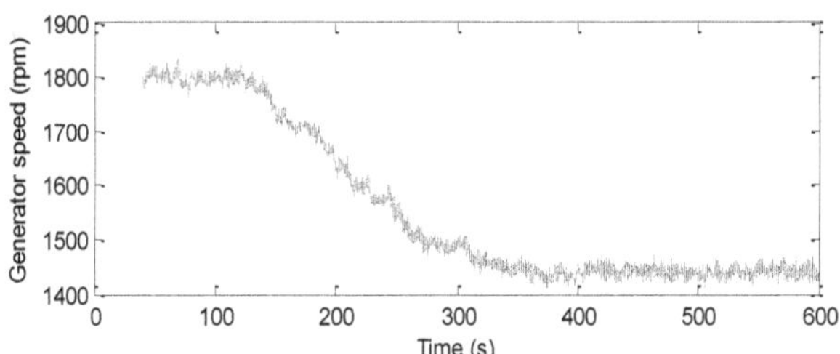

Figure 11. Measured generator speed for contingency controller.

The contingency controller has an observer or estimator of highly turbulent operating conditions. The observer uses measurements of the turbine rotor speed to estimate gusty conditions or conditions that cause rapid rotor accelerations suggesting extreme operating conditions that would not cause the turbine to shut down. For the illustrative example presented here, the generator was de-rated to 90% after a number of turbulent events were detected. It is assumed that the operating conditions transition from normal conditions to the turbulent wind conditions shown in Fig. (2). Furthermore, for this illustrative example, it is assumed that the operating conditions created by the turbulent wind are such that continued operation at rated speed could contribute to further damage of the turbine. The generator set-point is smoothly reduced by the contingency controller from rated speed to 90% of rated speed, thereby reducing transient behavior.

Figure 11 shows the generator speed for this illustrative example. Above-rated turbulent wind with IEC turbulence model C and a mean wind speed of 18 m/s is used to test the adaptive contingency controller. In practice, the operating conditions under which the turbine would be de-rated would be a design consideration depending on many factors, such as the wind resource at the turbine site, the desired capacity factor for the turbine, and the location of the turbine within a wind farm. The type and degree of damage and remaining useful life predictions from the PHM system would inform the contingency controller, enabling it to adjust turbine operation to achieve desired objectives.

7. CONCLUSION

We report here on first steps towards integrating structural health monitoring and contingency control for wind turbines to reduce loads on damaged blades. Ultimately, a trade-off between power capture and potential turbine damage must be made. In the study described in this paper, information about blade health, operator goals, and operating conditions, are linked in parametric form to determine when the contingency controller will de-rate the generator to mitigate further damage to turbine blades. A method for representing blade damage in a high fidelity simulation of a wind turbine is presented. Generator de-rating is employed in simulation with a contingency controller to show the integration of controls and structural health monitoring to reduce loads on damaged turbine blades.

ACKNOWLEDGEMENT

The authors would like to acknowledge the SSAT project under NASA's Aviation Safety Program and the University of Wyoming Wind Energy Research Center for their support of this work.

REFERENCES

Balaban E., Narasimhan S., Daigle M., Roychoudhury I., Sweet A., Bond C. & Gorospe G. (2013), Development of a Mobile Robot Test Platform and Methods for Validation of Prognostics-Enabled Decision Making Algorithms, International Journal of Prognostics and Health Management, Vol4(1) 006.

Butterfield, A Sheng, S, & Oyague, F. (Sept. 9-11, 2009), Wind Energy's New Role in Supplying the World's Energy: What Role will Structural Health Monitoring Play?, 7th International Workshop on Structural Health Monitoring, Stanford, CA.

Doebling, S.W., Farrar, C.R., Prime, M.B. & Shevitz, D.W. (May 1996), Damage identification and health monitoring of structural and mechanical systems from changes in their vibration characteristics: A literature review, Technical Report LA--13070-MS, Los Alamos National Lab., NM,.

Farrar, C.R. & Lieven (2007), N.A.J., Damage prognosis: the future of structural health monitoring, Phil. Trans. R. Soc. A, 365, pp. 623-632.

Fingersh, L.J. & Johnson, K.E. (Oct. 2002), Controls Advanced Research Turbine (CART) Commissioning and baseline data collection. National Renewable Energy Laboratory, NREL/TP-500-32879, Golden, CO.

Frost, S.A., Balas, M.J., Goebel, K, & Wright, A.D. (Jan. 2012), Adaptive contingency control: Wind turbine operation integrated with blade condition monitoring, Proceedings AIAA Aerospace Science Meeting, Wind Energy Symposium, Nashville, TN.

Frost, S.A., Balas, M.J., & Wright, A.D. (June 2011), Generator speed regulation in the presence of structural modes through adaptive control using residual mode filters, Mechatronics, 21(4): 660-667.

Frost, S.A., Goebel, K., Trinh, K.V., Balas, M.J., & Frost, A.M. (Mar. 2011), Integrating Systems Health Management with Adaptive Controls for a Utility-scale Wind Turbine, Proceedings AIAA Infotech@Aerospace Conference, St. Louis, MO.

Jonkman, J.M. & Buhl, M.L. (Aug. 2005), FAST user's guide, National Renewable Energy Laboratory, NREL/EL-500-38230, Golden, Colorado.

Laino, D.J. & Hansen, A.C. (Sept. 2001), User's guide to the computer software routines AeroDyn interface for ADAMS®. Salt Lake City, Utah: Windward Engineering, LC.

Litt, J.S., Parker, K.J., & Chatterjee, S. (2003), Adaptive Gas Turbine Engine Control for Deterioration Compensation due to Aging, Technical Report TM 2003-212607, NASA Glenn, Lewis, OH.

Lu, B., Li, Y., Wu, X., & Yang, Z. (June 24-26 2009), A review of recent advances in wind turbine condition monitoring and fault diagnosis, Proceedings IEEE Power Electronics and Machines in Wind Applications.

Manjock, A. (May 25, 2005), Evaluation report: Design codes FAST and ADAMS for load calculations of onshore wind turbines, Report No. 72042, Germanischer Lloyd WindEnergie GmbH, Hamburg, Germany.

Manwell, J.F., McGowan, J. G., & Rogers, A.L. (2009), Wind Energy Explained: Theory, Design and Application, 2nd ed., John Wiley & Sons Ltd.

Nelson, J.W., Cairns, D.S., & Riddle, T.W. (Jan. 2011), Manufacturing Defects Common to Composite Wind Turbine Blades: Effects of Defects, Proceedings AIAA Aerospace Science Meeting, Wind Energy Symposium, Orlando, FL.

NWTC Design Codes, FAST (2013) and TurbSim (2012), http://wind.nrel.gov/designcodes/simulators/fast/, http://wind.nrel.gov/designcodes/preprocessors/turbsim/, National Renewable Energy Lab., Golden, CO.

Rumsey, M.A. & Paquette, J.A. (2008), Structural health monitoring of wind turbine blades, in Proc. SPIE, vol. 6933, Smart Sensor Phenomena, Technology, Networks, and Systems, p. 69330E.

Shore, D. & Bodson, M. (2005), Flight Testing of a Reconfigurable Control System on an Unmanned Aircraft, AIAA Journal of Guidance, Control, and Dynamics, vol. 28, no. 4, pp. 696-707.

Stol K.A. (Sept. 2004), Geometry and structural properties for the Controls Advanced Research Turbine (CART) from model tuning. National Renewable Energy Laboratory, NREL/SR-500-32087, Golden, Colorado.

Tang, L., Kacprzynski, G.J., Goebel, K., Saxena, A., Saha, B., & Vachtsevanos, G. (Oct. 6-9, 2008), Prognostics-enhanced Automated Contingency Management for Advanced Autonomous Systems, International Conference on Prognostics and Health Management, Denver, CO.

Wahl, N.K., Mandell, J.F., & Samborsky, D.D. (2001), Spectrum fatigue lifetime and residual strength for fiberglass laminates, Doctoral dissertation, Montana State University, Bozeman.

Wright, A.D., Balas, M.J. & Fingersh, L.J. (2006), Testing state-space controls for the controls advanced research turbine, Transactions of the ASME. Journal of Solar Energy Engineering; 128(4): 506-515.

Yang, J. N., Jones, D. L., Yang, S. H. & Meskini, A. (July 1990), A Stiffness Degradation Model for Graphite /Epoxy Laminates, Journal of Composite Materials, Vol. 24, pp. 753-769.

Zhang, Y.M. & Jiang, J. (June 9-11, 2003), Bibliographical Review on Reconfigurable Fault-Tolerant Control Systems, in Proc. of the 5th IFAC Symposium on Fault Detection, Supervision and Safety of Technical Processes, Washington, D.C., USA, pp. 265-276.

A Two-Stage Diagnosis Framework for Wind Turbine Gearbox Condition Monitoring

Prasanna Tamilselvan[1], Pingfeng Wang[2, *], Shuangwen Sheng[3], and Janet M. Twomey[4]

[1,2,4]*Department of Industrial and Manufacturing Engineering, Wichita State University, Wichita, KS 67208, United States*
pxtamilselvan@wichita.edu;
pingfeng.wang@wichita.edu;
janet.twomey@wichita.edu;

[3]*National Renewable Energy Laboratory, Golden, CO 80401, United States*
shuangwen.sheng@nrel.gov

ABSTRACT

Advances in high performance sensing technologies enable the development of wind turbine condition monitoring system to diagnose and predict the system-wide effects of failure events. This paper presents a vibration-based two stage fault detection framework for failure diagnosis of rotating components in wind turbines. The proposed framework integrates an analytical defect detection method and a graphical verification method together to ensure the diagnosis efficiency and accuracy. The efficacy of the proposed methodology is demonstrated with a case study with the gearbox condition monitoring Round Robin study dataset provided by the National Renewable Energy Laboratory (NREL). The developed methodology successfully picked five faults out of seven in total with accurate severity levels without producing any false alarm in the blind analysis. The case study results indicated that the developed fault detection framework is effective for analyzing gear and bearing faults in wind turbine drive train system based upon system vibration characteristics.

1. INTRODUCTION

Maintaining wind turbines (WTs) in top operating condition ensures not only a continuous revenue generation but a reduction in electric power drawn from non-renewable and more polluting sources. Despite the large capital for establishing a wind farm, the operation cost of WTs is one of the primary contributors for the wind energy costs (Ebeling, 1997; Tamilselvan, Wang & Twomey, 2012; Tamilselvan, Wang & Wang, 2012; Tamilselvan, & Wang, 2013). The unexpected breakdowns can be prohibitively expensive since they immediately result in lost production, and poor customer satisfaction. The need for effective WT condition monitoring systems that enable accurate failure diagnosis in an early stage to facilitate optimum WT maintenance planning is reaching a critical stage. Advances in high performance sensing and signal processing technologies enable the development of WT health monitoring system and failure diagnosis tools applied on WT to detect, diagnose, and predict the system-wide effects of failure events.

Maintenance activities for WTs can be broadly classified into two categories namely corrective maintenance and preventive maintenance. Corrective maintenance is carried out after a failure event whereas preventive maintenance is done before the occurrence of a potential failure (Nielsen & Sorensen, 2010). Preventive maintenance can be further classified into scheduled maintenance and condition based maintenance (CBM). Scheduled maintenance is carried out as per on fixed scheduled times and it can be done in as minor and major scheduled maintenances. Some examples of minor scheduled maintenance in WTs include change of filters, lubrication etc. (Nilsson & Bertling, 2007). CBM is a form of preventive maintenance that involves continuous health monitoring of a WT unit. Currently the most common practice of maintenance activities in wind farms is scheduled maintenance. However with latest developments in the field of sensing and signal processing techniques, CBM has been gradually adopted into maintenance decision making of wind farms (Byon, Perez, Ntaimo, & Ding, 2010). In CBM, condition monitoring systems are installed on different system components, such as gearbox, bearing, drive train and generators, to record various sensory signals in order to determine physical states of these components. Different types of sensory signals can be used for condition monitoring purposes, such as vibration and electrical

* Corresponding Author. Tel: +1 316 978 5910; Fax: +1 316 978 3742.

signals. Usually CBM in WTs can be executed based upon vibration monitoring (Shi, Wang, Zhuo, & Liu, 2010, Gebraeel, Lawley, & Liu, 2002, Randall, 2011, Randall & Antoni, 2011, McFadden & Smith, 1984, Antoni, 2002, Harris, 2001, Lebold, McClintic, Campbell, Byington, & Maynard, 2000), oil analysis (Lu & Chu, 2010), or electrical signature analysis (Yang, Tavner, & Wilkinson, 2008). With the help of WT health information provided by condition monitoring systems, optimal O&M planning strategies can be ascertained to prevent system failures and improve turbine availability.

Vibration analysis is one of most commonly used mechanisms of condition monitoring in WTs (Shi, Wang, Zhuo, & Liu, 2010). It is mainly utilized to identify the present condition of WT components such as gearbox, drive train, bearings etc. and estimate their damage growths over the time. In the vibration analysis, vibration signals produced by the rotating components in WTs whose current health conditions need to be diagnosed are commonly analyzed either by broad band based methods or spectral line analysis methods (Lu & Chu, 2010). In the broad band analysis, parameters such as root mean square, peak value, or kurtosis are calculated based on the obtained output signals. Failure of the component can be estimated by the change observed in values of the above calculated parameters. The analysis using the spectral lines is based on the theory that each component exhibits a different frequency of vibration signatures. The frequencies vary for each component such as gear mesh, shaft harmonics or bearing harmonics. The failure of the component is said to occur if there is a measurable increase in the frequency of impulse signals for individual components.

Research on real-time failure diagnosis which interprets data acquired by smart sensors and utilizes these data streams in making critical decisions provides significant advancements so that early awareness of the WT health condition before unexpected failures are developed becomes possible (Tamilselvan, Wang & Jayaraman, 2012; Tamilselvan & Wang, 2012; Tamilselvan & Wang, 2013; Byon et al., 2010). Among many others, vibration-based health monitoring system that detects the faults of WT components based on the vibration signals produced by the rotating components during operation is one of the most vastly used mechanisms for WT condition monitoring. Although effective health diagnosis of WTs provides various benefits, such as improved reliability and reduced costs for turbine maintenance, analysis of massive heterogeneous vibration signals leading to accurate detection of failures of WT components in an early stage remains as a challenge problem.

This paper presents a vibration-based two-stage fault detection framework and integrates an analytical defect detection method and a graphical verification method together to ensure the efficiency and accuracy of failure diagnosis. The proposed methodology is demonstrated with NREL WT gearbox condition monitoring Round Robin study and the results are discussed. The rest of this chapter is organized as follows. Section 2 presents the proposed vibration-based condition monitoring framework. Section 3 introduces the Round Robin gearbox condition monitoring experiment and Section 4 details the data preprocessing. The analytical diagnosis methods and the graphical verification method are detailed in Sections 5 and 6 respectively. Section 7 reports the CM Round Robin study results and Section 8 briefly summarizes the work.

2. VIBRATION BASED HEALTH DIAGNOSIS FRAMEWORK

The framework for the proposed vibration-based two-stage health diagnosis is shown in Figure 1. The developed framework is composed of three essential modules: (i) data preprocessing for conversion of time domain vibration signal to frequency domain signal; (ii) analytical diagnosis module for the detection of defects in the rotating components using sideband and kurtosis evaluation as shown in the bottom left shaded box which include severity factor and matrix determination process; and (iii) graphical diagnosis module to determine the severity level of the defect.

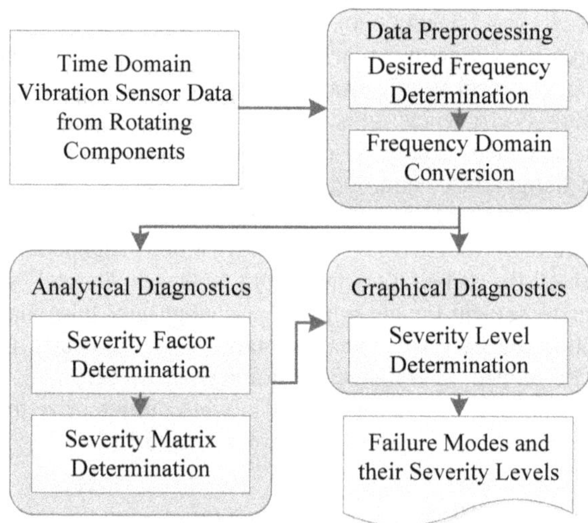

Figure 1. Vibration based two-stage health diagnosis framework

The time domain vibration sensory signal from the rotating components are preprocessed and converted into frequency domain signal for further condition monitoring analysis. The sideband and kurtosis based detection method is employed to analyze the frequency data analytically for detecting the failures of the rotating components. The results from the analytical diagnosis are used as inputs to the graphical diagnosis process. The failure modes and their severity levels are determined from the frequency domain signals by graphical diagnosis method.

2.1. Vibration Data Preprocessing

The preprocessing of vibration data involves three steps, as shown in Table 1. The primary step of the vibration analysis is the calculation of desired frequencies of the rotating components. For instance, the gear meshing frequencies for gears and the bearing frequencies such as Ball Passing Frequency Outer (BPFO), Ball Passing Frequency Inner (BPFI), and Ball Spinning Frequency (BSF) for bearings are desired frequencies of gears and bearings. The next step involves the identification of the relationship between the sensors and the components. Based on the identified relationship, the sensors related to different failure modes are segregated and their corresponding desired frequencies are noted. The final step in the preprocessing is to develop the frequency spectrum from the raw time domain vibration signal using the fast Fourier transformation (FFT) process for the desired sensors.

Step 1	Calculate Gear Meshing Frequency (GMF) for gears and bearing frequencies
Step 2	Determine relationship between sensors and components
Step 3	Develop FFT plot for desired sensors in each case

Table 1. Procedure for vibration data preprocessing

2.2. Analytical Diagnosis

The online analytical diagnosis approach identifies the defects or failures in the rotating components from the preprocessed frequency domain data. The developed analytical diagnosis method helps to narrow down the whole frequency spectrum to potential failure modes and their frequencies. The developed method with sideband and kurtosis based online defect detection process the frequency data analytically and the stepwise procedure is shown in Table 2.

The maximum amplitude values of the desired frequencies, the sidebands, and the kurtosis values for the sidebands are determined to calculate the severity factors to formulate the defect severity matrix. The failure modes and their severity levels from the frequency domain signals are determined by the defect severity matrix. The results from the analytical diagnosis are given as inputs to the graphical diagnosis process.

Step 1	Determine maximum amplitude values for sidebands and desired frequency
Step 2	Determine kurtosis values for sidebands
Step 3	Calculate severity factor 1, 2 and 3
Step 4	Formulate defect severity matrix

Table 2. The procedure of analytical diagnosis

2.2.1. Sideband and Kurtosis Analysis

The sidebands are indicators of the failure modes in the frequency spectrum of the rotating components based on their spread on both the sides of the desired frequency. The rising and inequality of the sidebands correspond to component defects, and moreover, the severity of the defect can be identified based on the frequency sideband features, as listed in Table 3.

The height and sharpness of the peak amplitudes in the frequency spectrum are measured by kurtosis. The spread of sidebands on either side of the desired frequency can be analyzed using kurtosis values. The differences in kurtosis values of both sidebands denote the inequality in the sidebands. The kurtosis ratio KR, is the ratio of the left side of jth, the desired frequency is K_{Lj}, to the right side of jth, the desired frequency K_{Rj} is as shown in Eq. 1 Similarly, the ratio of maximum amplitude of the sideband on left and right sides of the jth frequency is determined as AR shown in Eq. 1.

$$KR = \frac{K_{Lj}}{K_{Rj}}; \quad AR = \frac{A_{Lj}}{A_{Rj}} \tag{1}$$

Frequency sideband feature	Severity Level
Rising of sidebands around desired frequency	Low
Unequal sidebands on both sides	Medium
High sideband amplitude than frequency amplitude	High

Table 3. Sideband-based severity definition (Spectra Quest, 2006)

2.2.2. Severity Factors

The different failure modes and their severity are determined from the converted frequency domain signal through analytical sideband and kurtosis analysis. Table 3 shows the different severity levels based on the frequency sideband feature. The three severity factor metrics are developed for online defect detection and they are as follow.

The severity factor 1 (SF_1) ensures equal spread of the sidebands using the kurtosis ratio metric, as shown in Eq. 2. The threshold kurtosis ratio, KR_T, is considered to be 0.6. The value of SF_1<1 denotes the unequal spread of sidebands and vice versa. The severity factor 2 (SF_2) ensures equal maximum amplitude of sidebands on both sides of the desired frequency, as shown in Eq. 3. The threshold amplitude ratio, A_T, is considered to be 0.9. The value of SF_2< 1 denotes the unequal frequency amplitudes on both sides of the sidebands and vice versa.

$$SF_1 = \frac{\text{Min}\,(KR_j, KR_j^{-1})}{KR_T} \tag{2}$$

$$SF_2 = \frac{\text{Min}\,(AR_j, AR_j^{-1})}{A_T} \tag{3}$$

The severity factor 3 (SF_3) ensures that the maximum desired frequency amplitude is higher than the maximum amplitude of the sideband A_{max}, as shown in Eq. 4, where A_F is the maximum amplitude at the desired frequency. The value of $SF_3 < 1$ denotes the frequency amplitude of the sideband, A_{max}, which is higher compared to the desired frequency, A_{max}.

$$SF_3 = \frac{A_F}{\text{Max}\,(A_{Lj}, A_{Rj})} \tag{4}$$

The conditions SF1 ≤ 1, SF2 > 1, and SF3 > 1 show that the component has a low (L) severity defect. The severity factor characteristics of the medium (M) severity defect are SF1 ≤ 1, SF2 ≤ 1, and SF3 > 1 and SF1 > 1, SF2 ≤ 1, and SF3 > 1. Similarly, the high (H) severity defect conditions are SF1 ≤ 1, SF2 ≤ 1, and SF3 ≤ 1; SF1 > 1, SF2 ≤ 1, and SF3 ≤ 1; SF1 ≤ 1, SF2> 1, SF3 ≤ 1, and SF1 > 1, SF2 > 1, SF3 ≤ 1. If all three severity factors are greater than one, then the component has no defect (N). Based on these rules, the severity levels and the failure modes of the components are identified based on the each sensor. The procedure of assigning severity levels are clearly shown as a flowchart in Figure 2. The determined severity level will be assigned to the corresponding component u at level g through sensor m (S_{ugm}). Similarly, FFTs of different sensors are analyzed and their S_{ugm} are determined for all desired components.

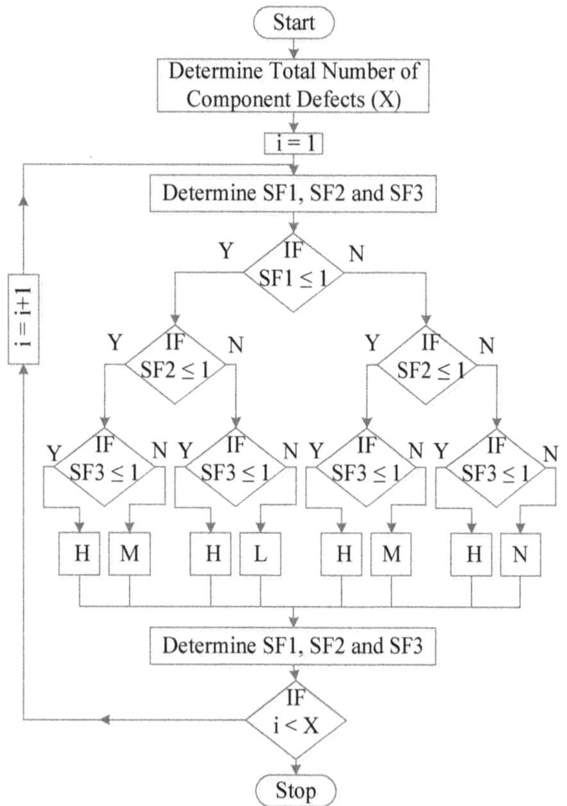

Figure 2. Flowchart of severity level assigning procedure

2.2.3. Severity Defect Matrix

The failure modes and their severity levels of the rotating components based on each sensor are identified with different severity metrics. However, the same defect of the rotating components can be identified by different sensors in and around the component location. Therefore, there is a need for developing a unified metric to make decisions on the failure mode and its severity level. The combination of results of all the components from the different sensors leads to the development of a defect severity matrix. The desired component matrix T is shown in the Eq. 5.

$$T = \begin{pmatrix} \text{Desired Component 1} \\ \text{Desired Component 2} \\ \vdots \\ \text{Desired Component } U \end{pmatrix} \tag{5}$$

The desired component matrix and the severity factor levels of all the components are utilized for developing a defect severity matrix. The severity ratio of component u at severity level g, S_{ug} is represented as Eq. 6, where g represents the different severity levels, such as low, medium, and high, and S_{ugm} represents the severity level of component u at level g through sensor m.

$$S_{ug} = \sum_{m=1}^{M} S_{ugm} \Big/ \sum_{g=1}^{3}\sum_{m=1}^{M} S_{ugm} \tag{6}$$

The defect severity matrix, DS, represents the defect component and its severity level in the matrix format as shown in Eq. 7, where rows of the matrix represent each desired component and columns represent the severity level of the components, such as low, medium and high. The analytical results are further fine-tuned using the graphical diagnosis process.

$$DS = \begin{pmatrix} S_{11} & S_{12} & S_{13} \\ \vdots & \ddots & \vdots \\ S_{U1} & S_{U2} & S_{U3} \end{pmatrix} \tag{7}$$

2.3. Graphical Diagnosis

The unified defect severity matrix results provide the initial insights about the component defects and their severity levels. There is the possibility of false identifications in the analytical methodology due to the overlap of different frequencies and their harmonic levels. Therefore, there is a need for verification of identified component defects graphically. The frequency spectrum of the predetermined component defects are verified graphically based on the sideband amplitudes and their spread. The developed two stage fault detection framework is demonstrated with NREL

Round-Robin gearbox condition monitoring in the next section.

3. GEARBOX DIAGNOSIS ROUND-ROBIN STUDY

The NREL Round-Robin gearbox reliability collaborative (GRC) test turbine drivetrain (Sheng et al., 2011, Sheng, 2012) is shown in Figure 3. The time domain vibration signals from the sensors placed on the gearbox are utilized for condition monitoring to determine the defects in gears and bearings of the GRC test gearbox.

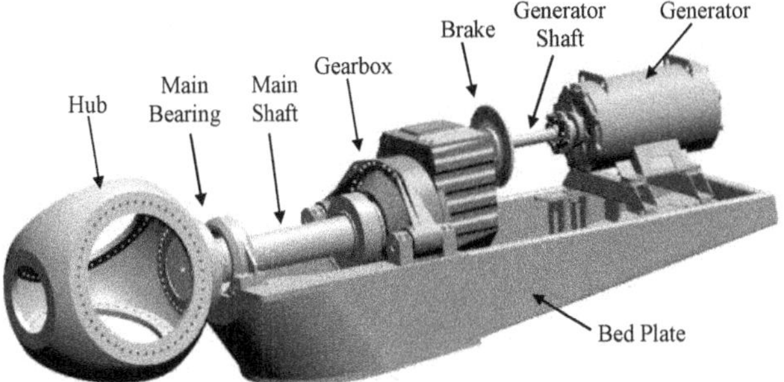

Figure 3. GRC Test Turbine Drivetrain (Sheng, 2012)

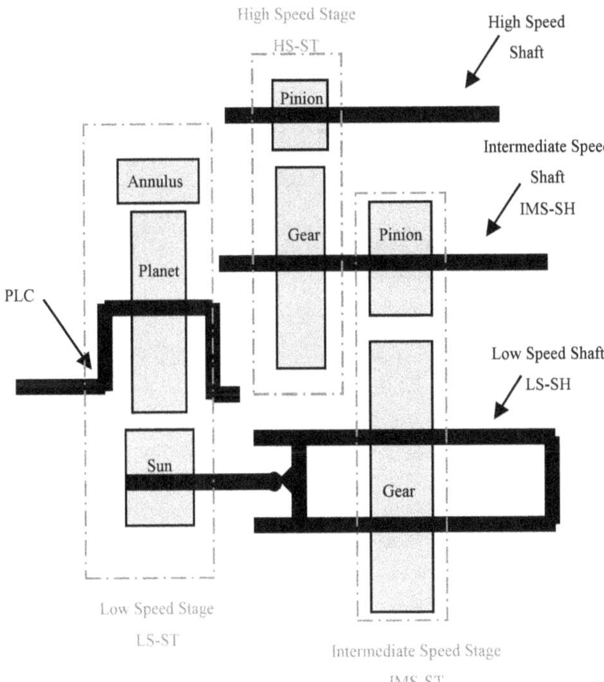

Figure 4. GRC gearbox internal nomenclature and abbreviations (Sheng, 2012)

The GRC gearbox had an overall gear ratio of 1:81.491. It has one planetary stage and two parallel stages namely a High Speed Stage (HS-ST) and an Intermediate Speed Stage (IS-ST) as shown in Figure 4. The main shaft was connected to planetary arm of the gearbox and the high speed pinion of the gearbox was geared to the generator. The experiment was conducted at two speed levels 1200 RPM and 1800 RPM.

Test case	Electric Power (% of rated)	Duration (min)	Speed (rpm)
CM_2a	25%	10	1200
CM_2b	25%	10	1800
CM_2c	50%	10	1800

Table 4: Test case data description

The three test cases were conducted at different power ratings and different speed levels as shown in the Table 4, in which HSS denotes the high speed shaft and LSS denotes the low speed shaft. The data of each sensor placed in the gearbox was collected as ten 1 minute dataset for each test case.

Figure 5. Test gearbox high speed stage gear damage (Sheng, 2012)

Damage #	Component / Location	Mode
1	HSS Gear Set	Scuffing
2	HSS Downwind Bearing (DWB)s	Overheating
3	ISS Gear Set	Fretting Corrosion, Scuffing and Polishing Wear
4	ISS Upwind Bearing (UWB)	Assembly damage, plastic deformation, scuffing and contact corrosion
5	ISS DWB	Assembly damage, plastic deformation and dents
6	Annulus/Ring Gear, or Sun Pinion	Scuffing and polishing and fretting corrosion
7	Planet Carrier UWB	Fretting Corrosion

Table 5: Actual Damages in the WT Gearbox (Sheng, 2012)

The condition monitoring of the gearbox was carried out by mounting 12 sensors at various locations around the gearbox. The vibration data was collected at 40 kHz per channel using a National Instruments PXI -4472B high speed data acquisition system (DAQ) and further details about the experimental setup can be found in (Sheng, 2012).The NREL provided the real damage results for the GRC gearbox (Errichello & Muller, 2012) and those damage selected for vibration analysis algorithm performance evaluation in the Round Robin study are listed in the Table 5. Figure 5 shows the real damage on high speed gear.

4. DATA PREPROCESSING

The Round-Robin study involves three speed stages: low speed (LS-ST), intermediate speed (IS-ST) and high speed (HS-ST). Among the total sensors from the Round-Robin gearbox, the desired sensors for the analysis of LS, IS and HS are AN 3 and AN 5 to AN 10.

The relationships between these sensors and the components are determined based on the location and proximity to the rotating components, as listed in the Table 6.

Sensor Name	Gear	Shaft	Bearing	Damage
AN3-Planet radial 180	Planet gear & Sun Pinion	Planet arm	Planet carrier UWB and DWB	Planet gear defect, Planetary arm bearing defect, planet bearing defect
AN5-LSS radial	IS-ST gear	LSS	LSS UWB and DWB	LSS bearings defect and ISS gear defect
AN6-ISS radial	IS-ST pinion & HS gear	ISS	ISS	ISS bearings defect, IS pinion defect, HS gear defect
AN7-HSS radial	HS-ST pinion	HSS	HSS	HSS bearings defect, HS pinion defect
AN8-HSS radial	HS-ST pinion	HSS	HSS UWB	HS pinion defect, HSS UWB defect
AN9-HSS rear radial	HS-ST pinion	HSS	HSS DWB	HS pinion defect, HSS DWB defect
AN10-Carrier Rear Radial	Planet gear & Sun Pinion	Planet arm	Planet UWB and DWB	Planet gear defect, planetary arm bearing defect

Table 6. Sensor and component relationship

Gear element	# of Teeth	Speed (rpm)	GMF (Hz)	SRF (Hz)
Ring gear	99	Fixed	NA	NA
Planet gear	39	14.74	29.45	0.25
Sun pinion	21	84.15		1.4025
LSS gear	82		115	
Intermediate pinion	23	300		5
ISS gear	88		440	
HSS pinion	22	1200		20

Table 7. Desired Gear Frequencies at 1200 rpm

Location	Type	Speed (rpm)	BPFI (Hz)	BPFO (Hz)	BSF (Hz)
Planet Carrier	UWB	14.74	6.11	5.44	2.11
	DWB		6.65	5.88	1.99
Planet	DWB & UWB	45.31	7.81	5.78	2.46
Low speed Shaft	UWB	84.15	42.19	37.75	13
	DWB		30.6	26.9	10.15
Inter mediate Shaft	UWB	300	49	35.9	15.6
	DWB		84.6	70.4	26
High Speed Shaft	UWB	1200	197	143	62.5
	DWB		228	172	66.5

Table 8. Desired Bearing Frequencies at 1200 rpm

The desired gear and bearing frequencies are determined and listed in the Tables 7 and 8 respectively, where UWB and DWB refer to the upwind bearing and downwind bearings respectively. The defects from the bearings and gears can be identified from their corresponding desired frequency amplitudes in the frequency spectrum. The FFT converts the time domain vibration signal into a frequency domain signal and helps in analyzing each desired frequency based on its amplitude and its harmonics.

5. ANALYTICAL DIAGNOSIS

The analytical diagnosis process identifies the bearing and gear defects in the GRC gearbox from the preprocessed frequency domain data. The developed analytical diagnosis method is utilized to determine the severity factors and defect matrix and confine the whole frequency spectrum into different potential failure modes. Based on the rules specified in Section 2.2 about the severity levels and the failure modes of the components, severity factors 1, 2 and 3 for the different gears and bearings of the GRC gearbox are determined. The severity factor analysis of sensor AN 6 for 2a case is listed in the Table 9.

Component	ISS gears	ISS UWB	ISS DWB	HSS UWB	HSS DWB
Desired Frequency	GMF	BPFI	BPFO	BPFO	BPFI
SF1	0.49	0.98	0.86	0.58	0.72
SF2	0.97	0.73	0.42	1.06	0.58
SF3	0.46	2.85	2.94	0.23	1.67
Low	0	0	0	0	0
Medium	0	1	1	0	1
High	1	0	0	1	0

Table 9. Severity factor analysis of sensor AN 6 for Case 2a

$$T = \begin{pmatrix} \text{Planet Gear and Sun Pinion} \\ \text{Planet Carrier upwind bearing} \\ \text{Planet Carrier downwind bearing} \\ \text{Planet up and downwind bearing} \\ \text{ISS Gear and ISS Pinion} \\ \text{HSS Gear and HSS Pinion} \\ \text{LSS upwind bearing} \\ \text{LSS downwind bearing} \\ \text{ISS upwind bearing} \\ \text{ISS downwind bearing} \\ \text{HSS upwind bearing} \\ \text{HSS downwind bearing} \end{pmatrix} \tag{8}$$

$$DS = \begin{pmatrix} 0 & 0 & 0 \\ 0 & 0 & 0 \\ 0 & 0 & 0 \\ 0 & 0 & 0 \\ 0.33 & 0.34 & 0.33 \\ 0.50 & 0.50 & 0 \\ 0 & 0 & 0 \\ 0 & 0 & 0 \\ 0.67 & 0.33 & 0 \\ 0.50 & 0.50 & 0 \\ 0.50 & 0.33 & 0.17 \\ 0.60 & 0.40 & 0 \end{pmatrix} \tag{9}$$

The desired component matrix T for the GRC gearbox is shown in the Eq. 8. The unified DS matrix of the Round-Robin study is determined by using the proposed severity factors and it is shown in the Eq. 9. The analytical diagnosis results indicated that there are no defects in the LSS UWB and DWBs. The values of DS matrix show that the IS gear and pinion each have a high severity defect; the HS gear and pinion each have a medium severity defect; and the bearings ISS upwind and downwind, and HSS downwind have a medium severity defect. The HSS upwind severity values show that it has a high severity defect. These defect results are used as an input to the graphical diagnosis process and the analytical results are further fine-tuned.

6. GRAPHICAL DIAGNOSIS

The frequency spectrum of the predetermined component defects are verified graphically based on the sideband amplitudes and their spread. The graphical diagnosis results are shown as frequency plots (Figures 6 - 10) for the identified damages.

Figure 6 shows the HSS gear GMF (660 Hz) and its sidebands of sensor AN7 HSS Radial. The unequal high sidebands on both sides with GMF maximum amplitude show that there is a high severity failure in HSS gear. Similarly in Figure 7, the second harmonic of the BPFO (172 Hz) of HSS DWB at (344 Hz) in sensor AN 6 ISS radial. The A_{max} of right sideband is almost two times the A_{max} of left sideband and moreover, the high amplitude of the right sideband is almost eight times the high amplitude of the desired frequency. These inferences from the figure prove that there is a high severity failure in the outer raceway. Since the sideband amplitudes are found in the second harmonic, there is a chance of misalignment of the bearing. Figure 8 shows the ISS gear GMF (115 Hz) and its sidebands of sensor AN5 LSS Radial. The rise of sidebands on both sides of GMF maximum amplitude show that the LSS gear is in early stage of failure.

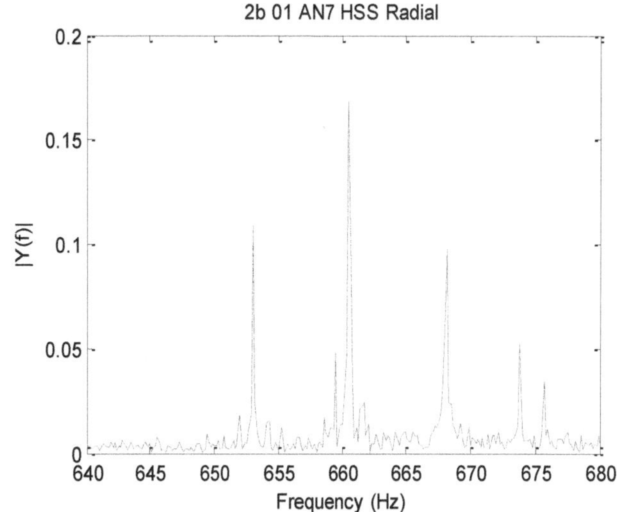

Figure 6. Damage 1 HSS Pinion

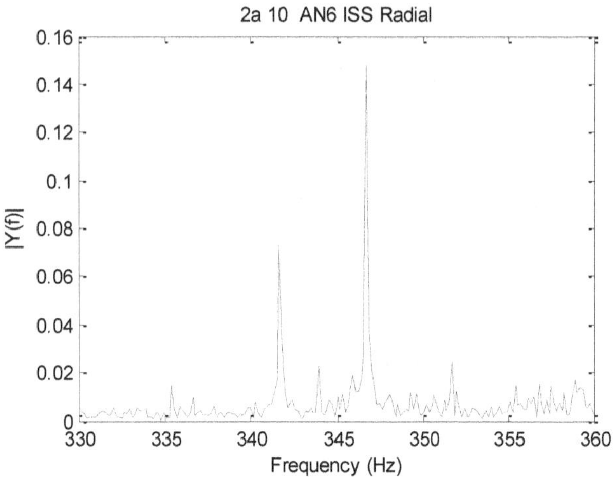

Figure 7. Damage 2 HSS DWB BPFO

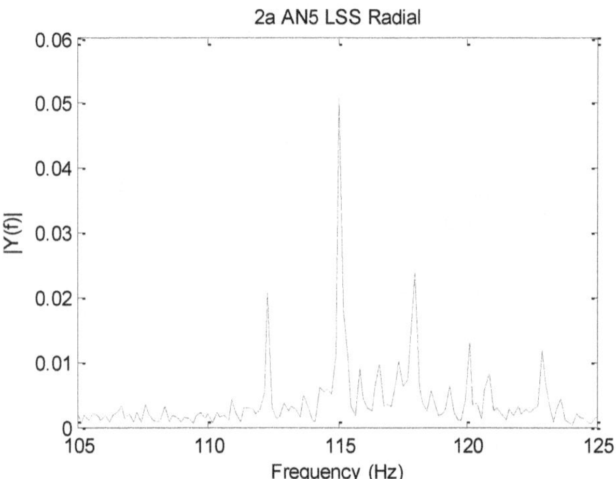

Figure 8. Damage 3 ISS Gear AN5 LSS Radial

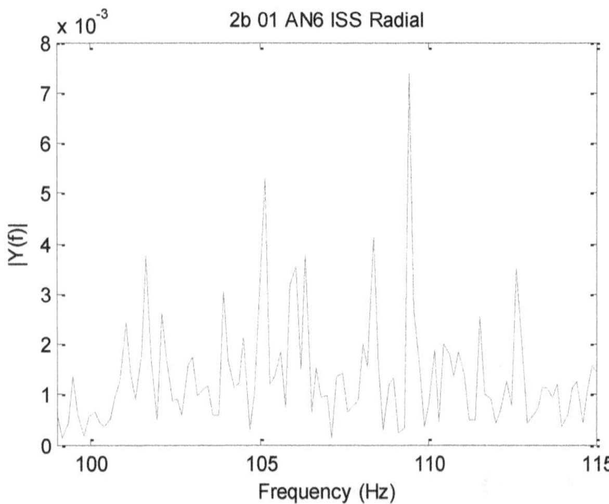

Figure 9. Damage 4 ISS UWB BPFI

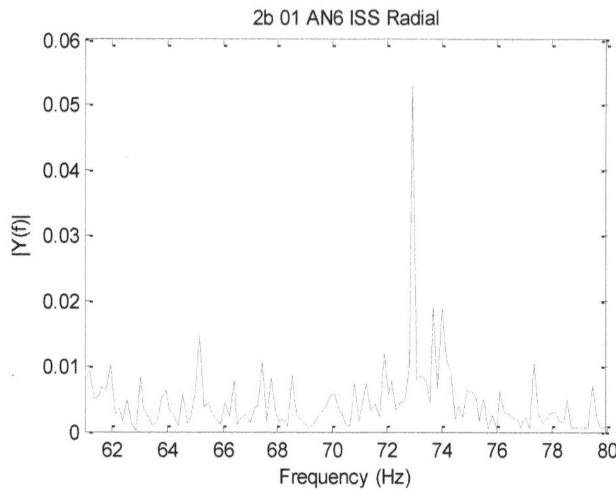

Figure 10. Damage 5 ISS DWB BPFO

The damage of ISS UWB is verified graphically using the frequency plot shown in Figure 9. The sensor value used is AN6 ISS radial and the BPFI is 53.8 Hz. There are rising of sideband amplitudes around the amplitude of BPFI in second harmonic frequency. The frequency plot clearly shows that there is early inner raceway failure and bearing misalignment. Similarly in Figure 10, the BPFO of ISS DWB at 73.7 Hz in sensor AN 6 ISS radial. There is rising of sideband amplitudes around one side of maximum amplitude of BPFO. These inferences from the figure prove that there is a high severity failure in the outer raceway. Thus, the component defects are identified graphically and the results are discussed in the next section.

7. DIAGNOSIS RESULTS

The results from the online analytical defect detection method are used as an input to the graphical diagnosis. The failure modes and their severity levels from the frequency domain signals are verified graphically, and the results are unified to the component level, with their corresponding severity levels, as shown in the Table 10.

Damage	Component	Mode	Severity
1	HSS pinion	Gear tooth failure of HSS pinion	High
2	HSS DWB	OR failure and bearing misalignment	High
3	ISS Gear	Early stages of gear failure	Low
4	ISS UWB	IR failure and bearing misalignment	Medium
5	ISS DWB	OR failure	High

Table 10. Gearbox fault diagnosis blind analysis results

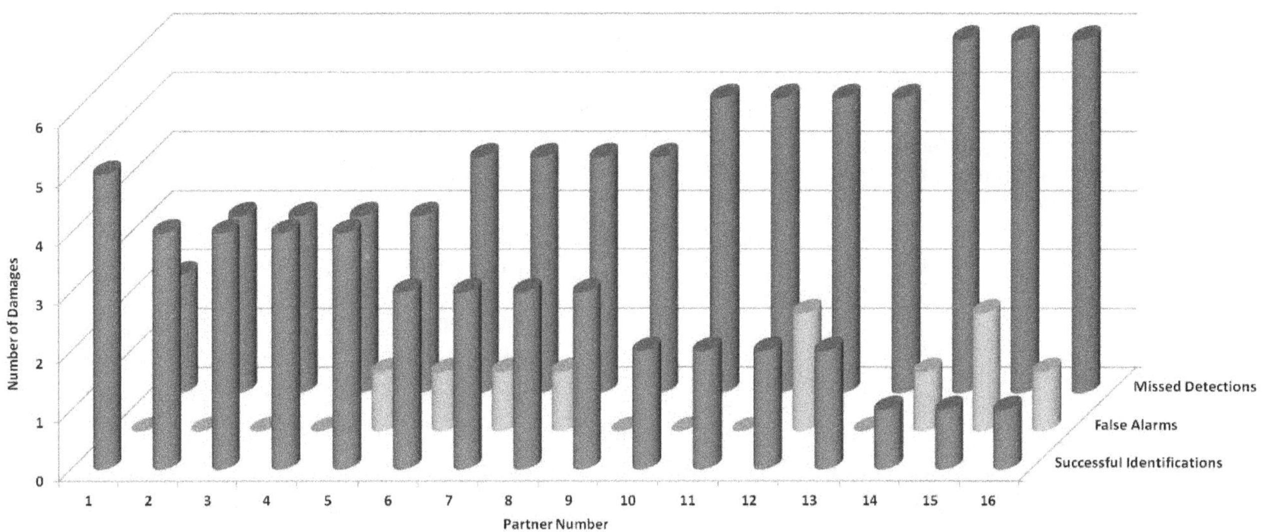

Figure 11. Comparison of Damage Results (Sheng, 2012)

The tabulated results were identified before the receiving knowledge of the actual failure modes from NREL. The possible number of failures that can be identified from the vibration analysis for this Round-Robin study is about seven as listed in Table 5. The developed two stage condition monitoring approach identified five failures and their severity. Moreover, the failures identified by the proposed vibration analysis approach do not have any false identification. The blind analysis results were submitted to NREL gearbox round robin competition. The proposed approach outperformed the other participants' approaches and the comparison results are shown in Figure 11. In the result comparison chart, our approach is masked as partner 1 which successfully picked up five faults with accurate severity levels without producing any false alarm in the blind analysis. Further details on the NREL gearbox round robin study can be found in the reference (Sheng, 2012).

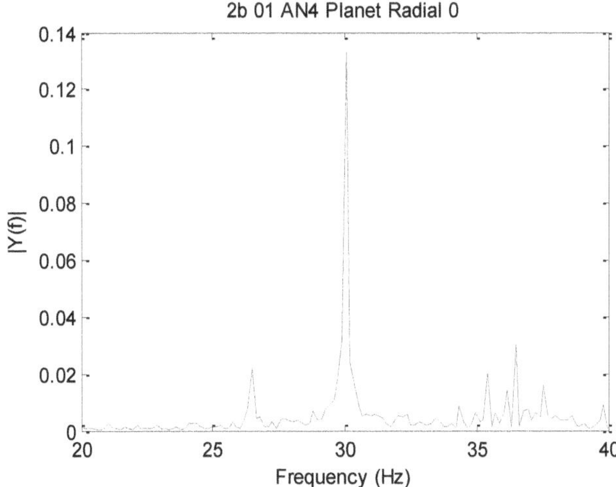

Figure 12. Damage 6 Planet Gear / Sun Pinion

After the blind analysis, the damage results are provided by NREL. The defects that were not identified in the blind analysis are defects in planet carrier upwind and Annulus/Sun pinion. These defects are verified graphically and able to identify the defect of Planet gear/sun pinion with low severity as shown in Figure 12 (due to high maximum amplitude at desired frequency and rise of sidebands around the desired frequency), but the defect of planet carrier upwind was not identified in Figure 13.

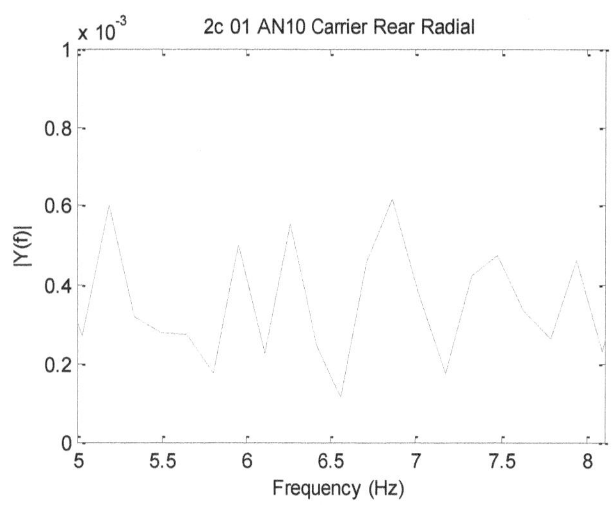

Figure 13. Damage 7 Planet Carrier UWB BPFO

Damage	Component	Mode	Severity
6	Planet Gear/Sun Pinion	Early stages of gear failure	Low

Table 11. Additional faults identified during post-analysis

The additional defect components detected during the post-result analysis is listed in the Table 11. Since the desired frequencies of the bearings planet upwind and downwind, and planet carrier upwind an downwind are less than 10 Hz, there is possibility of overlapping of sidebands of these desired frequencies. Therefore, it is not quite effective to trace the defect of planet carrier upwind through analytically or graphically.

8. CONCLUSION

This research showed that the developed vibration based two stage fault detection framework that integrates both analytical diagnosis and graphical diagnosis is quite effective for analyzing gear and bearing faults in WT transmissions, as proved by the CM Round-Robin study results. The proposed methodology picked up most faults (5 out of 7) with correct severity levels in the blind analysis, and more importantly our method does not produce any false alarm. Moreover, the post result analysis was able to identify one more fault.

During the study, we found that it is quite useful in identifying an initial set of potential failure modes using analytical diagnosis method, which will substantially reduce the work load in processing massive high frequency vibration data. With the preliminary results from the analytical diagnosis method, the graphical verification can be extremely useful to assure the correct diagnosis and avoid potential false identifications.

With lessons learned on this CM Round Robin study, the research on WT condition monitoring can be further carried out to failure prognostics through continuous condition monitoring and failure prediction, which will ultimately lead to automation of wind farm maintenance decision-making process to reduce the maintenance costs. Further study could also be developing a complex system design framework that can leveraging the results from this study to quantify the functionality, reliability, and cost/benefits of condition monitoring techniques and integrate them into a system-level WT design practice, which may serve a fundamental solution of enhancing reliability and reducing WT life cycle cost.

ACKNOWLEDGEMENT

This research is partially supported by National Science Foundation (CMMI-1200597) and Kansas NSF EPSCoR program and Wichita State University through the University Research Creative Project Award.

REFERENCES

Antoni, J. (2002). Differential Diagnosis of Gear and Bearing Faults. *Journal of Vibration and Acoustics*, vol. 124, no. 2, pp. 165-171.

Byon, E., Perez, E., Ntaimo, L., & Ding, Y. (2010). Simulation of Wind Farm Operations and Maintenance Using DEVS. Simulation, pp. 1-25.

Ebeling, C. E. (1997). *An Introduction to Reliability and Maintainability Engineering*. Long Grove, IL: Waveland.

Errichello, R., & Muller, J. (2012). *Gearbox Reliability Collaborative Gearbox 1 Failure Analysis Report.* NREL/SR-5000-530262. National Renewable Energy Laboratory, Golden, CO.

Gebraeel, N.; Lawley, M.; & Liu, R. (2002). Vibration-based condition monitoring of thrust bearings for maintenance management. *Intelligent Engineering System Through Artificial Neural Network*, vol. 12, pp. 543–551.

Harris, T.A. (2001). *Rolling Bearing Analysis*. 4th edition. New York, NY: John Wiley & Sons, York, pp. 993-1000.

Lebold, M., McClintic, K.., Campbell, R., Byington,C., & Maynard, K. (2000). Review of Vibration Analysis Methods for Gearbox Diagnostics and Prognostics. *Proceedings of the 54th Meeting of the Society for Machinery Failure Prevention Technology*, May 1-4, Virginia Beach, VA.

Lu, W., & Chu, F. (2010). Condition Monitoring and Fault Diagnostics of Wind Turbines. *Proceedings of Prognostics and Health Management Conference*, pp. 1-11.

McFadden, P.D., & Smith, J.D. (1984). Vibration Monitoring of Rolling Element Bearings by the High-Frequency Resonance Technique-A Review. *Tribology International*, vol. 17, no. 1, pp. 3-10.

Nielsen, J., & Sorensen, J. (2010). On Risk-Based Operation and Maintenance of Offshore Wind Turbine Components. *Reliability Engineering and System Safety*, vol. 96, no. 2011, pp. 218-229.

Nilsson, J., & Bertling, L. (2007). Maintenance Management of Wind Power Systems Using Condition Monitoring Systems- Life Cycle Cost Analysis for Two Case Studies. *IEEE Transactions on Energy Conversion*, vol. 22, no. 1, pp. 223-229.

Randall, R.B. (2011). *Vibration-based Condition Monitoring*. Hoboken, NJ: Wiley, 2011.

Randall, R.B., & Antoni, J. (2011). Rolling Element Bearing Diagnostics-A Tutorial. *Mechanical Systems and Signal Processing*, vol. 25, no. 2, pp. 485-520.

Shi, W., Wang, F., Zhuo, Y., & Liu, Y. (2010). Research on Operation Condition Classification Method for Vibration Monitoring of Wind Turbine. *Proceedings of Power and Energy Engineering Conference APPEEC*, Asia Pacific, pp. 1-6.

Sheng, S. (2012). *Wind Turbine Gearbox Condition Monitoring Round Robin Study–Vibration Analysis.* NREL/TP-5000-54530. Golden, CO: National Renewable Energy Laboratory, July, 2012.

Sheng, S., Link, L., LaCava, W., van Dam, J., McNiff, B., Veers, P., Keller, J., Butterfield, S., & Oyague, Francisco. (2011). *Wind Turbine Drivetrain Condition Monitoring During GRC Phase 1 and Phase 2 Testing.* NREL/TP-5000-52748. National Renewable Energy Laboratory, Golden, CO.

Spectra Quest Tech Note. (2006). *Analyzing Gearbox Degradation Using Time-Frequency Signature Analysis.*

Tamilselvan, P., Wang, P., & Twomey, J. (2012). Quantification of Economic and Environmental Benefits for Prognosis Informed Wind Farm Operation and Maintenance. *62nd Annual IIE Industrial Systems and Engineering Research Conference* (ISERC 2012), Orlando, FL, USA.

Tamilselvan, P., Wang, Y., & Wang, P. (2012). Optimization of Wind Turbines Operation and Maintenance Using Failure Prognosis. *IEEE 2012 Prognostics and Health Management* (PHM 2012), Denver, CO, USA.

Tamilselvan, P., Wang, P., & Jayaraman, R. (2012). Diagnostics with Unexampled Faulty States Using a Two-Fold Classification Method. *2012 IEEE International Conference on Prognostics and Health Management*, 18-21, June, Denver, CO, USA.

Tamilselvan, P., & Wang P. (2012). A Hybrid Inference Approach for Health Diagnostics with Unexampled Faulty States. AIAA 2012-1784, *53th AIAA/ASME/ASCE /AHS/ASC Structures, Structural Dynamics, and Materials Conference,* 23-26 April 2012, Honolulu, Hawaii, USA.

Tamilselvan, P., & Wang, P. (2013). Failure Diagnosis Using Deep Belief Learning based Health State Classification. *Reliability Engineering & System Safety*, vol. 115, no. 2013, pp. 124-135.

Yang, W., Tavner, P., & Wilkinson, M. (2008). Wind Turbine Condition Monitoring and Fault Diagnosis using Both Mechanical and Electrical Signatures. *Proceedings of IEEE international conference on Advanced Intelligent Mechatronics*, pp. 1296-1301.

BIOGRAPHIES

Prasanna Tamilselvan received his bachelor degree in Mechanical Engineering in 2007 from Anna University, India and his Master degree in 2010 in Industrial and Manufacturing Engineering from Wichita State University, USA. He is currently pursuing his doctorate in the Industrial and Manufacturing Engineering department at Wichita State University. He received IEEE PHM Best Paper Award in 2012. His expertise is in the areas of machine learning applications in failure diagnostics, prognostics and health management (PHM).

Dr. Pingfeng Wang is currently an assistant professor in the industrial and manufacturing engineering department at Wichita State University. Dr. Wang got his bachelor degree in Mechanical Engineering in 2001 from The University of Sci. & Tech. in Beijing, China; his Master degree in 2006 in Applied Mathematics from Tsinghua University in China, and his Ph.D. degree in Mechanical Engineering from the University of Maryland at College Park in 2010. His expertise is in the areas of reliability analysis and design for complex systems, structural health prognostics, and condition-based maintenance. He has written over 50 research articles in these areas, including one ASME Best Paper Award in 2008 and IEEE Best Paper Award in 2012. Since joining WSU in 2010, he has focused his research on developing methodologies to integrate the reliability based complex system design with health prognostics to improve the overall system resilience and sustainability.

Dr. Shuangwen (Shawn) Sheng is a senior engineer at National Renewable Energy Laboratory (NREL). He has B.S. and M.S. degrees both in electrical engineering and a Ph.D. in mechanical engineering. Shawn is currently leading wind turbine condition monitoring, failure database and wind plant operation & maintenance research at NREL. Shawn also has experience in mechanical and electrical system modeling and analysis, soft computing techniques, and automatic control. He has published his work in various journals, conference proceedings, and book chapters.

Janet M. Twomey received her BS, MS and PhD in industrial engineering from the University of Pittsburgh. Dr Twomey is currently a professor of Industrial and Manufacturing Engineering at Wichita State University, Wichita, Kansas. From 2001 to 2004 she held the position of program officer for Manufacturing Enterprise Systems, at the National Science Foundation (NSF). Prior going to NSF her research was in computational intelligence and sparse data. For this work she received an NSF CAREER Award. Dr. Twomey's current research applies life cycle thinking to analyze energy use in the industrial and commercial sectors.

A New Acoustic Emission Sensor Based Gear Fault Detection Approach

Yongzhi Qu[1], Eric Bechhoefer[2], David He[1], and Junda Zhu[1]

[1]*Department of Mechanical and Industrial Engineering, University of Illinois at Chicago, Chicago, IL, 60607, USA*
davidhe@uic.edu

[2]*NRG Systems, Hinesburg, VT, 05461, USA*
erb@nrgsystems.com

ABSTRACT

In order to reduce wind energy costs, prognostics and health management (PHM) of wind turbine is needed to reduce operations and maintenance cost of wind turbines. The major cost on wind turbine repairs is due to gearbox failure. Therefore, developing effective gearbox fault detection tools is important in the PHM of wind turbine. PHM system allows less costly maintenance because it can inform operators of needed repairs before a fault causes collateral damage happens to the gearbox. In this paper, a new acoustic emission (AE) sensor based gear fault detection approach is presented. This approach combines a heterodyne based frequency reduction technique with time synchronous average (TSA) and spectral kurtosis (SK) to process AE sensor signals and extract features as condition indictors for gear fault detection. Heterodyne techniques commonly used in communication are used to preprocess the AE signals before sampling. By heterodyning, the AE signal frequency is down shifted from MHz to below 50 kHz. This reduced AE signal sampling rate is comparable to that of vibration signals. The presented approach is validated using seeded gear tooth crack fault tests on a notational split torque gearbox. The approach presented in this paper is physics based and the validation results have showed that it could effectively detect the gear faults.

1. INTRODUCTION

The largest variable cost to owners and operators of wind turbines is unscheduled maintenance. PHM has been shown to be technique that can successfully reduce both scheduled and unscheduled maintenance. PHM system allows better maintenance practices as well as less costly maintenance because it can give indications and warnings prior to collateral damage occurring. In wind turbines, this might be the difference between an up tower maintenance effect or a down tower event, which requires a crane (a large fixed expense). Or, it could be the difference between refurbishing a gearbox instead of replacing it. Therefore the development of effective gearbox fault detection tools is important to the PHM of wind turbine. Currently, vibration is the most widely used tool in diagnosis of machine fault, such as: shaft, gears, and bearings. Common vibration sensors include accelerometer, displacement and velocity sensors. However, vibration signals have some drawbacks when it comes to detecting the incipient machine faults at low frequency. Accelerometers measure the second derivative of the displacement. For low frequency components, such as the carrier or planets, which operate below 2 Hz, even damage components may develop acceleration below the noise floor of the sensor. AE, on the other hand, does not measure acceleration and is not a function of displacement: it is independent of shaft rate. This has been observed in (Al-Ghamd & Mba, 2006), where early fault signatures were not present in vibration data, but was detected by AE. If these faults could be detected at an early stage, significant maintenance costs could be saved. In this paper, the development of a new AE sensor based gear fault detection approach is presented.

AE is commonly defined as transient elastic waves within a material, caused by the release of localized stress energy. It is produced by the sudden internal stress redistribution of material because of the changes in the internal structure of the material. Possible causes of these changes are crack initiation and growth, crack opening and closure, or pitting in various monolithic materials (gear, bearing material) or composite materials (concrete, fiberglass). Thus the ability to detect AE can be used to give diagnostics indications of component health. The challenges of using AE sensor include: the frequency of the output signals from AE sensor is generally high, even as high as several MHz. Thus a high sampling rate between 2MHz and 10MHz is normally needed for AE data collection. Other challenges include the high data volume, complicated feature of AE signals, which make the data processing highly difficult.

A number of methods have been developed for AE signal analysis, but few techniques have been successful in application. Most of the research into AE for machine condition monitoring focuses on time domain features, such as: peak, total energy, standard deviation, median, AE counts, root mean square (RMS) voltage and duration (Mba, 2003). These features all relate to the absolute energy levels of the measured signals. As the absolute energy could vary from one machine to another, or vary at different locations on the same machine, the effectiveness of these features may be compromised. Consequently, these features are not ideal for fault detection purpose.

Gao *et al.* (2011) proposed a wavelet transform based method to analyze AE signals, which could act as a supplement redundant method for vibration test. He and Li (2011) developed a data mining based method to classify the condition indicators derived from different AE data to detect fault. Later, Li and He (2012) introduced an EMD-based AE feature quantification method. In their work, successful detection of gear fault was achieved on AE data sampled at a rate as low as 500 kHz. Artificial neural networks (ANN) have also been adopted for AE signal classification. In (Pandya *et al.*, 2013), a supervised learning process was developed after EMD decomposition for bearing fault detection using AE signals.

In (Kilundu *et al.*, 2011), cyclostationarity analysis was compared with traditional envelope spectrum. It was proposed that the cyclic spectral correlation, a tool for characterizing cyclostationarity, was more efficient compared with envelope spectrum for bearing fault diagnosis. A comparison study between vibration and AE based on spectral kurtosis was reported in (Eftekharnejad *et al.* 2011). The conclusion was that by using AE features with spectral kurtosis, the fault could be detected at an earlier stage. Al-Balushi *et al.* (2002) developed an energy based feature, named energy index. By calculating the cumulative of the square root of the energy index, the tooth fault could be identified as high peak values.

There are still a number of issues in the reviewed methods. First, the AE data was collected at very high sampling frequency, typically 2~5MHz. Second, these methods tried to detect the gear faults using data driven approach rather than physics based approach. Data driven approaches normally rely on complicated computation algorithms such as EMD and wavelet analysis to compute the AE features.

This research is aimed to address the AE sensor based gear fault detection problem using physics based methods, similar to those techniques applied to vibration signal to analyze the AE data. For gear fault detection, it is common to use time synchronous average (TSA) to extract the gear signals from the raw signals. Generally, the signals collected from a gearbox contain broadband and non-synchronous noise, such as other shaft/gear pairs, or bearing tones. TSA is able to reduce the random and non-synchronized noise from other sources while enhancing the synchronous signals from the gear of interest. In order to perform TSA on the AE signals, high sampling frequency and large data volume need to be addressed. This requires a frequency demodulation/decimation technique be applied prior to data sampling and signal analysis. The heterodyne technique is proposed to demodulate the signals and down shift the signals to a low frequency range. After the heterodyne is applied, the AE signal can be sampled at frequencies comparable to that of vibration analysis. If a phase reference is available, the TSA could be generated.

The remainder of the paper is organized as follows. Section 2 illustrates in details the methodology. In Section 3, the setup of the experiments for the validation of the methodology is explained. Section 4 presents the analysis results of the experiments and illustrates how a fault is identified. Finally, Section 5 concludes the paper.

2. THE METHODOLOGY

The methodology will be illustrated in 4 parts. The first part discusses the heterodyne technique. TSA will be briefly reviewed in the second part. The third part explains how the kurtogram could be applied and idea of designing an optimal band pass filter from spectral kurtosis. Then, condition indicators for gearboxes diagnosis are discussed.

2.1. The Heterodyne Technique

Current AE signal processing steps are given in Figure 1.

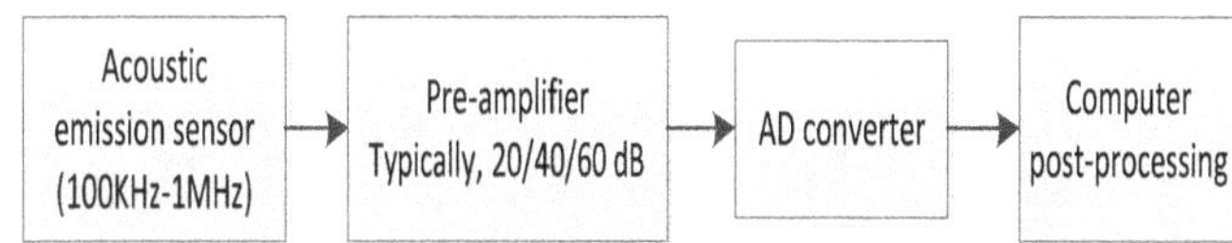

Figure 1. Traditional AE signal acquisition and preprocessing procedure

In a traditional AE signal processing procedure, all of the data is collected and stored to computer without any signal processing. There are two disadvantages associated with this procedure. First, it increases the data acquisition cost. Second, it relies on the computer to process the resulting large data set.

When taking a further look at an AE signal, one will find that the AE signal is virtually a carrier signal for the fault signal. The information of interest is related to the load signal, not the high frequency AE carrier signal. In order to get the load signal, demodulation process is required before sampling. As for rotational machine fault detection, the faults are mostly related to the rotational speed, which is generally in the low frequency range. Thus, the information of interest is related to the low frequency load signals, not the high frequency AE carrier signal. This low frequency load information is recovered through a demodulation process. The demodulation process is similar to information retrieval in an amplitude/phase modulated radio frequency signal. The carrier signal of a typical AM radio signal is several MHz, while the information modulated onto that signal is audio signal of a couple of kHz. After demodulating the carrier using an analog signal conditioning circuit, the acquisition system can then be sampled at audio frequency (10s of kHz). This signal processing can then be performed at lower cost with an analog circuit rather than a high speed analog to digital converter and the associated computation power required to process the large data set resulting from the high sample rate.

The AE signal demodulator implemented in this paper work similarly to a radio quadrature demodulator: shifting the carrier frequency to baseband, followed by low pass filtering. The technique applied here is called heterodyne. Mathematically, heterodyning is based on the trigonometric identity. For two signals with frequency f_1 and f_2, respectively, it could be written as

$$sin(2\pi * f_1 * t) * sin(2\pi * f_2 * t)$$
$$= \frac{1}{2}cos[2\pi(f_1 - f_2)] - \frac{1}{2}cos[2\pi(f_1 + f_2)] \quad (1)$$

where, f_1 is the carrier frequency, f_2 is the demodulator's reference input signal frequency. This process could be explained with a simple example.

For example, let $f_1 = 4$ Hz and $f_2 = 5$ Hz, note $y_1 = sin(2\pi * 4 * t)$ and $y_2 = sin(2\pi * 5 * t)$. Take their multiplication as $Y = y_1.* y_2$, as shown in Figure 2.

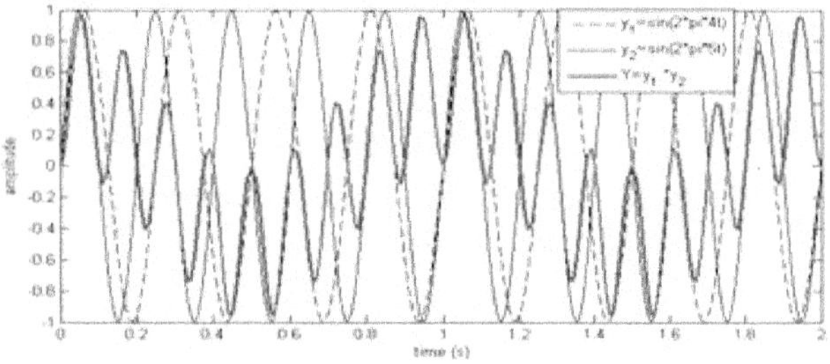

Figure 2. The multiplication of two sinusoid signals

The modulated signal is then low pass filtered to reject the high frequency image at frequency $(f_1 + f_2)$, as shown in Figure 3.

A detailed discussion of the heterodyne technique applied on the raw AE signal is given in the following. In general, amplitude modulation is the major modulation form for AE signal. Although, frequency modulation and phase modulation could present in the AE signal potentially, they are considered trivial and will not be discussed here. The amplitude modulation function is given in Eq. (2).

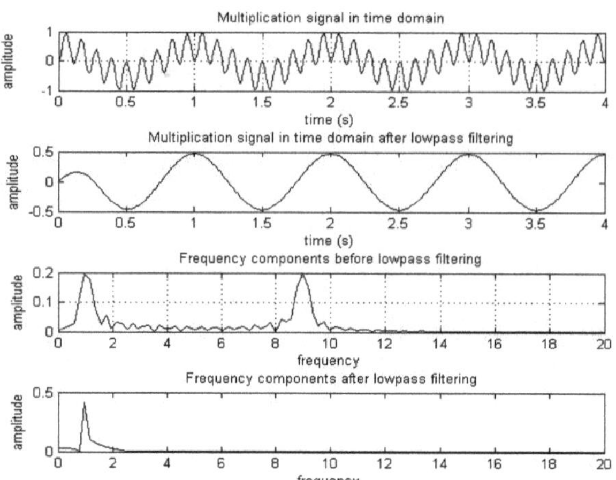

Figure 3. The extraction of the heterodyned signal by frequency domain filtering

$$U_a = (U_m + mx)cos\,\omega_c t \quad (2)$$

where, U_m is the carrier signal amplitude, ω_c is the carrier signal frequency, m is the modulation coefficient. x is the modulated signal, note as

$$x = X_m cos\Omega t \quad (3)$$

Then, with heterodyne technique, the modulated signal will be multiplied with a unit amplitude reference signal $\cos(\omega_c t)$. The result is given in the following.

For the amplitude modulation signal,

$$U_o = (U_m + mx)\,cos\,\omega_c t\,cos\,\omega_c t$$
$$= (U_m + mx)\left[\frac{1}{2} + \frac{1}{2}cos(2\omega_c t)\right] \quad (4)$$

Then substitute Eq. (3) in into Eq. (4), it gives:

$$U_o = \frac{1}{2}U_m + \frac{1}{2}X_m cos\Omega t + \frac{1}{2}U_m cos(2\omega_c t)$$
$$+ \frac{1}{4}mX_m[cos(2\omega_c + \Omega)t + cos(2\omega_c - \Omega) \quad (5)$$

Since U_m does not contain any useful information related with the modulated signal, it could be set as 0, or removed by de-trending. From Eq. (5), it can be seen that only the modulated signal will be left after low pass filtering, where the high frequency components around frequency $2\omega_c$ will be removed.

The diagram of the proposed down sampling system using heterodyne is shown in Figure 4.

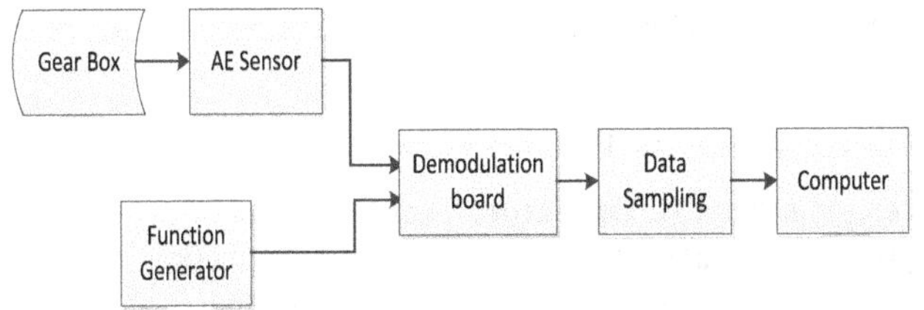

Figure 4. Proposed AE signal acquisition and preprocessing procedure

By adding a demodulation step, it could achieve the purpose of reducing the signal frequency to 10s of kHz. This is close to the frequency range of general vibration signals. Any data acquisition board with a low sampling rate could be able to sample the pre-processed AE data.

2.2. AE Signal TSA

TSA has been widely used in processing the vibration signals for rotating machine fault diagnosis (McFadden & Toozhy, 2000; Bonnardot *et al.*, 2005). The idea of TSA is to use the ensemble average of a raw signal over certain number of revolutions in order to enhanced signal of interest with less noise from other sources. For a function $x(t)$, digitized at a sampling interval nT, resulting in sampling in samples $x(nT)$. Denoting the averaged period by mT, TSA is given as (Braun, 1975):

$$y(nT) = \frac{1}{N}\sum_{r=0}^{N-1} x(nT - rmT) \qquad (6)$$

More details about TSA could be found in (McFadden, 1987).

The successful application of TSA on vibration signal analysis provides the possibility of using it to process AE signals. Basically, two types of TSA algorithms are available in literature, i.e., TSA with tachometer, and tachometer less TSA. In comparison with TSA with tachometer, tachometer less TSA needs to estimate the angular information from the vibration data. For slow speed variation cases, time domain feature like gear meshing information could be used. However, tachometer less TSA will introduce more phase reference errors and thus have less accuracy than TSA with tachometer. In this work, TSA with tachometer is applied.

Despite of the popular application of TSA to vibration signal analysis, application of TSA to AE signal processing for gear fault diagnosis has not been reported in the literature. The complicated feature and large data volume of AE signals make it unrealistic to perform TSA algorithm directly on AE data with an on-line condition monitoring system. In this paper, the authors explore the application of TSA to AE signal analysis.

TSA enables the direct comparison of the acoustic signals produced by each tooth on the same gear over one revolution. TSA for gear diagnosis generally computes the acoustic signals of a single shaft revolution. After TSA is calculated, basically all kind of fault detection condition indicators can be evaluated on the TSA signal.

2.3. Spectral kurtosis (SK) and optimal band pass filter

The spectral kurtosis was proposed by Dwyer (1983), as a statistical tool that can be used to identify the non-Gaussian components in a signal as well as their location in the frequency domain. A more formal definition of SK was provided in (Capdevielle, 1996) from the perspective of higher-order statistics. By Capdevielle's definition, SK is the normalised fourth-order cumulant of the Fourier transform and can be used as a measure of distance of a process from Gaussianity. Therefore, it can act as a measure of the peakiness of the probability density function of the process at a frequency of f. However, SK did not draw much attention from the researchers until it was revisited and further developed by Antoni (2006). The SK of a signal $x(t)$ is defined as the energy-normalized fourth-order spectral cumulant as:

$$K_x(f) = \frac{S_{4x}(x)}{S_{2x}^2(f)} - 2 \qquad (7)$$

where $S_{nx}(f) = \langle |X(t,f)|^n \rangle$, $\langle\cdot\rangle$ stands for the time averaging operator, $X(t,f)$ is the complex envelop of signal $x(t)$.

$X(t,f)$ can be estimated by any time-frequency analysis methods, such as: short time Fourier transform (STFT), the filter bank method, Wigner-Ville distribution, and wavelet package..

Take STFT for example, the STFT of signal $x(t)$ discretely sampled as $X(n)$ is defined as:

$$X_w(kP,f) = \sum_{n=-\infty}^{\infty} X(n)w(n-kP)e^{-j2\pi nf} \qquad (8)$$

where, $w(n)$ is a positive analysis window, P is a given temporal step.

As noted, the SK is suitable for identifying the peakiness of a signal with regard to frequency. It is able to extract non-stationary event in the signal. In general, the vibration signals measured from rotating machinery is considered as stationary. However, an AE signal is considered non-stationary. Gear signals can be classified as cyclostationary process. As indicated in (Antoni, 2006), the signals from rotating machinery can be resynchronized with a phase reference and then form a non-stationary signal with a periodic statistical structure. It is therefore conditionally non-stationary, which is suitable to use SK for fault detection.

In order to estimate SK, the kurtogram was proposed by Antoni and Randall (2006). A kurtogram is a three dimension graph which gives the kurtosis value for different frequency and different window size. Window size N_w is an important parameter because it directly affects the spectral resolution of the SK. A short N_w will yield high SK value, but too short a N_w will also lose some details and reduce the frequency resolution. Therefore, both the frequency and N_w should be optimized while the maximum SK can be identified. Since a kurtogram can identify the optimal frequency range and optimal window size where the signal displays the maximum peakiness, it is very useful for filter design. After the frequency line where the maximum SK is obtained, several filter methods could be applied to extract an enhanced SNR signal, such as Wiener filter, matched filter and band pass filter (Antoni & Randall 2006).

For optimal band pass filter, the objective is to find: (1) the central frequency $\boldsymbol{f_c}$ and (2) the bandwidth $\boldsymbol{B_f}$ of the filter which maximizes peakiness on the filtered signal. For fault detection problem, in order to recover the impulse associated with a faulty signature, a band pass filter which is used to maximize the kurtosis of the envelope of the filtered signal. As demonstrated in (Antoni & Randall 2006), this problem is strictly equivalent to finding the frequency $\boldsymbol{f}$ and the window length $\boldsymbol{N_w}$ that maximises the STFT-based SK over all possible combinations. The optimal central frequency $\boldsymbol{f_c}$ and bandwidth $\boldsymbol{B_f}$ of the band pass filter are determined as those values which jointly maximize the kurtogram. Therefore, both the center frequency $\boldsymbol{f_c}$ and window length $\boldsymbol{N_w}$ could be identified by using kurtogram. By doing this, the best compromise between maintaining the highest possible signal to noise ratio and extracting the impulse like signature of the fault is achieved.

2.4. Condition indicators for gearboxes diagnosis

Many vibration based condition indicators for gear fault detection have been reported in literature. Most of the condition indicators deal with the data distribution, such as peakiness, amplitude level, deviation from the mean and so on. A major difference between these condition indicators lies in the signal from which they are calculated. Generally four types of signals are used for computation, i.e., raw signals, time synchronous average signals, residual signasl and difference signals (Večeř *et al.* 2005; Lebold *et al.* 2000). A residual signal is generally defined as a synchronous averaged signal with the shaft, gear mesh, and their associated harmonic frequencies removed (Zakrajsek, 1993). The difference signal is defined by further removing the first order sidebands from the residual signal (e.g. the distinction between the residual and difference signals is the first order sidebands). For this filtering process, the spectrum values corresponding to these features are set to zero and the inverse Fourier transform is performed to convert it back to the time domain. However, these definitions are not strict. Also in practice, different filtering methods of performing the above mentioned process will give different results.

Other operations on the TSA include:

Teager's Energy Operator: Teager's energy operator is a type of residual of the autocorrelation function (Kaiser, 1990; Teager, 1992). For a nominal gear, the predominant vibration is gear mesh. Surface disturbances, scuffing, and etc., generate small higher frequency values which are not removed by autocorrelation. The CIs of the EO are the standard statistics of the EO vector. The mathematics formula is as follows:

$$\psi[x_i] = x_i^2 - x_{i-1} \cdot x_{i+1} \quad (9)$$

where $\psi[x_i]$ is the ith element in EO, x_i is the ith element of x.

Statistics are performed on the analysis, which include:

RMS: The root mean square (RMS) for a discretize sampled signal is defined as:

$$x_{rms} = \sqrt{\frac{1}{N}\sum_{i=1}^{N}(x_i{}^2)} \quad (10)$$

where x_{rms} is the root mean square value of data set x, x_i is the *i*-th element of x, N is the length of data set x.

From the definition of RMS, it is easy to understand that the RMS may not increase greatly with isolated peaks in the signal, and consequently it is not sensitive to incipient tooth crack or initial failure. Its value will increase as the speed and load increase.

Crest factor: Before crest factor could be defined, peak value must be understood. Peak value generally refers to the maximum value in the collected data. The crest factor then could be given in Eq. (11)

$$CF = \frac{|x|_{peak}}{x_{rms}} \quad (11)$$

where CF is the crest factor, $|x|_{peak}$ is the peak amplitude in data x, x_{rms} is the RMS.

This parameter is more sensitive to initial gear fault, such as one tooth crack. Since the RMS will not change in incipient fault, but the crest factor should see an increase.

Kurtosis: kurtosis describe how peaky or how smooth of the amplitude of data set x. If a signal contains sharp peaks with high values generated by a fault in the gearbox, it is expected that its distribution function will be sharper. Thus, the kurtosis of the fault signal should be higher than that of the healthy signal. The function of kurtosis is given below,

$$Kurt = \frac{N\sum_{i=1}^{N}(x_i - \bar{x})^4}{(\sum_{i=1}^{N}(x_i - \bar{x})^2)^2} \quad (12)$$

where *Kurt* is the kurtosis of data set x, x_i is the i-th element of x, N is the length of data set x.

It is worth to mention that for any normal distribution, the kurtosis value is 3. This could be easily verified by the moment generating function.

Some other gear fault algorithms are functions of operations, such as:

FM4: The FM4 parameter is simply the kurtosis of the difference signal. It is assumed that a healthy gearbox difference signal should display a Gaussian amplitude distribution, while a damaged gearbox will produce some high peak value which does not conform to Gaussian distribution.

$$FM4 = \frac{N\sum_{i=1}^{N}(d_i - \bar{d})^4}{(\sum_{i=1}^{N}(d_i - \bar{d})^2)^2} \quad (13)$$

where d_i is the i-th element of the difference signal, N is the length of difference signal.

NA4: NA4 is an improved version of FM4. NA4 is based on the argument that sideband signal contains the fault related information. So, the NA is calculated based on the residual signal which keeps the sideband while removing other meshing components. Also, NA4 takes an average value of the variance. The NA4 formula is given as:

$$NA4 = \frac{N\sum_{i=1}^{N}(r_i - \bar{r})^4}{(\frac{1}{M}\sum_{j=1}^{M}\sum_{i=1}^{N}(r_{ij} - \bar{r}_j)^2)^2} \quad (14)$$

where r_i is the i-th data point in the residual signal, r_{ij} is the i-th data point of the j-th group of residual signal, M is number of the data group of TSA residual signal, N is the number of data point in one TSA residual signal.

Figure 5 shows the overall process of computing the condition indictors using the presented approach.

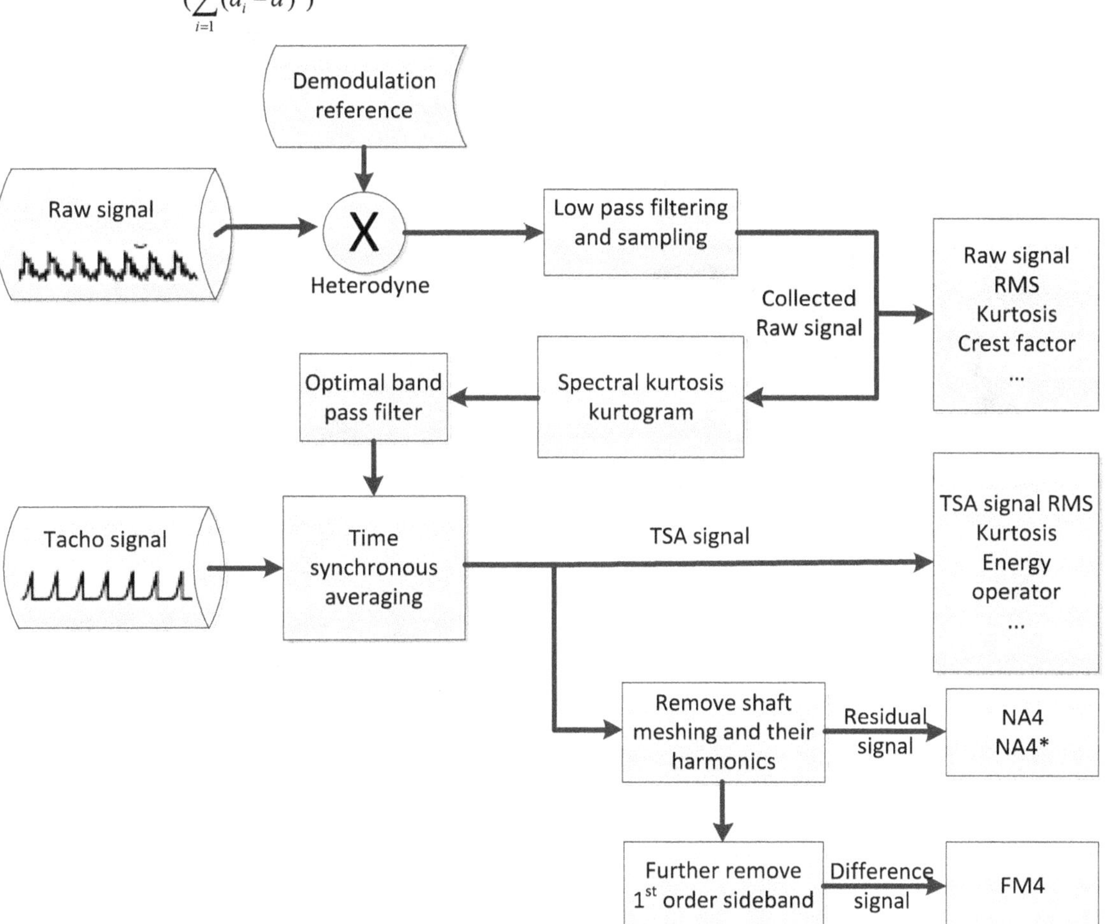

Figure 5. The overall process of computing the condition indicators

3. EXPERIMENTAL SETUP

In this section, the experiment setup for validating the AE sensor based gear fault detection approach is explained. In Figure 6, the demodulation board (Analog devices - AD8339) and sampling devices (NI-DAQ 6211) are shown. The demodulation board performed the multiplication of sensor signals and reference signals. It is an analog device and much more affordable than a high sampling rate board. It takes two inputs, one from the AE sensor, and the other from function generator as reference signal. The basic principle of AD8339 could be explained by Gilbert cell mixers. In electronics, the Gilbert cell is commonly used as an analog multiplier and frequency mixer. The output current of this circuit is an accurate multiplication of the base currents of the both inputs. According to Eq. (1) it could convert the signal to baseband and twice the carrier frequency. The output of the demodulation board goes to the sampling board and the high frequency component is filtered out. NI-DAQ 6211 is a low frequency data acquisition device, with a sample frequency up to 250kS/s. Before data acquisition, another task was to determine the frequency of the reference signal for demodulation. The objective was to down shift the AE signal frequency as low as possible. In order to remove the carrier frequency, the reference signal frequency needs to be as close to the AE carrier frequency as possible. Thus, the next step was to identify the AE sensor response frequency.

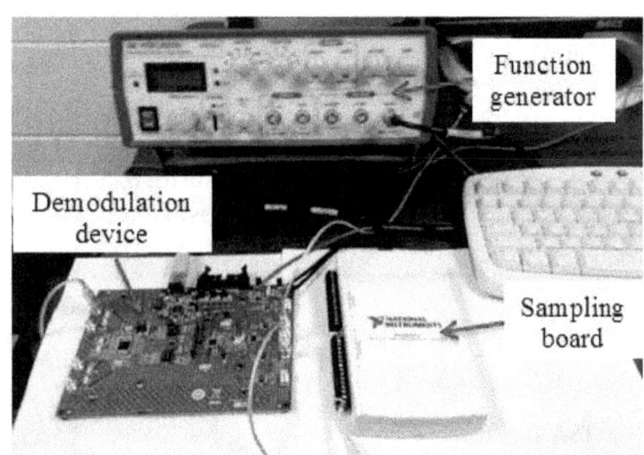

Figure 6. Demodulation device and data sampling board

Each AE sensor has its specific frequency response range, which further depends on the system being sensed and the sensor itself. With reference to the AE sensor user manual, a coarse range of the sensor response frequency is given. In order to identify a more accurate AE sensor response frequency, a function generator with sweep function was used to test the system and record the output. With a wide range of sweep frequency signal as the reference signal to demodulation board, the demodulation result varies accordingly. As mentioned before, the energy impact information of AE sensor is carried by a high frequency modulated signal. If the AE signal is successfully demodulated, most of the energy related signal will be shifted to below 50kHz, and thus it will be captured by the low frequency sampling board, which is set at 100kHz sampling rate. Otherwise most energy related information will still dwell in high frequency and therefore would be lost after low pass filtering. Using swept frequency input, one can analyze the energy envelope of the signal at different frequencies. Then the AE sensor response frequency is identified in the demodulated AE signal having the largest energy level. By this method, the accurate frequency range of the sensor output could be found. In the conducted experiment, 400 kHz was identified as the AE carrier signal center frequency. This frequency was then used as the demodulation reference frequency.

Seeded gear tooth crack fault tests were conducted on a notational two-stage split torque gearbox. The gearbox and the AE sensor location on the gearbox are shown in Figure 7.

Figure 7. The notational split torque gearbox and the AE sensor location

As shown in Figure 8, the notational split torque gearbox has an intermediate stage which could split the torque and change the transmission ratio between input and output shaft.

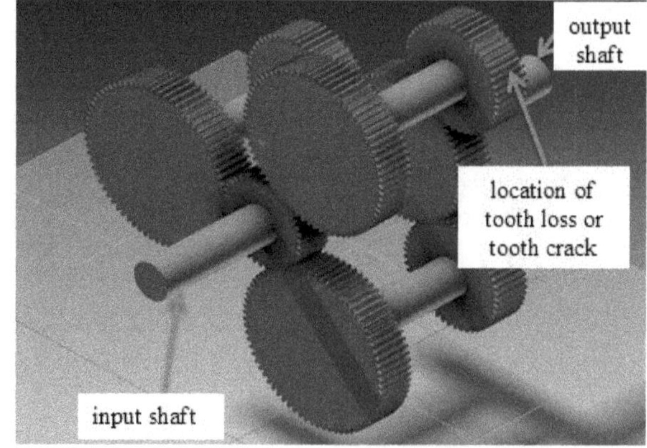

Figure 8. The structure of the notational split torque gearbox

For the faulty gearbox, one of the intermediate gears with 48 teeth was damaged by cutting the root of a gear tooth with a depth equal to half width of the gear tooth by EDM (electric discharge machining) with a wire of 0.5 mm diameter, to simulate the root crack damage in real applications. The seeded tooth crack is shown in Figure 9.

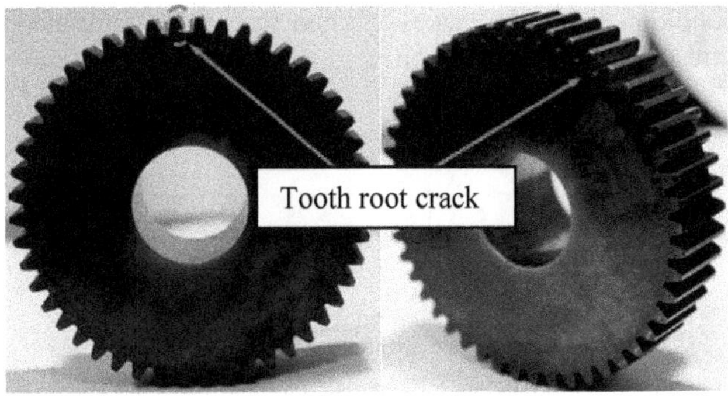

Figure 9. Seeded tooth root crack

Since this is a speed reduction gearbox, the input side and the output side have a 2.4 times speed reduction ratio. The corresponding output shaft speed and intermediate shaft (faulty gear shaft) speed are provided in Table 1.

Table 1. Output shaft speed corresponding to input shaft speed

Input shaft speed (Hz)	10	20	30	40	50	60
Faulty gear shaft frequency	5.56	11.1	16.7	22.2	27.8	33.3
Output shaft speed	4.17	8.33	12.5	16.7	20.8	25

For signal acquisition, Labview signal express software was used. During the experiments, continuous AE signals were collected. The data sampling rate was set to 100 kHz for all the tests. There was no torque load during the test. The gearbox input shaft speed is running from 10Hz-60Hz with 10Hz interval. For each speed, 5 data sets are collected. It should be noted that with a load, the faulty gear feature will be larger due to the increased impact on the gear. It is hypothesized that with a higher load, the faulty feature in AE could be more easily detected. In zero loading experiments, the identification of gear fault is more challenging than loaded cases.

4. RESULTS

The AE data after heterodyning were collected by a low sampling rate device, with the sampling rate fixed at 100 kHz. Additionally, the tachometer signals were collected together with the AE data from the main input shaft, which were used to perform the TSA calculation. Two collected examples of faulty AE data and healthy AE data are shown in Figure 10. For simplicity, the tachometer signals are not shown in the figure.

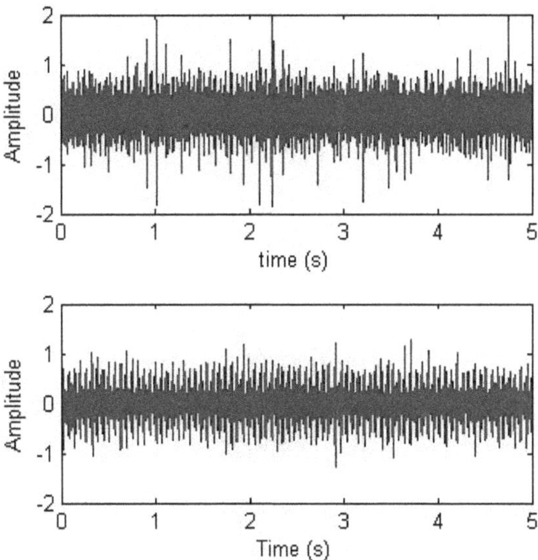

Figure 10. Healthy (upper) and faulty AE signals (lower) collected with heterodyne

Before performing TSA, the data was first analyzed using kurtogram. Figure 11 shows an example of kurtogram for a faulty signal at 30Hz. Based on the kurtogram, the center frequency and bandwidth where the spectral kurtosis is maximized could be identified. An optimal band pass filter was designed to filter the signal with regard to the corresponding frequency range.

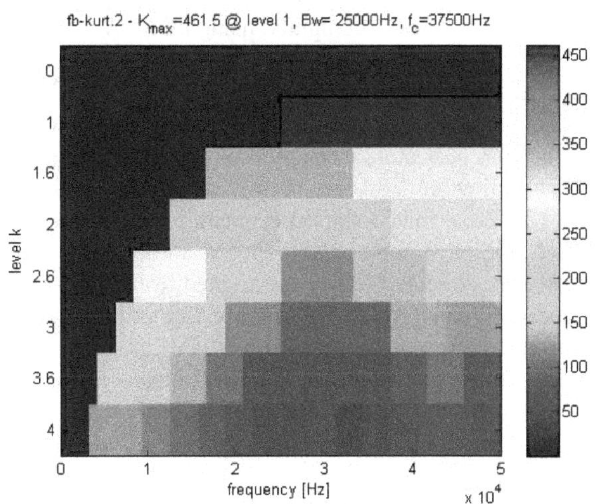

Figure 11. The kurtogram of spectral kurtosis for a faulty signal at 30Hz input shaft speed (The center frequency identified from this kurtogram is 37500Hz, with a band width of 25000Hz)

After the band pass filter, the TSA was computed using the filtered data. Since the raw signal was filtered before TSA, to maintain the phase unchanged, a zero-phase filter was used. About 260 averages were taken for each group of data. The TSA signal at 30Hz is shown in Figure 12.

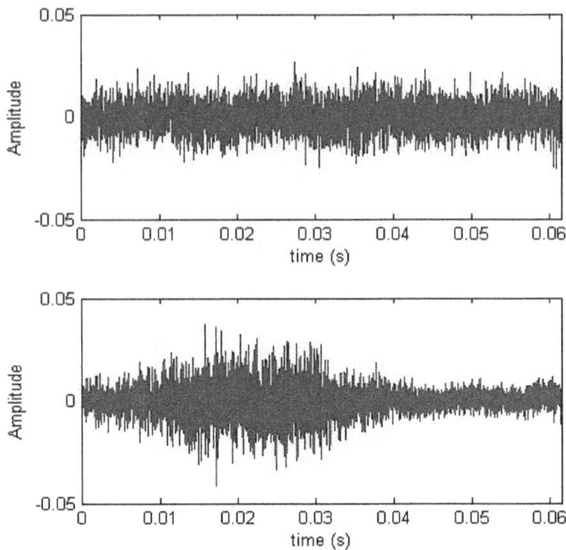

Figure 12. Healthy signal TSA (upper) and faulty signal TSA (lower)

In order to test the effectiveness and sensitiveness of different condition indicators, the following groups of condition indicators were compared:

(a) RMS, P2P, kurtosis and crest factor of the raw data.

(b) RMS, P2P, kurtosis and crest factor of the TSA data. In addition, condition indicators FM4 and NA4 were also computed using the TSA data.

(c) RMS, P2P, kurtosis and crest factor for the Teager's energy operator of the TSA.

The condition indicators calculated on the raw data without band pass filter or TSA are shown in Figure 13 through Figure 16. Note that in the experiment, 5 sets of data were collected for each input shaft speed: a total of 30 data samples were collected. They are aligned from low speed to high speed. It can be seen from Figure 13 that the raw RMS cannot separate the faulty gear from the healthy one. As the speed of the gearbox increases, the RMS increase gradually. From Figure 14, one can see that health signals have slightly larger P2P values than the faulty signals. This could be caused by the random noises, either from the misalignment of the gearbox or from the sensors. In Figure 15, it shows that the healthy signals have larger kurtosis values than the faulty signals. Based on the raw kurtosis, it is possible to separate the faulty signals from the healthy ones when the input speed is lower than 40 Hz. However at a speed higher than 40 Hz, it is impossible to distinguish healthy signals from faulty signals. Similarly in Figure 16, the healthy crest factors have larger amplitudes than faulty ones.

Based on the results of the raw data condition indicators, it is impossible to separate the health signals from the faulty ones. Also, the fact that the condition indicators of raw healthy signals have larger amplitude than the faulty condition indicators makes it impractical to set fault alarm threshold in real application.

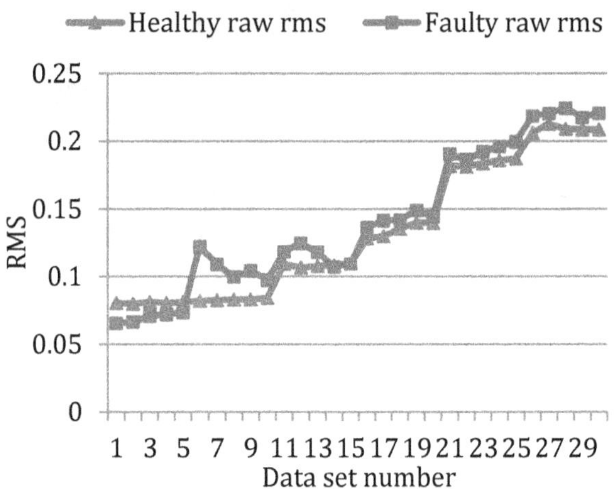

Figure 13. Raw data RMS

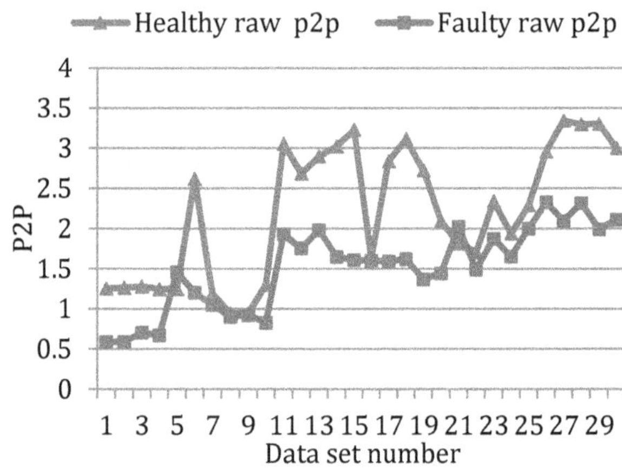

Figure 14. Raw data P2P

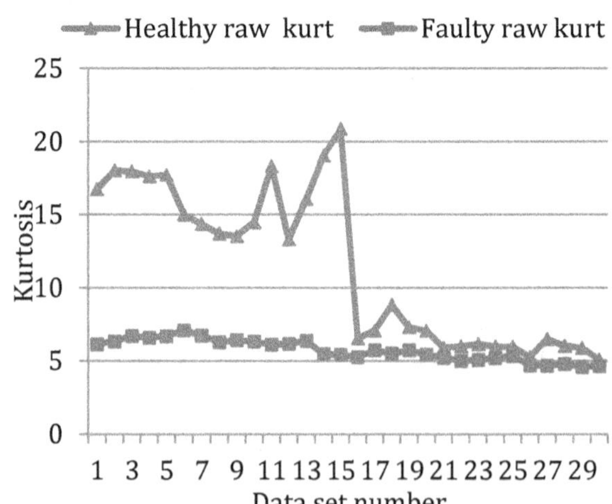

Figure 15. Raw data kurtosis

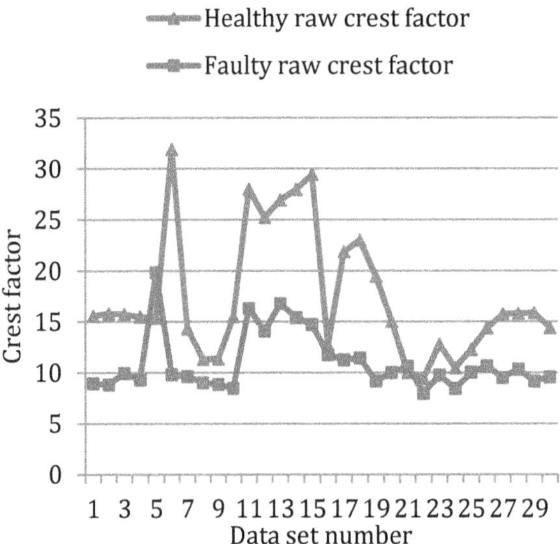

Figure 16. Raw data crest factor

In order to minimize the random noises and enhance the faulty features hidden in the raw signals, SK based filter followed by TSA was performed on the raw data.

The plots of the condition indicators computed using TSA are provided in Figure 17 through Figure 22.

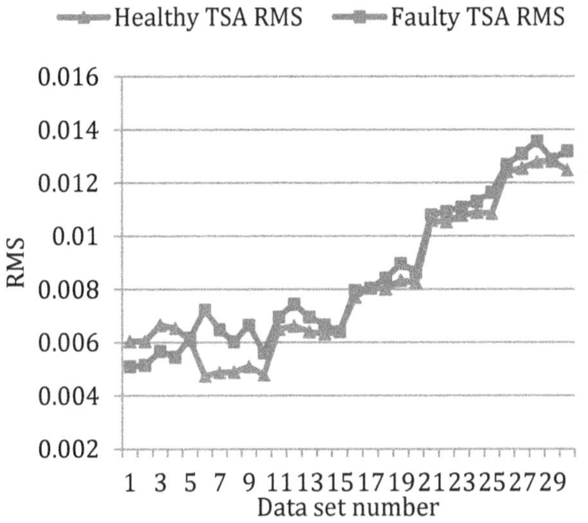

Figure 17. TSA RMS of the healthy data and faulty data

From Figure 17, one can see that the behavior of the TSA RMS is similar to that of the raw RMS in Figure 13. As the input shaft speed increases, the RMS increases for both health and faulty gears. Since both the RMS of the health signals and the RMS of the faulty signals overlap over the entire testing conditions, it is impossible to separate the gear fault using TSA RMS.

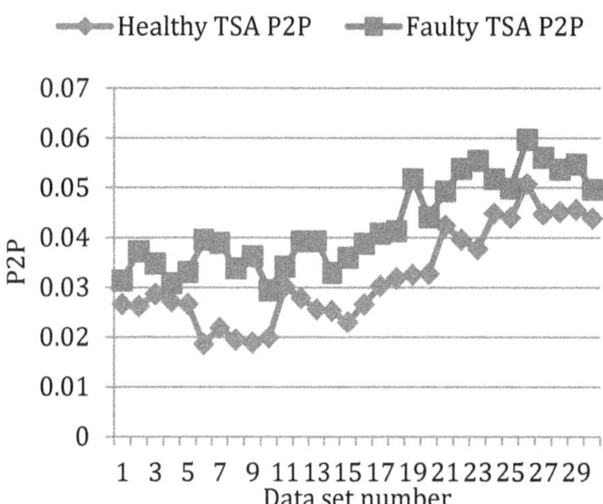

Figure 18. TSA P2P of the healthy data and faulty data

In comparison with P2P of the raw signals, one can see that the P2P of the faulty TSA signals are all larger than that of the healthy TSA signals under each individual shaft input speed, as shown in Figure 18. This confirms that the random noise is removed by TSA while the faulty features in the tooth crack condition are significantly enhanced.

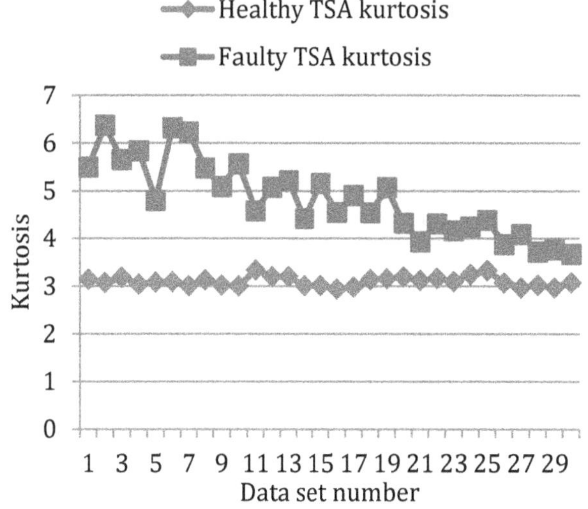

Figure 19. TSA kurtosis of the health data and tooth crack data

From Figure 19, it can be seen that the TSA kurtosis values of the health data remain almost constant around 3. As mentioned before, for any Gaussian distribution, the kurtosis is exactly calculated as 3. This concludes that the health gear TSA satisfied the Gaussian distribution. It means the amplitude of the AE impact waves generated by each tooth meshing complies with Gaussian distribution as expected. On the other hand, the TSA kurtosis of the faulty data is all above 3.6. This simply illustrates behavior of the faulty signal patterns.

Since kurtosis is non-quantitative value, it does not depend on the absolute amplitude. Kurtosis can serve as a reliable condition indicator for gearbox fault detection under variable loads and speeds.

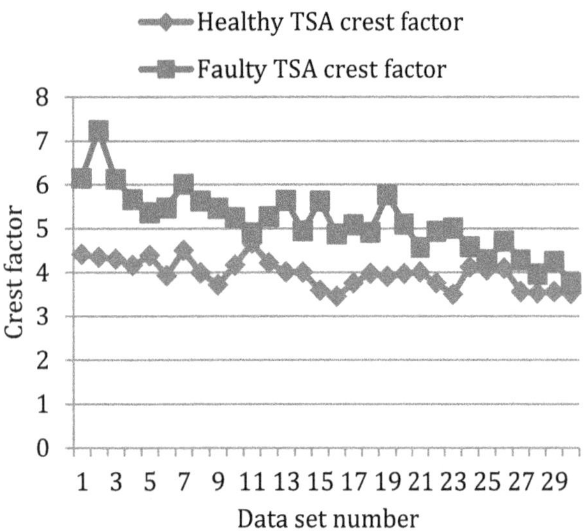

Figure 20. TSA crest factor of the healthy data and faulty data

The crest factor shows the statistics of the peak and the mean amplitude ratio. As show in Figure 20, all of the faulty TSA crest factors are larger than their health counterparts. Even though the differences of the crest factors between the healthy and faulty signals shown in Figure 20 are not as significant as those shown in Figure 19, the TSA crest factor still can be used as an effective condition indicator for detecting the gear fault.

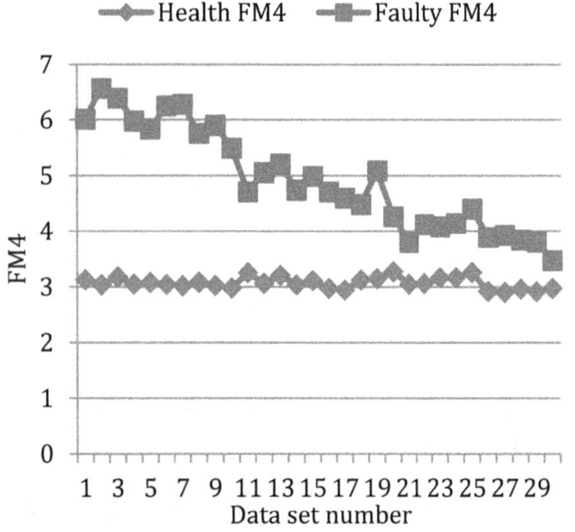

Figure 21. TSA FM4 of the health data and tooth crack data

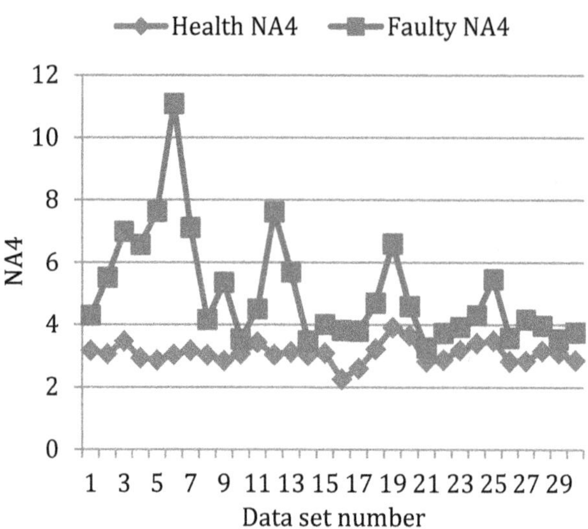

Figure 22. TSA NA4 of health and tooth crack data

Figure 21 and Figure 22 show the plots of the FM4 and NA4 condition indicators, respectively. FM4 is the difference signal kurtosis while NA4 is the residual signal kurtosis. From Figure 21, the healthy data FM4 identified itself as near Gaussian distribution. The faulty FM4 is larger which indicates fault feature. For NA4 in Figure 22, the condition indicator has more fluctuation because the variance was averaged across each 5 sets of data at each operational speed.

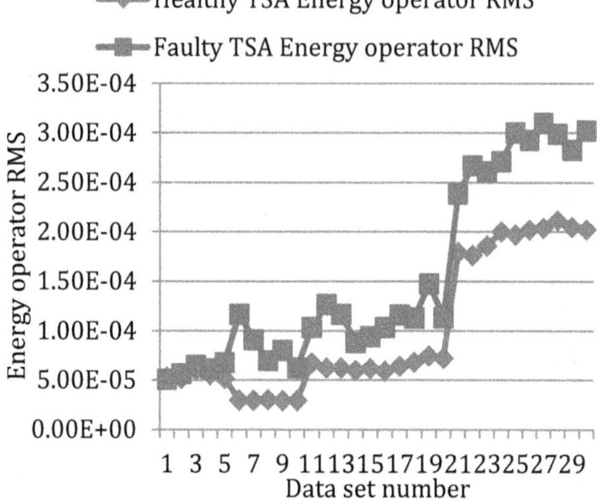

Figure 23. Energy operator RMS of healthy and tooth crack TSA data

From the above shown results of the condition indicators computed on the raw signals and TSA signals, RMS cannot separate the healthy signals from the faulty signals. Surprisingly, when taking the energy operator, one could actually see that faulty signal RMS values clearly separate themselves from the healthy signal RMS values as the speed increases, as shown in Figure 23.

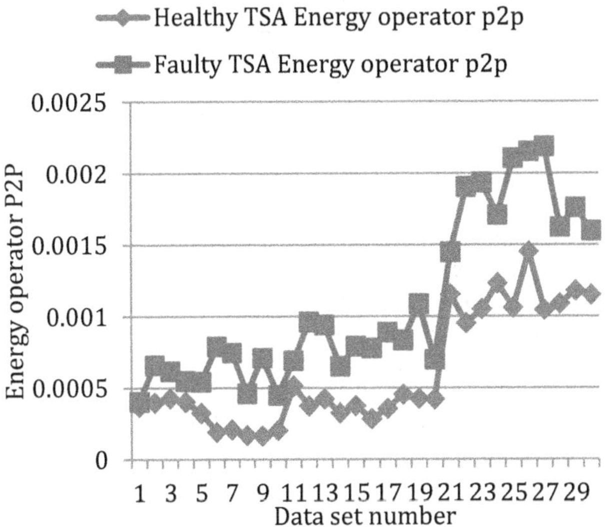

Figure 24. Energy operator P2P of healthy and tooth crack TSA data

Figure 24 shows the P2P values of the TSA energy operators. The P2P condition indicator could roughly separate the faulty signals from the healthy signals.

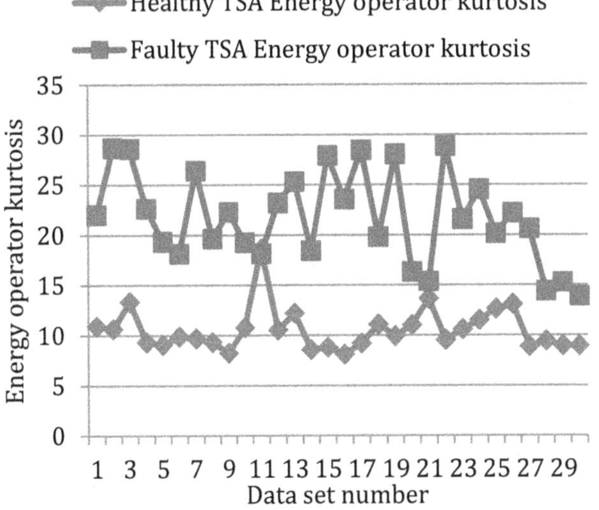

Figure 25. Energy operator kurtosis of healthy and tooth crack TSA data

Figure 25 shows the energy operator for both cases. Energy operator by definition is another kurtosis based condition indicator. From the plot, it is easy to see that energy operator can basically separate the healthy gear from the faulty gear.

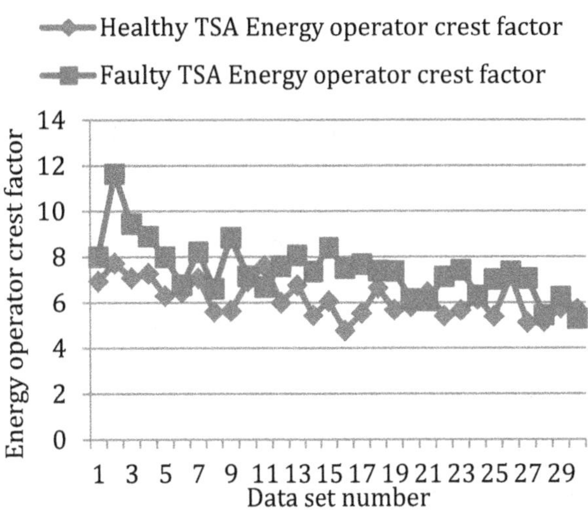

Figure 26. Energy operator crest factor of healthy and tooth crack TSA data

Figure 26 plots the TSA energy operator crest factor. It could be seen from the plot that the faulty signal crest factors are roughly larger than the healthy ones. However, there are some overlaps in a few samples. Based on the experiment results, crest factors are less reliable than kurtosis in term of tooth crack detection.

It was observed that as the gearbox speed increase, the kurtosis based condition indicators generally decrease. Two reasons might count for these behaviors. The first reason is that when the gearbox input shaft speed increases, the variance value in the denominator of Eq. (12) – Eq. (14) increases rapidly, which results in a decrease of the kurtosis based condition indicators. The second reason is that when an incipient fault is presented, the fault features are relatively small. When the gearbox is operating at a high speed, the normal gear meshing has similar impact amplitude as the fault feature, which overwhelms the fault feature. As noted, an AE signal tends to display more Gaussian type characteristics. For gear fault detection using AE signals, this might be a shortcoming at high shaft rates because AE signal amplitudes increase significantly with speed. The high amplitude of the normal gear meshing impact signals might overwhelm the incipient fault features.

Conversely, in the high shaft speed conditions, the RMS and P2P of the energy operator give clear indication for the faulty signals. The energy operator RMS and P2P of the fault signals increase much faster than the healthy signals. So it is possible to use these condition indicators to compensate the performance degradation of kurtosis based condition indicators.

One solution to this problem is to substitute the denominator with the variance of a gearbox in good condition, which then leads to the condition indicator NA4* (Lebold *et al.*, 2000). NA4* is more suitable for continuous monitoring.

In the case of naturally grown fault, it is convenient to use the variance when the gearbox is new and under good condition. The evolution of the fault growth will be easily observed. On the other hand, while this phenomenon is not exactly as desired in a high speed operational condition (1000 RPM or higher), it may benefit the detection of gearbox fault operated at lower speed (within 1000 RPM).

In summary, the condition indicators on the raw AE data did not convey much useful information for fault detection. The SK filter and TSA greatly enhanced the fault features. Most of the condition indicators on the TSA could clearly separate the faulty condition signals and healthy signals: the TSA kurtosis and FM4 worked the best in this tooth crack fault detection experiment. Furthermore, the Teager's energy operator based condition indicators could successfully separate the faulty signals from the healthy gear signals. Teager's energy operator had significant improvement on the RMS condition indicator. In addition, Teager's energy operator improved the separation between healthy TSA kurtosis and faulty TSA kurtosis.

5. CONCLUSIONS

In order to reduce wind energy costs, PHM of wind turbine is needed to reduce the operations and maintenance costs associated with running a wind farm. One of the major costs on wind turbine repairs is due to gearbox failure. Therefore, developing effective gearbox fault detection tools is important to the PHM of wind turbine.

In this paper, a new AE sensor based gear fault detection approach was presented. This new approach combines a heterodyne based frequency reduction technique with TSA and spectral kurtosis to process AE sensor signals and extract features as condition indictors for gear fault detection. Heterodyne technique commonly used in communication is first employed to preprocess the AE signals before sampling. By heterodyning, the AE signal frequency is down shifted from several hundred kHz to below 50 kHz. This reduced AE signal sampling rate is comparable to that of vibration signals.

The approach presented in this paper is physics based. The presented approach was validated using seeded gear tooth crack fault tests on a notational split torque gearbox. Condition indicators, such as RMS, P2P, kurtosis, and crest factor, were computed from the raw signals, TSA signals, and Teager's energy operator signals, separately. The results showed that the condition indicators computed on the TSA signals and Teager's energy operators could effectively separate the faulty signals from the healthy signals. Among all the condition indicators tested, kurtosis related condition indicators, like, TSA kurtosis, FM4, and Teager's energy operator kurtosis, have showed the best performance of detecting the gear tooth crack for all the testing conditions.

REFERENCES

Al-Balushi K.R., SamantaB. (2002), "Gear fault diagnosis using energy-based features of acoustic emission signals", Proceedings of the Institution of Mechanical Engineers. Part I: Journal of Systems and Control Engineering, Vol.216, No. 3, pp. 249 – 263.

Al-Ghamd, A.M. and Mba, D. (2006), "A comparative experimental study on the use of acoustic emission and vibration analysis for bearing defect identification and estimation of defect size", *Mechanical Systems and Signal Processing*, Vol. 20, No. 7, pp. 1537 – 1571.

Antoni J. (2006), "The spectral kurtosis: a useful tool for characterizing non-stationary signals", *Mechanical Systems and Signal Processing*, Vol. 20, No. 2, pp. 282–307.

Antoni J. and Randall R.B. (2006), "The spectral kurtosis: application to the vibratory surveillance and diagnostics of rotating machines", *Mechanical Systems and Signal Processing*, Vol. 20, No. 2, pp. 308–331.

Bonnardot F., Badaoui M. El, Randall R.B., Daniere J., and Guillet F. (2005), "Use of the acceleration signal of a gearbox in order to perform angular resampling (with limited speed fluctuation) ", *Mechanical Systems and Signal Processing*, Vol. 19, No. 4, pp. 766 - 785.

Braun S. (1975), "The exaction of periodic waveforms by time domain averaging", *Acustica*, Vol. 32, No.2, pp. 69 - 77.

Capdevielle V., Servie`re C., and Lacoume J.-L. (1996), Blind separation of wide-band sources: application to rotating machine signals, Proceedings of the Eighth European Signal Processing Conference, vol. 3, pp. 2085–2088.

Dwyer R.F. (1983) "Detection of non-Gaussian signals by frequency domain kurtosis estimation", *Proceedings of the International Conference on Acoustic, Speech, and Signal Processing*, Vol. 8, pp. 607–610.

Eftekharnejad B., Carrasco M.R., Charnley B., and Mba D. (2011), "The application of spectral kurtosis on Acoustic Emission and vibrations from a defective bearing", *Mechanical Systems and Signal Processing*, Vol. 25, No. 1, pp. 266 - 284.

Gao L., Zai F., Su S., Wang H., Chen P., and Liu L., (2011), "Study and Application of Acoustic Emission Testing in Fault Diagnosis of Low-Speed Heavy-Duty Gears", *Sensor*, Vol.11, No.1, pp. 599 – 611.

He D., Li R., Zhu J., and Zade M. (2011), "Data Mining Based Full Ceramic Bearing Fault Diagnostic System Using AE Sensors", *IEEE Transactions on Neural Network*, Vol. 22, No. 12, pp. 2022 – 2031.

Kaiser J. F. (1990), "On Teager's Energy Algorithm and Its Generalization to Continuous Signals", Proc. 4th IEEE Digital Signal Processing Workshop, Mohonk (New Palts), NY.

Kilundu B., Chiementin X., Duez J. and Mba D., (2011), "Cyclostationarity of Acoustic Emissions (AE) for monitoring bearing defects", *Mechanical Systems and Signal Processing*, Vol. 25, No. 6, pp.2061 - 2072.

Lebold, M.; McClintic, K.; Campbell, R.; Byington, C.; Maynard, K. (2000), Review of Vibration Analysis Methods for Gearbox Diagnostics and Prognostics, *Proceedings of the 54th Meeting of the Society for Machinery Failure Prevention Technology*, Virginia Beach, VA, May 1-4, pp. 623-634.

Li R. and He D. (2012), "Rotational Machine Health Monitoring and Fault Detection Using EMD-Based Acoustic Emission Feature Quantification", *IEEE Transactions on Instrumentation and Measurement* , Vol. 61, No.4 , pp. 990 – 1001.

Mba D. (2003), "Acoustic Emissions and monitoring bearing health", *Tribology Transactions*, Vol.46, No.3, pp. 447 – 451.

McFadden P.D. (1987), "A revised model for the extraction of periodic waveforms by time domain averaging", *Mechanical Systems and Signal Processing*, Vol.1, No.1, pp. 83 – 95.

McFadden P.D. , Toozhy M.M. (2000), "Application of synchronous averaging to vibration monitoring of rolling element bearings", *Mechanical Systems and Signal Processing*, Vol. 14, No. 6, pp.891 - 906.

Pandya D. H., , Upadhyay S.H., and Harsha S.P., (2013), "Fault diagnosis of rolling element bearing with intrinsic mode function of acoustic emission data using APF-KNN", *Expert Systems with Applications*, http://dx.doi.org/10.1016/j.eswa.2013.01.033

Teager H. M. and Teager S. M. (1992), "Evidence for Nonlinear Sound Production Mechanisms in the Vocal Tract", in Speech Production and Speech Symp. Time-Frequency and Time-Scale Analysis, Victoria, British Columbia, Canada, pp. 345-348.

Večeř P., Kreidl M. and Šmíd R. (2005), "Condition Indicators for Gearbox Condition Monitoring Systems", *Acta Polytechnica*, Vol. 45, No. 6, pp. 35 - 43.

Zakrajsek J.J., Townsend D.P., Decker H.J. (1993), "An analysis of gear fault detection methods as applied to pitting fatigue failure data", Technical Report NASA TM-105950, AVSCOM TR-92-C-035, *NASA and the US Army Aviation Systems Command.*

BIOGRAPHIES

Yongzhi Qu received his B.S. in Measurement and Control and M.S. in Measurement and Testing from Wuhan University of Technology, China. He is a PhD candidate in the Department of Mechanical and Industrial Engineering at The University of Illinois Chicago. His research interests include: rotational machinery health monitoring and fault diagnosis, especially with acoustic emission sensors, embedded system design and resources allocation and scheduling optimization.

Eric Bechhoefer received his B.S. in Biology from the University of Michigan, his M.S. in Operations Research from the Naval Postgraduate School, and a Ph.D. in General Engineering from Kennedy Western University. His is a former Naval Aviator who has worked extensively on condition based maintenance, rotor track and balance, vibration analysis of rotating machinery and fault detection in electronic systems. Dr. Bechhoefer is a board member of the Prognostics Health Management Society, and a member of the IEEE Reliability Society.

David He received his B.S. degree in metallurgical engineering from Shanghai University of Technology, China, MBA degree from The University of Northern Iowa, and Ph.D. degree in industrial engineering from The University of Iowa in 1994. Dr. He is a Professor and Director of the Intelligent Systems Modeling & Development Laboratory in the Department of Mechanical and Industrial Engineering at The University of Illinois-Chicago. Dr. He's research areas include: machinery health monitoring, diagnosis and prognosis, complex systems failure analysis, quality and reliability engineering, and manufacturing systems design, modeling, scheduling and planning.

Junda Zhu received his B.S. degree in Mechanical Engineering from Northeastern University, Shenyang, China, and M.S. degree in Mechanical Engineering from The University of Illinois at Chicago in 2009. He is a Ph.D. candidate at the Department of Mechanical and Industrial Engineering. His current research interests include lubrication oil condition monitoring and degradation simulation and analysis, rotational machinery health monitoring, diagnosis and prognosis with vibration or acoustic emission based signal processing techniques, physics/data driven based machine failure modeling, CAD and FEA.

An Integrated Framework of Drivetrain Degradation Assessment and Fault Localization for Offshore Wind Turbines

Wenyu Zhao[1], David Siegel[1], Jay Lee[1], and Liying Su[2]

[1]*NSF I/UCRC for Intelligent Maintenance Systems, University of Cincinnati, Cincinnati, Ohio, 45221-0072, USA*
zhaowy@mail.uc.edu
siegeldn@mail.uc.edu
jay.lee@uc.edu

[2]*Sinovel Wind Group Co., Ltd., Beijing, 100872, China*
suliyingamily@163.com

ABSTRACT

As wind energy proliferates in onshore and offshore applications, it has become significantly important to predict wind turbine downtime and maintain operation uptime to ensure maximal yield. Two types of data systems have been widely adopted for monitoring turbine health condition: supervisory control and data acquisition (SCADA) and condition monitoring system (CMS). Provided that research and development have focused on advancing analytical techniques based on these systems independently, an intelligent model that associates information from both systems is necessary and beneficial. In this paper, a systematic framework is designed to integrate CMS and SCADA data and assess drivetrain degradation over its lifecycle. Information reference and advanced feature extraction techniques are employed to procure heterogeneous health indicators. A pattern recognition algorithm is used to model baseline behavior and measure deviation of current behavior, where a Self-organizing Map (SOM) and minimum quantization error (MQE) method is selected to achieve degradation assessment. Eventually, the computation and ranking of component contribution to the detected degradation offers component-level fault localization. When validated and automated by various applications, the approach is able to incorporate diverse data resources and output actionable information to advise predictive maintenance with precise fault information. The approach is validated on a 3 MW offshore turbine, where an incipient fault is detected well before existing system shuts down the unit. A radar chart is used to illustrate the fault localization result.

1. INTRODUCTION

With the rapid increase in the adoption of wind power for renewable energy generation, wind farm development and wind capacity installation have seen extensive growth. As Global Wind Energy Council (2012) pointed out, global capacity has reached 237 GW in 2011 and is projected to achieve 759 GW, which is more than three times current capacity, by the year of 2020.

Meadows (2011) shows that a 1975 MW offshore capacity has been installed in Europe, whereas a 135 MW capacity is available in China; for the United States, the forecast of offshore capacity is 10 GW by 2020 and 54 GW by 2030. However, availability and reliability of offshore turbines are imposing challenges for productive and efficient offshore wind farms.

A comprehensive report by National Renewable Energy Laboratory (2010) provided similar insight that: U.S. offshore wind power has great potential of supporting a considerable percentage of electricity needs; while the improvement of reliability through condition monitoring is one of major technology trends that will greatly support operations and maintenance for turbines both onshore and offshore.

CMS has been an emerging technology for monitoring turbine health status and diagnosing component failures. A study by LeBlanc and Graves (2011) shows that, the application of CMS is rising despite initial doubt of its capability. In certain offshore wind farm guidelines, CMS is even mandatory for turbine monitoring (GL Renewables Certification, 2012). A framework of CMS is provided in the study as well, where typical requirements of sensor locations are shown. The benefit of adopting CMS is discussed and justified, based on failure rates of key components and related cost. It proves that, on average, predicting one gearbox failure can clearly justify the budget of deploying a CMS on the turbine system. In addition, McMilan and Ault (2007) investigated Markov model between drivetrain components and quantified revenue and risk of condition monitoring, based on reliability data of the components.

As detailed by ISET (2005), CMS utilizes various types of communication infrastructure to transfer real time sensor data and control information to data centers, where servers are used to host and process the data. Another study (Amirat, Benbouzid, Al-Ahmar, Bensaker and Turri, 2009) also provided a review of data collection schemes for the electrical system, blade and drivetrain condition monitoring.

In literature, much research is being conducted for condition monitoring of wind turbines based on data infrastructure. Lu, Li, Wu and Yang (2009) gave a diagnosis review of the gearbox, bearing generator, power electronics, rotor, blades and overall system with condition monitoring techniques including vibration, torque, oil debris, temperature, acoustic emission and electric current & power analyses. Hameed, Hong, Cho, Ahn and Song (2009) provided a related review of fault detection methods for global and subsystem levels based on CMS. Crabtree, Feng and Tavner (2010) developed a multivariate approach that combines vibration and oil debris analysis for detecting gearbox failure at an early stage. Entezami (2010) proposed an overview and approach to connect the control system with turbine condition monitoring. Sheng and Veers (2011) described the gearbox reliability collaborative research at the National Renewable Energy Laboratory, where a fully instrumented drivetrain test bed is built for generating lab test data.

Furthermore, SCADA system is also frequently used for monitoring wind turbine condition. Commonly used variables in different SCADA systems are shown in Table 1.

Category	Variable Examples
Ambient	Temperature, wind direction, wind speed
Blades	Pitch angle
Controller	Hub temperature, Ground temperature.
Gear	Gear bearing temperature, oil temperature
Generator	Bearing temperature, rotation speed
Grid	Production voltage, current, power factor
Hydraulic	Hydraulic oil temperature
Nacelle	Direction, temperature
Production	Average power, accumulated power
Rotor	Rotation speed
System	Logs of active alarms, turbine state
Hour Counter	Service hours

Table 1. Commonly Used SCADA Variables

In a study by Qiu, Feng, Tavner, Richardson, Erdos and Chen (2012), SCADA data from an onshore turbine is used for alarm analysis and probability-based reliability modeling. SCADA data is also suitable for evaluating turbine power generation performance, which is complicated by the dynamic environment parameters and operation conditions (Lapira, Siegel, Zhao, Brisset, Su, Wang, AbuAli and Lee, 2011).

In most of the available literature, CMS and SCADA systems are used separately for condition monitoring purposes, mainly due to the issue of data availability in certain research activities. Moreover, the majority of tools and techniques are developed and validated on lab-scale test beds. To address such issues, a degradation assessment framework is proposed to integrate CMS data and SCADA variables for the evaluation of drivetrain degradation. Although usually used for a different purpose than a CMS, SCADA provides operational information that can assist the screening and processing of CMS data. In addition, some SCADA systems can provide variables that can serve as health indicators of drivetrain components. Previous research, including Qiue et al. (2009) and Edwin, Theo, Henk, Luc, Xiang and Simon (2008), show that SCADA variables can be used for fault detection at early stage, especially through analyzing temperature measurement from drivetrain components. The framework is eventually validated with an offshore turbine drivetrain.

The remainder of the paper is organized as following: Section 2 describes the methodology of integrating SCADA system data and CMS data, extracting and selecting features, assessing drivetrain degradation and identifying fault location; Section 3 demonstrates an application of the methodology in monitoring the drivetrain for a 3 MW offshore turbine, as well as a monitoring platform prototype with visualization tools; Section 4 discusses the conclusion of presented work, and plan for future development and validation; acknowledgement and references ensue as the last portion of paper.

2. METHODOLOGY

The overall framework integrates selected information from both SCADA and CMS systems, provides global degradation assessment of the drivetrain, and identifies faulty component(s) when fault detection is determined to be positive. The systematic methodology is shown in Figure 1.

SCADA variables that are related with wind turbine operation are initially used to assist in deciding if individual CMS data instances can represent the true degradation condition for drivetrain. CMS data from all sensors in retained instances are then processed by a set of feature extraction tools, while SCADA variables that indicate drivetrain conditions are selected. CMS features and

SCADA variables are then concatenated and input to a degradation assessment method, to evaluate how the overall condition differs from a baseline. When degradation is significant and fault detection is confirmed, the location of the fault is decided based on each component's contribution to the overall degradation. Eventually, the analytical results are visually presented.

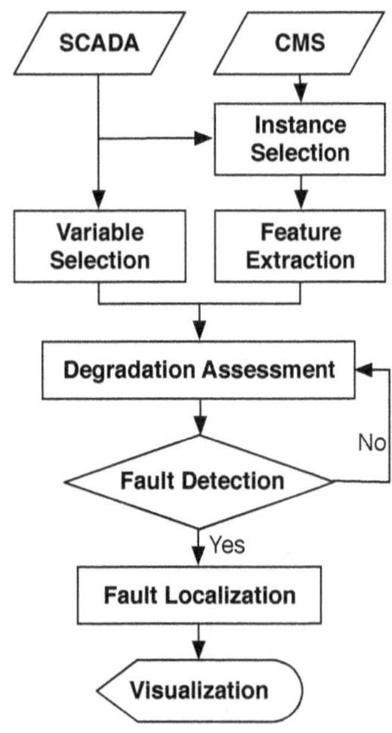

Figure 1. The integrated framework for drivetrain degradation assessment and fault localization

The rationale and techniques for each step are described as follows.

2.1. CMS Instance Selection

In applications of condition monitoring systems for wind turbines, it is a common practice to configure the sampling scheme as a routine program, so that several seconds or minutes worth of vibration waveforms with high sampling frequency from all instrumented sensors are acquired and stored at certain time instances throughout a day. This strategy is due to the limitation of proper infrastructure for data acquisition and transferring, and the concern of computational capacity for large-scale wind farms. In a few cases, vibration data is processed and only its features will be archived for further investigation; nevertheless the feature extraction is usually still time-based with a static period.

For a condition monitoring system that utilizes multiple accelerometers to measure the vibratory behavior of the drivetrain, it is important to decipher the quality of vibration data before actually processing the data. If the instance of data waveforms does not characterize the drivetrain's true health at the time of acquisition, including such instance in later analysis will generate false health information and affect decision making for maintenance. With rule-based criteria learned from wind farm operation and the control mechanism, irrelevant CMS data instances should be discarded based on SCADA measurements and only meaningful instances are kept for subsequent analysis. For example, if it is detected that the rotor speed has been zero for certain duration and there has been no rotation for the drivetrain, CMS data instance collected within this duration is determined to not contain vibration information that can be used for degradation assessment. Such instances should be rejected prior to further analysis.

2.2. SCADA Variable Selection

For majority of SCADA systems, some variables, measured by sensors within close proximity to the drivetrain, are also incorporated. They are valuable additional indicators for deciding the degradation condition of the drivetrain and its components, and sometimes can provide incipient failure detection with superior performance (Feng, Qiu, Crabtree, Long and Tavner, 2011). Examples of these variables include temperature readings of the rotor, gearbox and generator, as well as the gearbox oil pressure.

A heuristic method is used to select SCADA variables based on variable name and measurement location. Given that SCADA data is typically recorded more frequently than CMS data in wind power applications, SCADA records are selected only when a retained CMS instance exists at matching time interval.

2.3. CMS Data Feature Extraction

For health assessment and diagnosis, values of SCADA variables can be directly used as health indicators, whereas for CMS data features are normally computed to reduce its dimension and obtain representative indicators. A toolbox of signal processing techniques for vibration-based wind turbine monitoring has been designed and developed (Siegel, Zhao, Lapira, AbuAli and Lee, 2013), to extract features corresponding with key drivetrain components such as bearings, shafts, and gears respectively.

2.3.1. Time Domain Features

Time domain features provide statistical measures of a variable. Three commonly used features for vibration analysis are root mean square (RMS), kurtosis and crest factor (Lebold, MacClintic, Campbell, Byington and Maynard, 2000). For a data vector $X_i, i = 1,2,3 \dots N$, these features are defined as:

$$RMS = \sqrt{\frac{\sum_{i=1}^{N} {X_i}^2}{N}} \tag{1}$$

$$Kurtosis = \frac{\sum_{i=1}^{N}(X_i - \bar{X})^4}{N} \Big/ \left[\frac{\sum_{i=1}^{N}(X_i - \bar{X})^2}{N}\right]^2 \quad (2)$$

$$Crest\ Factor = max(|X_i|)/RMS \quad (3)$$

RMS is calculated as the vector's Euclidean norm, divided by the square root of vector length, as shown in Eq. (1). It represents the magnitude, or energy, of the vibration signal. A high RMS value can indicate indefinite or severe damages. Kurtosis, computed as Eq. (2), is the ratio between the vector's fourth moment about the mean and square of its second moment about the mean. It is essentially a measure of signal peakedness, which normally increases when damage causes impulses and unevenness in data. Crest factor can be used to detect high-amplitude impacts, when such impacts generate large signal impulse and increase the ratio between maximum value and RMS of the signal indicated by Eq. (3).

2.3.2. Spectral Kurtosis Filtering

To monitor damages in a complex mechanical system like a drivetrain, it is necessary to detect impulsive vibration behavior stimulated by defective gears or bearings. Time domain features of raw data can fulfill the task to a certain degree, but the impulsive behavior is often obscured by additive noise from irrelevant vibration resources. Therefore, band-pass filters in the frequency domain are applied to preserve the most impulsive frequency content and de-noise the signal.

Spectral kurtosis filtering (SKF) is a technique to optimize the configuration of band-pass filter for noise reduction. Based on time-frequency analysis results, it adopts the kurtosis computation from the time domain analysis to seek the most impulsive frequency band.

As developed by Antoni (2006) for non-stationary signal analysis, the short-time Fourier transform (STFT) of the signal is first calculated and denoted as $H(t,f)$. For each frequency index decided by the STFT, the kurtosis of its amplitude over discrete time is calculated as Eq. (4):

$$SK_X(f) = \frac{\langle H^4(t,f)\rangle}{\langle H^2(t,f)\rangle^2} - 2 \quad (4)$$

A statistical threshold S_α is computed (Antoni and Randall, 2006) to decide the significance of spectral kurtosis given level of significance α:

$$S_\alpha = u_{1-\alpha}\frac{2}{\sqrt{K}} \quad (5)$$

where $u_{1-\alpha}$ is the quantile with significance level α, and K is the number of time windows in STFT analysis. If the SK value for certain frequency is higher than the threshold, a Wiener filter is multiplied with the frequency spectrum of the original signal, where the multiplier is square root of the SK value. Therefore, the frequency content that is originally impulsive with high SK level is further amplified, whereas the other content is attenuated. The signal is eventually transformed back to a time series for extracting time domain features.

An example for effect of spectral kurtosis filtering is shown in Figure 2, where filtered data (bottom plot in red) apparently accentuate impulsive behavior more than raw data (top plot in blue). To quantify the difference, time domain kurtosis values before and after the filtering are 5 and 20.4 respectively; crest factors are 6.1 and 16.3 respectively.

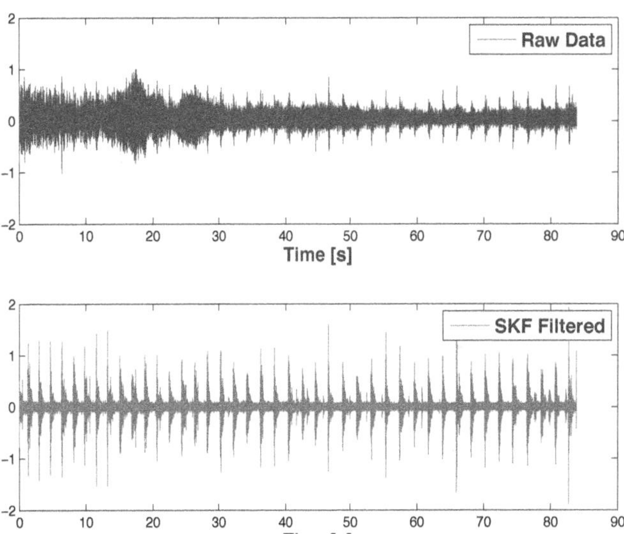

Figure 2. An example of spectral kurtosis filtering result

Spectral kurtosis filtering discovers the inherent dynamics of vibration spectrum, and automatically de-noises the signal without prior knowledge or visual inspection for band selection. In research with similar objectives, Barszcz and Randall (2009) previously investigated using spectral kurtosis for tooth crack detection in wind turbine planetary gears.

2.3.3. Envelope Analysis

As rolling element bearings and gear wheels rotate around their shafts in the gearbox system, bearing damages and gear defects often cause multiple impacts per each revolution and excite the resonant vibration of the entire structure. In vibration data, it results in an amplitude modulation phenomenon, where the structural resonance is the high-frequency carrier wave and the component fault frequency is the low-frequency envelope that modulates the waveform in time domain.

In frequency domain, modulation behavior is represented with concentrated high-amplitude peaks around resonance frequency where fault frequencies exist as sidebands spaced on both sides of resonance. To extract signatures related with specific faults, demodulation is performed with following steps:

- A band-pass filter is designed to filter vibration data around the excited resonance frequency. The design could be achieved by modal analysis, observation of spectrum for collected vibration data, or aforementioned spectral kurtosis filtering technique.
- Envelope analysis of data that is processed with the band-pass filter. A frequently used method is Empirical Mode Decomposition (EMD), which is an iterative filtering process shown as Figure 3 (Peng, Tse and Chu, 2005).

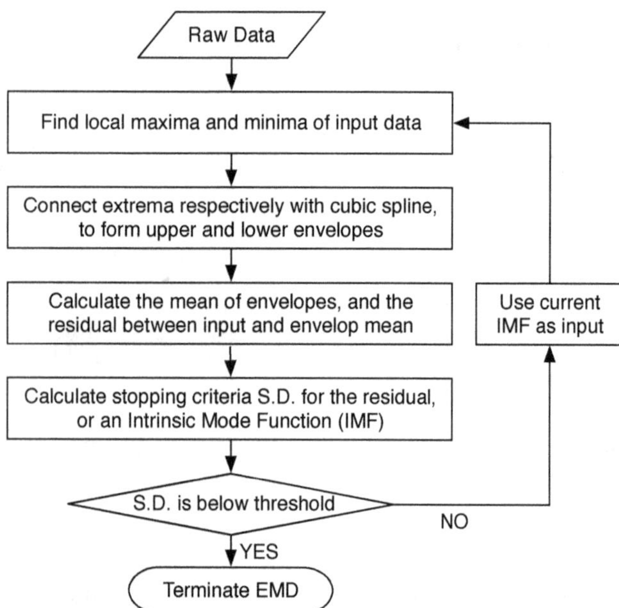

Figure 3. Empirical Mode Decomposition (EMD) process

- The method of extracting envelopes with local extrema has disadvantages of possible overshoots and breakpoints when a cubic spline is applied. To overcome such disadvantages, a Hilbert transform method is adopted (Liu, Riemenschneider and Xu, 2006) to find the upper envelope of signal $x(t)$, by finding the principal value (PV) $y(t)$ with Eq. (6) first, and analytic signal $z(t)$ with Eq. (7). The envelope is eventually the absolute value of analytic signal $z(t)$.

$$y(t) = \frac{1}{\pi} PV \int_{-\infty}^{\infty} \frac{x(\tau)}{t-\tau} d\tau \tag{6}$$

$$z(t) = x(t) + iy(t) \tag{7}$$

- After finding the envelope of band-pass filtered data with Hilbert transform, several feature extraction methods can be used to further analyze the envelope as signature of component defects. For example, bearing fault frequencies including ball pass frequency inner-race (BPFI), ball pass frequency outer-race (BPFO), ball fault frequency (BFF) and fundamental train frequency (FTF) for faults on bearing inner race, outer race, roller element and cage respectively. Furthermore, time domain statistics can also be indicators for defects, such as RMS and crest factor.

2.3.4. Wavelet Energy Analysis

The use of wavelets for time-frequency analysis as a method for automated feature extraction has seen a growing interest in the area of condition monitoring (Peng and Chu, 2004). For CMS vibration analysis, the focus is how to use wavelet analysis for feature extraction, and thus the discussion on the background of time-frequency analysis and continuous wavelet transform is omitted here. The wavelet decomposition and the wavelet packet decomposition are the more commonly used algorithms for feature extraction purposes, particularly in wind turbine monitoring area as well (Yang, Tavner and Wilkinson, 2008). The wavelet decomposition applies a filtering operation in which the signal is divided into an approximation signal (low frequency) and a detail signal (high frequency). The approximation signal consists of frequency content from 0 to approximately 1/4 of the Nyquist frequency (F_{nyq}), while the detail signal consists of frequency content from 1/4 of the Nyquist frequency to 1/2 of the Nyquist frequency. This represents the decomposition at the first level, and the approximation signal is further decomposed to a specified number of levels. The selection of the mother wavelet influences the filtering result, wherein higher coefficients and values can be obtained when the mother wavelet function is a closer match to the original signal (Jiang, Tang, Qin and Liu, 2011).

In general, the frequency content at level n for the approximation signal is given by Eq. (8) and the frequency content for the detail signal at level n is given by Eq. (9).

$$0 \le f \le \frac{F_{nyq}}{2^{n+1}} \tag{8}$$

$$\frac{F_{nyq}}{2^{n+1}} \le f \le \frac{F_{nyq}}{2^{n}} \tag{9}$$

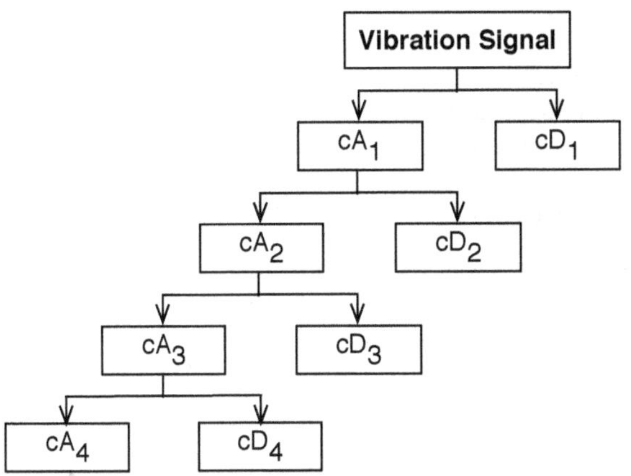

Figure 4. Wavelet Decomposition Diagram (Level 4)

An example wavelet decomposition diagram for level 4 is illustrated in Figure 4, where the approximation signal at each level is further decomposed. The wavelet decomposition only decomposes approximation signals, while the wavelet packet decomposition decomposes both approximation and detail signals.

Mechanical fault signatures for bearing, shaft, and gear components create amplitude and frequency modulation effects. In addition, these faults can excite the structural resonances of the system. Monitoring the vibration changes in different frequency bands from a baseline state is a way to monitor the overall health state of the drivetrain. The energy extracted at each node of the wavelet decomposition is used for monitoring the changes in the vibration in different frequency bands. The wavelet decomposition energy at each node can be calculated using Eq. (10), which consists of the squared summation of the coefficients at that particular node for the N coefficients. The wavelet energy vibration feature is a normalized frequency band vibration value, in which the vibration at each node is normalized by the total energy in the vibration signal and the feature is given as a percentage value. An example calculation of the wavelet energy feature for the approximation signal for a level 4 decomposition is provided in Eq. (11). In this equation, the coefficients for the approximation signal are denoted as w_{a4}, and the detail coefficients are denoted as w_{d1}, w_{d2}, w_{d3}, and w_{d4} respectively. For this example of level 4 decomposition, 5 wavelet features would be extracted, since a feature would be extracted at each level for the detail signals, and a feature is extracted for the level 4 approximation signal.

$$Energy\ A_4 = \sum_{i=1}^{N} {w_{a4i}}^2 \tag{10}$$

$$\frac{100 \times \sum w_{a4}^2}{\sum w_{d1}^2 + \sum w_{d2}^2 + \sum w_{d3}^2 + \sum w_{d4}^2 + \sum w_{a4}^2} \tag{11}$$

2.4. Drivetrain Overall Degradation Assessment

Upon the completion of SCADA variable selection and CMS feature extraction, the set of features from all sensors and selected SCADA variables are used to evaluate drivetrain degradation as explained in subsequent sections of the paper. Degradation assessment estimates present drivetrain condition by comparing a feature distribution model with a known healthy condition as the baseline model. As operation conditions change for the wind turbine, the drivetrain work regime varies over time and affects its response even under comparable health condition. Therefore, the features are assumed to comprise distributions from multiple models, and a modeling method that can learn and represent data with a mixture model is preferred.

Moreover, a distance metric is used to quantify degradation by measuring the dissimilarity between present features and the baseline. A threshold can be defined as an unacceptable level for the distance, and fault detection can be confirmed when the distance measure exceeds the threshold.

In this study, a Self-organizing Maps (SOM) approach is used for degradation assessment (Kohonen, 1990). Being a type of artificial neural network, SOM is able to automatically discover signal patterns and organize signals to create spatial separation between clusters. When used for unsupervised learning tasks where the data labels are not available for classification, SOM can cluster data instances so that inter-cluster distance is high and intra-cluster distance is low.

To train a SOM, a 2D map is initialized with m neurons corresponding with n input vectors x:

$$x = [x_1, x_2, \ldots, x_n]^T \tag{12}$$

Each neuron has a weight vector that has the same dimension n of an input vector:

$$w_j = [w_{j1}, w_{j2}, \ldots, w_{jn}]^T, j = 1,2, \ldots, m \tag{13}$$

For each of the input vectors, the Euclidean distance between the particular input vector and all weight vectors are calculated. The weight vector with smallest distance, hence highest similarity is chosen as the Best Matching Unit (BMU), w_c, for that input vector, as shown in Eq. (14).

$$\|x_i - w_c\| = \min_j\{\|x_i - w_j\|\} \tag{14}$$

After the first iteration of finding the BMUs, the values of weight vectors are updated so that each BMU is topologically closer to the input vector. The updated is computed as Eq. (15):

$$w_j(t+1) = w_j(t) + \alpha(t) h_{j,w_c}(t) \left(x - w_j(t)\right) \tag{15}$$

where t denotes the iteration step; h_{j,w_c} denotes the topological neighborhood kernel centered around the BMU, which is typically chosen as Gaussian function; and $\alpha(t)$ denotes the learning rate, which is monotonically decreasing with the training iteration. Through this competitive learning process where weight vectors that are closer to input space get updated with higher weight, the map of weight vectors eventually converge to a certain number of clusters.

Minimum quantization error (MQE) is the distance metric for SOM method (Yu and Wang, 2009), computed as the Euclidean distance between an input vector and its BMU, as shown in Eq. (16). Therefore the training of SOM can be viewed as the process of minimizing the average MQE for input vectors and achieving the optimal map structure. For a testing process, where the present degradation condition is assessed, features are used as input vectors for the trained map as they are collected. The MQE value is calculated for each feature vector against its BMU in the trained SOM map, which can be found in the "codebook" of the SOM. The larger the MQE value is, the more severe the degradation is.

$$MQE = \|x - w_{BMU}\| \tag{16}$$

2.5. Fault Localization

As drivetrain degradation grows and becomes significant, the MQE value is expected to exceed its prescribed threshold and trigger an alarm for fault. It is desirable to locate the fault at component level, so that specific advice can be provided for deciding which component is at a more critical condition and needs to be repaired.

With SOM-MQE technique being used, fault localization is achieved by computing MQE contribution of features and variables from each component. MQE is essentially the Euclidean norm of the vector subtraction between an input vector and its BMU, as shown in Eq. (17), where e_i is the difference for the ith feature among all k features.

$$MQE = \sqrt{e_1^2 + e_2^2 + \cdots + e_k^2} \tag{17}$$

The features can be grouped by drivetrain components based on their contextual information. CMS vibration features are grouped based on location of the sensor that generated the feature, whereas SCADA variables are grouped based on variable names. For each component, its contribution to MQE value is calculated as Eq. (18), where e_i are features of the same component.

$$Contribution = \frac{\sum e_i^2}{MQE^2} \tag{18}$$

3. CASE STUDY

To further validate and implement the aforementioned methodology, a case study based on an offshore wind turbine is conducted.

3.1. Description of Turbine and Data

The test bed is a 3 MW wind turbine. A split torque, three-stage planetary gearbox is used to connect the rotor and the generator on the drivetrain. Schematics of the drivetrain, as well as locations of vibration sensors, are shown in Figure 5.

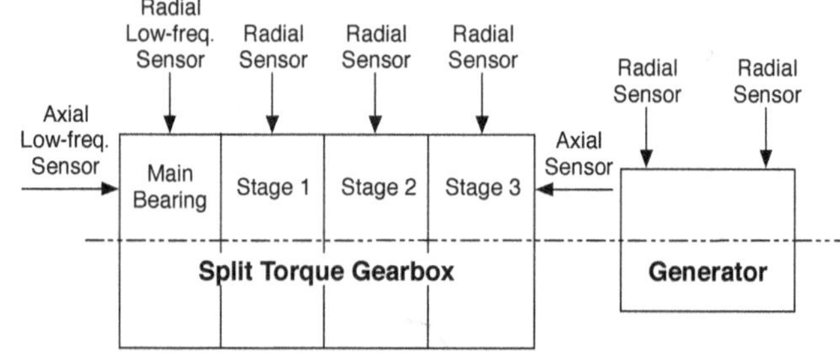

Figure 5. Schematics of test bed drivetrain

The first two stages each consist of a sun gear, planetary gears, a planet carrier and a ring gear, whereas the third stage is a parallel stage with a pair of gears and an output pinion. The input shaft of the gearbox drives ring gear of the first stage and planet carrier of the second stage simultaneously with identical rotation speed. For the first stage, planet carrier does not rotate, thus planetary gears only rotate on their own axes without rolling relative with its sun gear. The sun gear, driven by the ring gear through rotation of planetary gears, connects with the ring gear of the second stage. Therefore all of ring gear, planetary gears and planetary carrier rotate for the second stage, to drive the rotation of its sun gear. Then the sun gear of second stage connects with the third stage, and outputs rotation that drives the generator. Parameters of the gears are listed in Table 2, where CW means the rotation is clockwise and CCW means counterclockwise. The computation of overall gear ratio results in 76.64, which is the nominal ratio between rotation speeds of generator and rotor shaft.

Stage	Gear	No. Of Tooth	Rotation Direction
1st	Sun	66	CW
	Planet (8)	37	CW
	Ring	142	CCW
2nd	Sun	30	CCW
	Planet (4)	62	CCW
	Ring	154	CW
3rd	Input	116	CCW
	Pinion	26	CW

Table 2. List of known gear parameters

In total, eight (8) accelerometers are installed along the drivetrain (Figure 5), with two on the main bearing, four on the gearbox and two on generator. The sampling rate is 6250 Hz for all sensors.

The condition monitoring system is configured to collect vibration data from all accelerometers synchronously once per day at midnight, with a few exceptions that data is collected at a different time of the day or data is collected more than once per day. Data duration for all eight channels of each collected instance is around 85 seconds.

SCADA data is also available for the turbine unit, where statistics of mean, standard deviation, maximum and minimum for over a hundred variables are recorded every ten (10) minutes. The total duration of both CMS and SCADA data is fifteen (15) months. MATLAB® is used for developing all techniques described in previous Section and generating results for this case study.

3.2. Analysis and Results

A filtering algorithm (Grubbs, 1969) that detects outlier observations in a time series is used to reject drastic outliers for each SCADA variable in advance. In this algorithm, the null hypothesis is defined as there is no outlier in a distribution, whereas alternative hypothesis is defined as there is at lease one outlier in the distribution. For any given sample of the distribution, X_i, a Grubbs' test statistics G is generated as Eq. (19), where $\bar{X}$ is the distribution's mean value and σ is the distribution's standard deviation.

$$G = |\bar{X} - X_i|/\sigma \tag{19}$$

A critical value Z is computed as shown in Eq. (20), where N is distribution sample size and t is critical value of the t-distribution with $N-2$ degrees of freedom and $\alpha/(2N)$ of significance level. If $G > Z$, X_i is determined to be an outlier and null hypothesis is rejected.

$$Z = \frac{N-1}{\sqrt{N}} \sqrt{\frac{t^2}{N-2+t^2}} \tag{20}$$

For each SCADA variable, its maximum and minimum values are tested with a significance level of 0.05, and rejected if they are determined to be outliers. Extrema values of the filtered distribution will be tested repeatedly until no outlier is detected.

SCADA records with apparent timestamp error are rejected as well.

For each available CMS data instance, a SCADA record with a matching timestamp is selected as the reference for deciding whether the CMS instance should be discarded or kept. In the case when there is no SCADA record with exact same timestamp for a particular CMS instance since data sampling between the two systems may not be synchronized in most instances, the first SCADA record sampled right after the CMS instance is chosen as the reference.

Four variables in reference SCADA records are then examined with following rules:

- Rotor speed average [rpm] is higher than 0;
- Generator speed average [rpm] is higher than 0;
- Average active power [kw] is higher than 0;
- Average wind speed [m/s] is higher than cut-in wind speed, which in this case is 2.

The corresponding CMS instance is retained when the drivetrain is operational, which is indicated, in most cases, by all four aforementioned rules being met.

Reference SCADA records of retained CMS instances are kept for future analysis. In these records, four variables are selected for degradation assessment:

- Rotor bearing temperature average;
- Gearbox stage 1 temperature average;
- Gearbox stage 2 temperature average;
- Gearbox stage 3 temperature average.

The selected temperature readings are then subtracted by the variable *Environment temperature average*, to offset the seasonal effect on the absolute reading of the variables.

For feature extraction of each channel of CMS vibration data, four categories of features are extracted:

- RMS, kurtosis and crest factor are time domain features;
- Spectral kurtosis is used to filter vibration waveform, with a beforehand STFT of window size 256 samples and 80 percent of overlap. RMS, peak-to-peak and kurtosis values are extracted from the filtered time domain signal;
- Envelope analysis is used to demodulate the signal around resonance frequencies, where resonance frequency is often found at the interval between 1000 Hz and 1600 Hz. Five features are extracted from the demodulated signal: RMS of envelope, RMS of band-pass filtered data, maximum peak of envelope spectrum in low frequency range, frequency index of the peak, and crest factor of envelope spectrum. In this specific case study, bearing configuration parameters that are indispensable for bearing fault frequency calculation are not available. Therefore fault frequencies are not considered in this example.
- Wavelet energy analysis is conducted with Daubechies 4 wavelet, and five features, including four energy bands for four levels of detail signal and one energy band for level 4 approximation signal, are extracted.

In this case study, features are extracted only from sensor 1 to sensor 6 since the gearbox system is of higher interest and it is more suitable to study analytical methods that are specific for the generator independently.

In total, 96 CMS features are extracted with 16 features per each accelerometer. With selected SCADA variables, 100 health indicators are included in SOM-MQE calculation for degradation assessment. For this case study, a Monte Carlo based statistical method is (Bechhoefer, He & Dempsey, 2011) adopted to generate a threshold for triggering fault detection when MQE value exceeds the threshold. To be conservative and avoid users' disbelief due to false alarms, a probability of false alarm (PFA) is set as 10^{-6}. Gaussian distribution and Rayleigh distribution are fit with the first 30 MQE instances , where Rayleigh distribution has a lower negative log likelihood value and is a better fit for the instances. The same test is conducted for varying number of instances (25 – 35) and Rayleigh distribution consistently outperforms Gaussian distribution, therefore the distribution is assumed to be Rayleigh in this case. Threshold values are calculated based on the different numbers of training instances, and the average is taken to be the eventual threshold.

The reminder of the historical data is used for testing the SOM model. Each instance is input to the trained SOM, to generate a MQE value. Eventually, the trend of MQE (in dB) progression over time is shown in Figure 6.

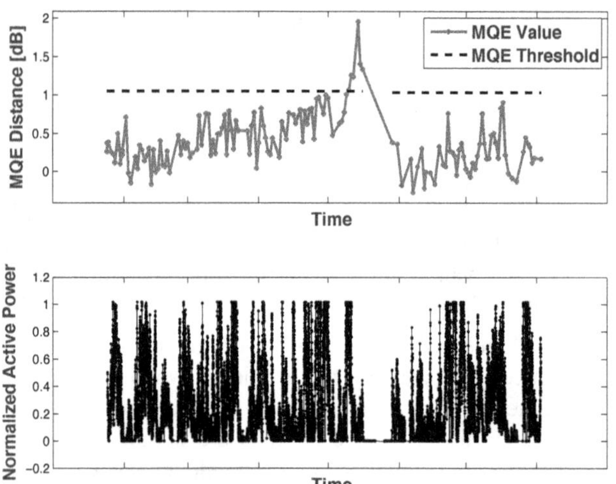

Figure 6. SOM-MQE result for degradation assessment

As observed from the SOM-MQE result, there is a short duration in the middle of the history when MQE value noticeably exceeded the MQE threshold. Due to data confidentiality issue, the exact dates cannot be revealed. However, it can be found from SCADA variable *Average active power [kW]* that the turbine unit was producing zero power for a two-week duration. It was probably triggered by simply monitoring the level of SCADA variables and pausing the turbine operation due to certain alerts.

In comparison, the MQE excess occurred about five days before the operation pause. The result shows that SOM-MQE is capable of detecting drivetrain anomaly at an early stage.

After the wind turbine resumes operation, the SOM model is re-trained since there might have been component replacement and drivetrain behavior should be compared with a new baseline. As shown in Figure 6, the new SOM model has a new MQE threshold for fault detection as well.

For fault localization, features are grouped based on sensor locations in the schematics and SCADA variable names. The components here are denoted as component 1, 2, 3 and 4, where actual component names are omitted as proprietary information. When MQE exceeds its threshold and fault is detected, the contribution of MQE increase is calculated for each component based on Eq. (18), which results in 0.9, 0.05, 0.02 and 0.03 for the occurrence of the major downtime. Therefore the critical component in this case is decided to be component 1.

A radar chart is created to view component criticality simultaneously (Figure 7). In this chart, each axis represents the contribution of each component to MQE abnormality. The closer the data point is to the center, the smaller the contribution is.

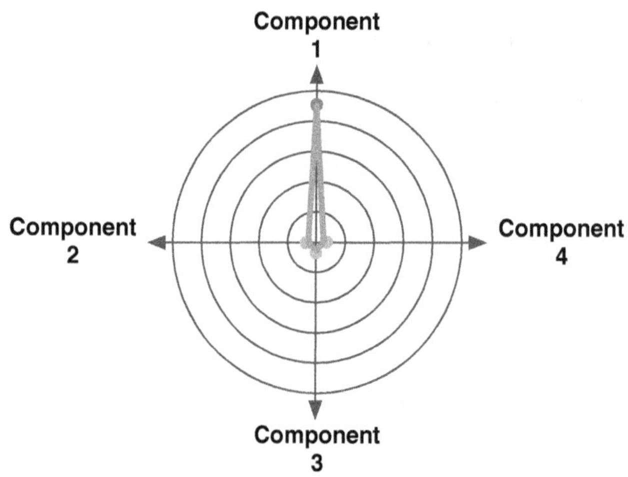

Figure 7. Radar chart for fault localization

3.3. Visualization

To apply the developed tools for large-scale wind farms, a monitoring platform prototype is established for data management, visualization, analysis and fault reporting. The software modules include a) main interface, which directs user to different analytical modules: b) data organization, which sorts data instances and convert them to a compatible format; c) signal visualization and filtering, which provides visual observation of raw signal and configuration of data filtering; d) feature extraction, which enables feature configuration and extraction for various feature types, as well as SCADA variable selection; and e) degradation assessment and fault localization. The main module is able to invoke any other modules for particular tasks, and results from a former module are archived so that a latter module can use them as input. When the analytical methods are improved and validated with more turbine units and data, the platform can operate in a systematic manner to analyze raw SCADA and CMS data and to provide users with direct health degradation information.

4. CONCLUSION AND FUTURE WORK

In this paper, an integrated methodology of degradation assessment and fault localization for wind turbine drivetrain components is presented. The result of the methodology is achieved by combining input from SCADA system and CMS, and validated with a planetary gearbox system of an offshore wind turbine.

Besides using selected SCADA variables, a few feature extraction methods are employed to extract health indicators from CMS vibration data. The methods include time domain features, spectral kurtosis filtering, envelope analysis and wavelet energy analysis. A Self-organizing Map and minimum quantization error approach is adopted to evaluate the degradation condition of drivetrain, and contribution calculation is used to decide the location of defect on the drivetrain. In the case study, an incipient defect is detected and located before detection by the existing system, indicating the potential of predictive monitoring with the presented methods.

In terms of future work, there are a few items to be considered for improving the methods and applications

- Regarding wavelet transform for feature extraction, the selection of mother wavelet function is crucial for obtaining the optimal decomposition results. Rather than depending on experience, preference or visual inspection, an intelligent method can be designed and validated for monitoring either onshore or offshore turbine drivetrains. In literature, there has been interesting and valuable investigation for reference (Rafiee, Tse, Harifi and Sadeghi, 2009).
- There are other multimodal methods that are applicable for degradation assessment. For example, Gaussian mixture model (GMM) estimates data distribution as linear combination of multivariate Gaussian distribution components. A L2 distance metric can be used to measure degradation.
- In the case study, rotation speed for CMS vibration data is missing. As a result, there are some techniques that could not be evaluated on the test bed, such as time synchronous average, especially for the variable speed transmission system of wind turbines (Zhang, Wen and Wu, 2012). If rotational speed can be made available in the future, perhaps through a tachometer signal, more in-depth analysis can be conducted to investigate fault diagnosis methods.
- As bearing specifications are not available for the presented case study, bearing failure frequencies are not inspected and potential bearing faults are not explored. Adding bearing-specific knowledge in the future will enable bearing diagnosis for different stages of drivetrain with given sensors.
- The CMS instance selection method, which is proposed and implemented in this paper, can be incorporated with control of data sampling for CMS. CMS can refer to SCADA variables to evaluate if certain time duration is suitable for vibration data acquisition. An adaptive sampling mechanism can be developed to ensure CMS data quality, improve computation efficiency, and enhance degradation model accuracy.

ACKNOWLEDGEMENT

The authors would like to acknowledge the collaborators at the Sinovel Wind Group, especially Mr. Wei Hu, Mr. Feng Wang and Dr. Yubin Zhu, for their support in terms of condition monitoring system expertise, wind turbine operation experience and providing data for the case study. The authors would like to acknowledge the National Science Foundation (NSF) for its support of research and project activities at Industry/University Cooperative Research Center for Intelligent Maintenance Systems (IMS) at the University of Cincinnati, as well as IMS industry members for offering applied research opportunities that proves to be fundamental to this specific area. The authors would like to acknowledge Ms. Christina Lucas for reviewing the manuscript.

APPENDIX

Screenshots of software modules discussed in Section 3.3 are included in this appendix.

a) Main interface

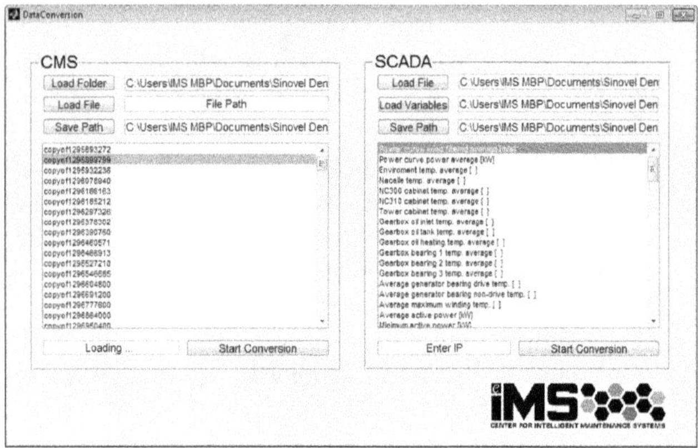

b) Data organization module

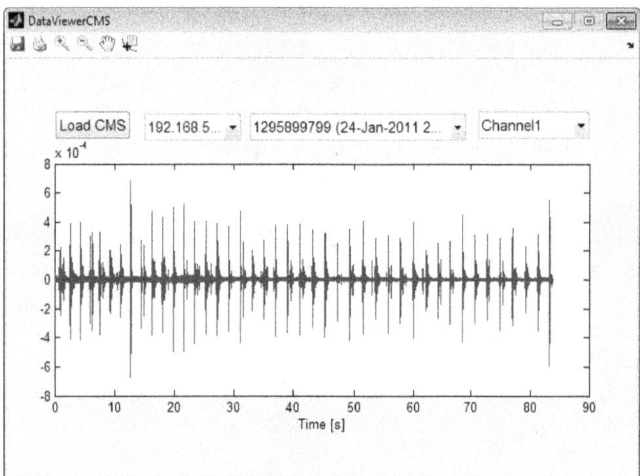

c) CMS data visualization module

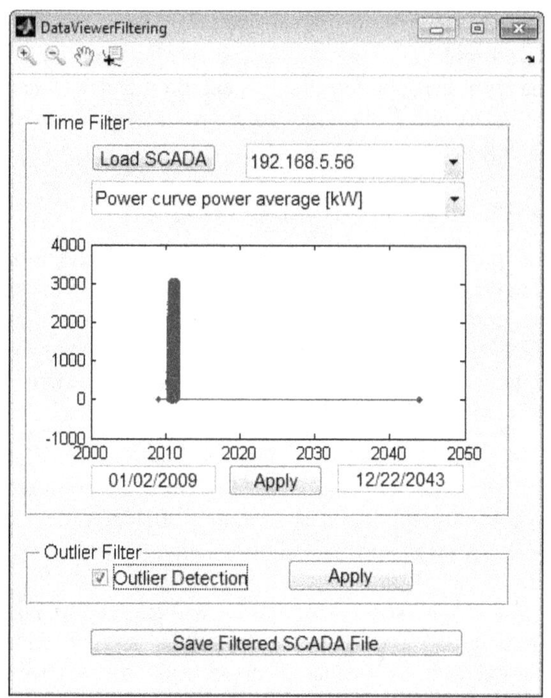

d) SCADA variable filtering module

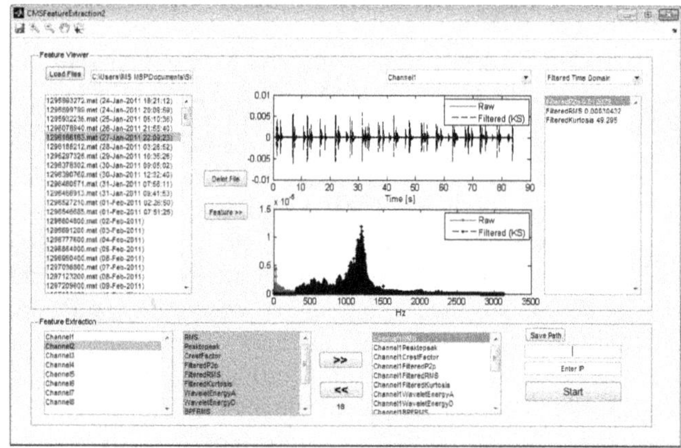

e) CMS feature extraction module

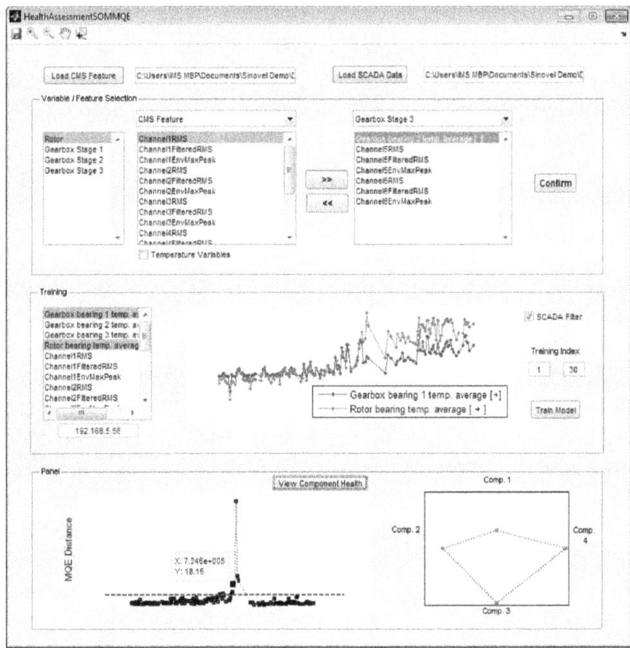

f) Degradation assessment and fault localization module

Figure 8. Prototype of wind farm monitoring platform

REFERENCES

Amirat, Y., Benbouzid, M. E. H., Al-Ahmar, E., Bensaker, B., & Turri, S. (2009). A brief status on condition monitoring and fault diagnosis in wind energy conversion systems. *Renewable and Sustainable Energy Reviews*, 13 (9), pp. 2629-2636. doi:10.1016/j.rser.2009.06.031

Antoni, J. (2006). The spectral kurtosis: a useful tool for characterising non-stationary signals. *Mechanical Sysetms and Signal Processing*, 20 (2), pp. 282-307. doi:10.1016/j.ymssp.2004.09.001

Barszcz, T. & Randall, R. (2009). Application of spectral kurtosis for detection of a tooth crack in the planetary gear of a wind turbine. *Mechanical Systems and Signal Processing*, 23 (4), pp. 1352-1365. doi:10.1015/j.ymssp.2008.07.019

Bechhoefer, E., He, D., & Dempsey, P. (2011). Gear health threshold setting based on a probability of false alarm. *Annual Conference of the Prognostics and Health Management Society, September 25 – 20, Montreal, Canada.*

Crabtree, C. J., Feng, Y., & Tavner, P. J. (2010). Detecting incipient wind turbine gearbox failure: A signal analysis method for on-line condition monitoring. *Scientific Track Proceedings of European Wind Energy Conference*, April 20-23, Warsaw, Poland.

Edwin, W., Theo, V., Henk, B., Luc, R., Xiang, J., & Simon, W. (2008). Assessment of condition monitoring techniques for offshore wind farms. *Journal of Solar Energy Engineering, 130 (3), pp. 0310041-0310049.*

Entezami, M. (2010). *Wind turbine condition monitoring system*. Doctoral dissertation report. University of Birmingham, Birmingham, United Kindom. http://www.sampaolesi-impiantielettrici.com/documenti/eolico/Wind_Turbines_Condition_Monitoring_Systems_University_of_Birmingham_2010.pdf

Feng, Y., Qiu, Y., Crabtree, C. J., Long, H., & Tavner, P. J. (2011). Use of SCADA and CMS signals for failure detection and diagnosis of a wind turbine gearbox. *European Wind Energy Conference & Exhibition*, March 14-17, Brussels, Belgium.

Fraunhofer-Institute for Wind Energy and Energy System Technology. (2005). Final report: Advanced maintenance and repair for offshore wind farms using fault prediction and condition monitoring techniques. http://ec.europa.eu/energy/renewables/wind_energy/doc/offshore.pdf

Global Wind Energy Council. (2012). *Global Wind Energy Outlook 2012*. http://www.gwec.net/wp-content/uploads/2012/11/GWEO_2012_lowRes.pdf

Grubbs, F. E. (1969). Procedures for detecting outlying observations in samples. *Technometrics*. 11 (1), pp. 1-21. doi:10.1080/00401706.1969.10490657

Hameed, Z., Hong, Y. S., Cho, Y. M., Ahn, S. H., & Song, C. K. (2009). Condition monitoring and fault detection of wind turbines and related algorithms: a review. *Renewable and Sustainable Energy Reviews*, 13 (1), pp. 1-39. doi:10.1016/j.rser.2007.05.008

Jiang, Y., Tang, B., Qin, Y., & Liu, W. (2011). Feature extraction method of wind turbine based on adaptive Morlet wavelet and SVD. *Renewable Energy*, 36 (8), pp. 2146-2153. doi:10.1016/j.renene.2011.01.009

Kohonen, T. (1990). The self-organizing map. *Proceedings of the IEEE*, 78 (9), pp. 1464-1480.

Lapira, E., Siegel, D., Zhao, W., Brisset, D., Su, J., Wang, C., AbuAli, M., & Lee, J. (2011). A systematic framework for wind turbine health assessment under dynamic operating conditions. *Proceedings of the 24th International Congress on Condition Monitoring and Diagnostics Engineering Management*, (1-9), May 30 – June 1, Stavanger, Norway.

LeBlanc, M., & Graves, A. (2011). Condition monitoring systems: trends and cost benefits. *NREL Wind Turbine Condition Monitoring Workshop*, September 19-20, Broomfield, CO. http://www.nrel.gov/wind/pdfs/day1_sessioni_04_garradhassan_leblanc.pdf

Lebold, M., McClintic, K., Campbell, R., Byington, C., & Maynard, K. (2000). Review of vibration analysis methods for gearbox diagnostics and prognostics. *Proceedings of the 54th Meeting of the Society for Machinery Failure Prevention Technology* (623-634), May 1-4, Virginia Beach, VA, USA

Liu, B., Riemenschneider, S., & Xu, Y. (2006). Gearbox fault diagnosis using empirical mode decomposition and Hilbert spectrum. *Mechanical Systems and Signal Processing*, 20 (3), pp. 718-734. doi:10.1016/j.ymssp.2005.02.003

Lu, B., Li, Y., Wu, X., & Yang, Z. (2009). A review of recent advances in wind turbine condition monitoring and fault diagnosis. *Proceedings of IEEE Power Electronics and Machines in Wind Applications,* (1-7), June 24-26, Lincoln, NE.

McMillan, D., & Ault, G. W. (2007). Quantification of condition monitoring benefit for offshore wind turbines. *Wind Engineering*, 31 (4), pp. 267-285. doi: 10.1260/030952407783123060

Meadows, B. (2011). Offshore wind O&M challenges. *NREL Wind Turbine Condition Monitoring Workshop*, September 19-20, Broomfield, CO. http://www.nrel.gov/wind/pdfs/day1_sessioni_05_nrel_meadows.pdf

Musial, W., & Ram, B. (2010). Large-scale offshore wind power in the United States: Assessment of opportunities and barriers. National Renewable Energy Laboratory Report No. TP-500-40745.

Peng, Z. K., & Chu, F. L. (2004) Application of the wavelet transform in machine condition monitoring and fault diagnostics: a review with bibliography. *Mechanical Systems and Signal Processing*, 18 (2), pp. 199-221. doi:10.1016/S0888-3270(03)00075-X

Peng, Z. K., Tse, P. W., & Chu, F. L. (2005). An improved Hilbert-Huang transform and its application in vibration signal analysis. *Journal of Sound and Vibration*, 286 (1-2), pp. 187-205. doi: 10.1016/j.jsv.2004.10.005

Qiu, Y., Feng, Y., Tavner, P., Richardson, P., Erdos, G., & Chen, B. (2012). Wind turbine SCADA alarm analysis for improving reliability. *Wind Energy*, 15 (8), pp. 951-966. doi: 10.1002/we.513

Rafiee, J., Tse, P. W., Harifi, A., & Sadeghi, M. H. (2009). A novel technique for selecting mother wavelet function using an intelligent fault diagnosis system. *Expert Systems with Applications*, 36 (3), pp. 4862-4875. doi:10.1016/j.eswa.2008.05.052

Sheng, S., & Veers, P. (2011). Wind turbine drivetrain condition monitoring – An overview. *Machinery Failure Prevention Technology: The Applied Systems Health Management Conference*, May 10-12, Virginia Beach, VA.

Siegel, D., Zhao, W., Lapira, E., AbuAli, M., & Lee, J. (2013). A comparative study on vibration-based condition monitoring algorithms for wind turbine drive trains. *Wind Energy*, doi:10.1002/we.1585

Yang, W., Tavner, P. J., & Wilkinson, M. R. (2008). Condition monitoring and fault diagnosis of a wind turbine synchronous generator drive train. *IET Renewable Power Generation*, 3 (1), pp. 1-11. doi: 10.1049/iet-rpg:20080006

Yu, J., & Wang, S. (2009). Using minimum quantization error chart for the monitoring of process states in multivariate manufacturing processes. *Computers & Industrial Engineering*, 57 (4), pp. 1300-1312. doi:10.1016/j.cie.2009.06.009

Zhang, X., Wen, G., & Wu, T. (2012). A new time synchronous average method for variable speed operating condition gearbox. *Journal of Vibroengineering*, 14 (4), pp. 1766-1774.

BIOGRAPHIES

Wenyu Zhao is currently a Ph.D. researcher at Center for Intelligent Maintenance Systems (IMS) at the University of Cincinnati, and a graduate student with the School of Dynamic Systems. He has conducted research and industry projects in area of Prognostics and Health Management (PHM) for areas including rotary machinery, renewable energies, hydraulic system and vehicle system. Prior to joining IMS, he received his B.S.E in Mechanical Engineering from Shanghai Jiao Tong University (SJTU). His work experience includes internships at Parker Hannifin Corp. and Metal Industry Research and Development Centre. He won 2nd place in 2009 PHM Data Challenge. His current research interest is to develop a unified framework and methodology for prognostics of an interconnected system.

David Siegel is currently a Ph.D. student in Mechanical Engineering at the University of Cincinnati, and a research assistant for the Center for Intelligent Maintenance Systems. Related work experience in the field of prognostics and health management include internships at General Electric Aviation, and at the U.S Army Research Lab. He is also a two-time winner of the prognostics and health management data challenge. His current research focus is on component-level prognostic methods, as well as health monitoring algorithms for systems operating under multiple operating regimes.

Jay Lee is an Ohio Eminent Scholar and L.W. Scott Alter Chair Professor in Advanced Manufacturing at the University of Cincinnati. He is the founding director of National Science Foundation (NSF) Industry/University Co-operative Research Center (I/UCRC) for Intelligent Maintenance Systems (IMS), which is a multi-campus NSF Center of Excellence between the University of Cincinnati (lead institution), the University of Michigan, and the Missouri University of Science and Technology. His current research areas include autonomic computing, embedded IT and smart prognostics technologies for industrial and healthcare systems, cloud-based prognostics as a service, self-aware sensor and system, and dominant design tools for product and service innovation. He is a Fellow of ASME, SME, and International Society of Engineering Asset Management (ISEAM).

A Fuzzy-FMEA Risk Assessment Approach for Offshore Wind Turbines

F. Dinmohammadi[1], M. Shafiee[2]*

[1] *Renewable Energy Organization of Iran (SANA), Tehran, Iran*
[2] *School of Applied Sciences, Cranfield University, Cranfield, Bedfordshire, MK43 0AL, United Kingdom*

mahsha@chalmers.se

ABSTRACT

Failure Mode and Effects Analysis (FMEA) has been extensively used by wind turbine assembly manufacturers for risk and reliability analysis. However, several limitations are associated with its implementation in offshore wind farms: (i) the failure data gathered from SCADA system is often missing or unreliable, and hence, the assessment information of the three risk factors (i.e., severity, occurrence, and fault detection) are mainly based on experts' knowledge; (ii) it is rather difficult for experts to precisely evaluate the risk factors; (iii) the relative importance among the risk factors is not taken into consideration, and hence, the results may not necessarily represent the true risk priorities; and etc. To overcome these drawbacks and improve the effectiveness of the traditional FMEA, we develop a fuzzy-FMEA approach for risk and failure mode analysis in offshore wind turbine systems. The information obtained from the experts is expressed using fuzzy linguistics terms, and a grey theory analysis is proposed to incorporate the relative importance of the risk factors into the determination of risk priority of failure modes. The proposed approach is applied to an offshore wind turbine system with sixteen mechanical, electrical and auxiliary assemblies, and the results are compared with the traditional FMEA.

1. INTRODUCTION

Offshore wind energy has experienced an extensive and worldwide growth during the past several years. For instance, of the 9,616 MW installed wind energy capacity in the EU in 2011, 866 MW (i.e., 9%) was offshore, which increased the EU's offshore wind power capacity to 3,810 MW—less than one percent of the total electricity demand (EWEA, 2012). Certain forecasts indicate that the share of offshore wind power in EU's electricity demand will reach up to 14% by 2030.

Comparing with onshore wind power, offshore winds tend to flow at higher speeds, thus it allows turbines to produce more electricity (Bilgili, Yasar and Simsek, 2011). However, a wind power system located on sea comes with higher failure rate, lower reliability, and higher operation and maintenance (O&M) costs. So, with the development of wind farms in remote areas, the need for efficient tool to identify and then limit or avoid risk of failures is of increasing importance.

Failure Mode and Effects Analysis (FMEA) has been extensively used by wind turbine assembly manufacturers for analyzing, evaluating, and prioritizing the potential failure modes (Andrawus, 2008). FMEA is a structured, bottom-up approach that starts with potential/known failure modes at one level and investigates the effect on the next sub-system level (Kumar and Kumar, 2005). Hence, a complete FMEA analysis of a system often spans all the levels in the hierarchy from bottom to top (see Fig. 1).

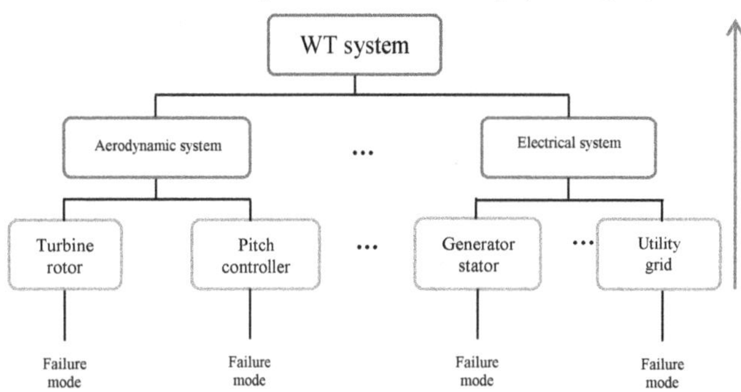

Fig. 1– Hierarchical structure of a typical wind turbine system.

A failure mode is defined as the way in which a component, subsystem or system could potentially fail to perform its desired function. Examples of failure modes in wind turbine systems are: material fatigue, deterioration, deformation, strips, fracture, detachment, blockage, misalignment, collapse, and etc. (Tavner, Xiang and Spinato, 2007).

A failure cause is defined as a weakness that may result in a failure. Typical causes of failures in wind turbine systems are: using incorrect material, poor welding, corrosion, assembly error, calibration error, over stressing, overheating, icing, maintenance fault, forming of cracks, being out of balance, connection failure, and etc.

The failure modes are usually detected through visual inspection, *online* condition monitoring techniques – such as oil analysis and ultrasonic testing (for more see Márquez, Tobias, Pérez and Papaelias, 2012), and time-based preventive maintenance actions. For each identified failure mode, their ultimate effects need to be determined by a cross-functional team which is usually formed by specialists from various functions (e.g., design, operation and maintenance, and power production). A failure effect is defined as the result of a failure mode on the function of the system as perceived by the user. Some of the effects of a failure in wind turbine systems are loss of electricity production, poor power quality to the grid, and a significant audible noise. Also, the effects of a failure in one component can be the cause of a failure mode in another component.

As outlined by Pillay and Wang (2003), the process for carrying out an FMEA can be divided into several steps as shown in Fig. 2.

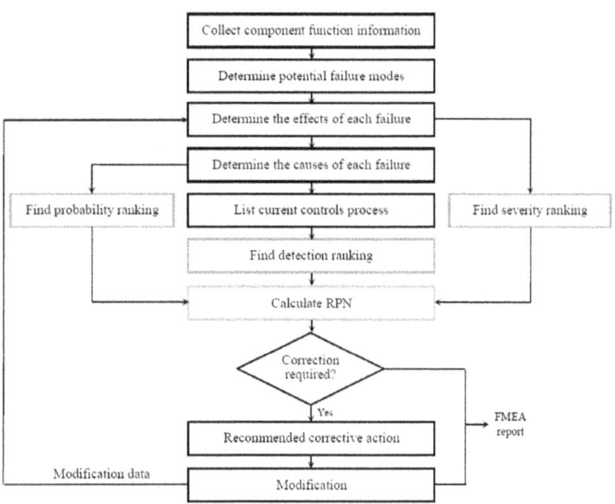

Fig 2– FMEA process.

Basically, each failure mode in the FMEA technique is evaluated by three factors as severity (S), likelihood of occurrence (O), and the difficulty of detection (D). A number between 1 and 10 (with 1 being the best and 10 being the worst case) is given for each of the three factors, and a risk-priority-number (RPN) is obtained, which is RPN = $S \times O \times D$. The RPN value helps the FMEA team to identify the components or subsystems that need the priority actions for improvement. Depending on the wind farm manager's decision, different criteria are used to trigger the improvement actions. For instance, action could be required if the overall RPN exceeds a predefined threshold, or for the highest RPN regardless of a threshold. Finally, at the last step, some hardware, software or design modifications are made in the system to minimize the failure effects.

Even though FMEA is probably the most popular tool for reliability and failure mode analysis in wind turbine systems, several limitations are associated with its implementation in offshore wind farms:

(i) The failure data gathered from inspectors, vibration sensors, and the SCADA system is often missing or unreliable. Hence, the assessment information of three risk factors (severity, occurrence, and detection) is mainly based on experts' knowledge and expertise;
(ii) Comparing with onshore wind power, the history of offshore wind power generation is fairly recent. Hence, it is difficult or even impossible for experts to precisely evaluate the three risk factors S, O and D. The risk factors are often expressed in a linguistic way (such as 'likely' , 'important', 'very high' and etc);
(iii) In the traditional FMEA methodology, the three risk factors are assumed to have the same importance (Braglia, 2000). However, it is observed that many O&M experts give more preference to the 'fault detection' factor.

So, the results of the traditional FMEA methodology may not necessarily represent the true risk priorities in offshore wind turbine systems, and this can entail a waste of resources and time.

To overcome the above drawbacks and improve the effectiveness of the traditional FMEA methodology, we develop a fuzzy-FMEA approach to determine the effects of failure on offshore wind turbine systems. Firstly, a fuzzy inference approach is considered to represent the assessment information using linguistic terms. Then, by using the weight vector of three risk factors, a grey theory analysis is proposed to rank the failure modes. To our knowledge, this paper is the first attempt to make the traditional FMEA methodology more applicable for offshore wind turbine systems, especially when the failure data is unavailable or unreliable.

The rest of this paper is organized as follows. In Section 2, we give a brief overview of FMEA methodology so as to set the background for the main contribution of the paper. Section 3 describes the wind turbine system considered in this paper. In section 4, the proposed fuzzy approach which utilizes the fuzzy IF–THEN rules and grey relation analysis. Finally, in section 5, the results obtained from the proposed approach are compared with the traditional FMEA.

2. FMEA: AN OVERVIEW

FMEA as a formal system analysis methodology was first proposed by NASA in 1963 for their obvious reliability requirements. Then, it was adopted and implemented by

Ford Motor in 1977 (Gilchrist, 1993). Since then, it has become a powerful tool extensively used for risk and reliability analysis of systems in a wide range of industries, including automotive, construction, aerospace, nuclear, and electro-technical.

2.1. FMEA in Wind Turbines

A brief review of the literature shows that only a few researchers have worked on improving the traditional FMEA methodology to make it more practical for wind turbine systems. Arabian-Hoseynabadi, Oraee and Tavner (2010) presented a design-stage FMEA methodology for prioritization of failures in a 2-MW wind turbine system (named as R80) within the RELIAWIND project. The authors' methodology used four-point scales for severity rating (Table 1), occurrence rating (Table 2), and detection of a failure (Table 3) to represent the risk of the 64 possible severity–occurrence–detection combinations.

Table 1– Severity rating scale for wind turbine FMEA.

Scale #	Description	Criteria
1	Category IV (minor)	Electricity can be generated but urgent repair is required.
2	Category III (marginal)	Reduction in ability to generate electricity.
3	Category II (critical)	Loss of ability to generate electricity.
4	Category I (catastrophic)	Major damage to the Turbine as a capital installation.

Table 2– Occurrence rating scale for wind turbine FMEA.

Scale #	Description	Criteria
1	Level E (extremely unlikely)	A single failure mode probability of occurrence is less than 0.001.
2	Level D (remote)	A single failure mode probability of occurrence is more than 0.001 but less than 0.01.
3	Level C (occasional)	A single failure mode probability of occurrence is more than 0.01 but less than 0.10.
5	Level A (frequent)	A single failure mode probability of occurrence is greater than 0.10.

Table 3– Detection rating scale for wind turbine FMEA.

Scale #	Description	Criteria
1	Almost certain	Current monitoring methods almost always will defect the failure.
4	High	Good likelihood current monitoring methods will detect the failure.
7	Low	Low likelihood current monitoring methods will defect the failure.
10	Almost impossible	No known monitoring methods available to detect the failure.

From the scales that they assign to the three risk factors, the following results can be concluded:

(i) The proposed methodology gives importance weights of (0.21, 0.26, 0.53) to (S, O, D). This implies that their methodology gives more preference to the fault detection factor.

(ii) From the existing sixty-four combinations, only thirty-nine different RPN values can be obtained, which they are heavily distributed at the bottom of the scale from 1 to 100 (see Fig. 3).

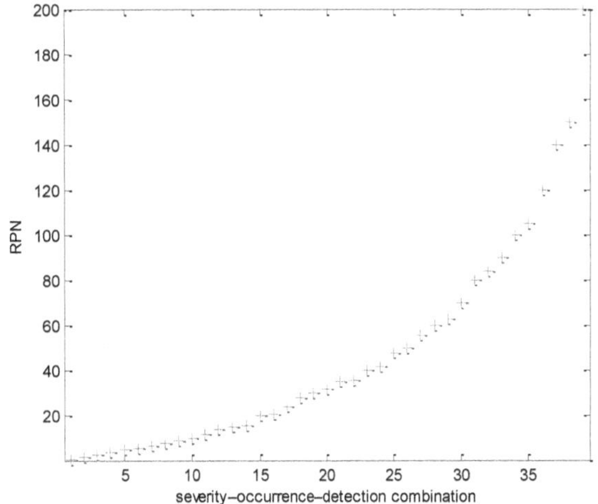

Fig 3–The RPN values for the sixty-four 'occurrence–severity–detection' combinations

Kahrobaee and Asgarpoor (2011) proposed a cost-priority-number (CPN) approach, in which the system's risk is calculated as $C \times P \times N$, where C is the cost consequences of each failure, P is the probability of occurrence, and N is the possibility of not detection. This approach has been recently extended in Dinmohammadi and Shafiee (2013) by incorporating all the costs associated with each failure (corrective replacement, spare parts, transportation, manpower, and production loss) in calculating the CPN value. Also, a quantitative study is carried out on two the same type of onshore and offshore wind turbines, and some useful comparisons are made.

2.2. Fuzzy FMEA

Fuzzy logic is a tool for transforming the vagueness of human feeling and recognition into a mathematical formula. It also provides meaningful representation of measurement for uncertainties and vague concepts expressed in natural language. In line with this, there has been a growing trend in FMEA literature to use fuzzy linguistic terms for describing the three risk factors S, O, and D. Readers can refer to Yang, Bonsall, and Wang (2008); Keskin and Özkan (2009); Gargama and Chaturvedi (2011) as good sources of fuzzy-FMEA approach. Most of the existing studies in the fuzzy FMEA literature have concerned with the fuzzy rule-base approach by using 'If–Then' rules. Fig. 4 shows an overall view of the fuzzy rule base technique, in which there are three major steps to carry out the assessment (Chin, Chan and Yang, 2008):

(i) *Fuzzification* process uses linguistic variables to convert the three risk factors S, O and D into the fuzzy representations. Using the linguistic variables and their definitions, ranking three risk factors can be made in a scale basis. These inputs are then fuzzified to determine the degree of membership in each input class.

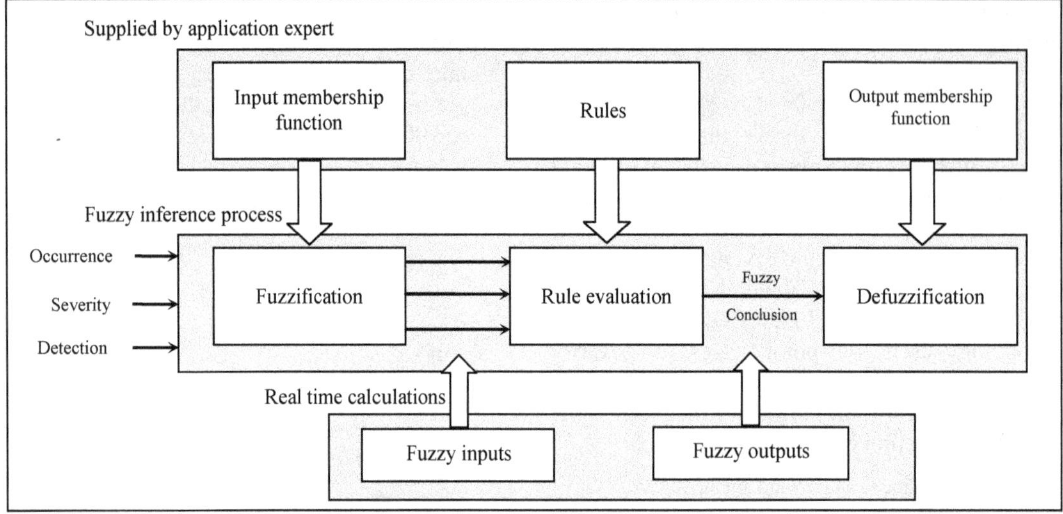

Fig 4–Overall view of the fuzzy-FMEA approach

(ii) *Rule evaluation* consists of the expert knowledge about the interactions between various failure modes and effect that is represented in the form of fuzzy if–then rules. Such rules are usually more conveniently formulated in linguistic terms than in numerical terms. The outputs of the fuzzy inference system are variously named as 'riskiness', 'critically failure mode', 'priority for attention', and 'fuzzy RPN' in the fuzzy FMEA studies.

(iii) *Defuzzification* process creates a crisp ranking from the fuzzy RPN to give the prioritization level for the failure modes.

3. WIND TURBINE SYSTEM CONSIDERED IN THIS STUDY

Nowadays, many kinds of wind turbine systems compete in the market. According to Li and Chen (2008), wind turbines can be categorized by their generator, gearbox, and their power converter types.

Fixed speed wind turbines which operate with constant speed 'Danish concept' were produced until the late 1990s with the power ratings below 1-MW. They used a multi-stage gearbox, and a standard squirrel-cage induction generator directly connected to the grid through a transformer. From the late 1990s, fully variable speed wind turbines were introduced in wind power industry. The first generation of fully variable speed wind turbines (with power ratings approximately 1-MW) used a multi-stage gearbox, a relatively low-cost standard wound rotor induction generator, and a power electronic converter feeding the rotor (Carlin, Laxson and Muljadi, 2003). The doubly fed induction generator (DFIG) technology is currently the most widely used in the wind turbine industry because of its low investment cost and good energy yield (Muller, Deicke and De Doncker, 2002). Since 1991, there have also been variable speed wind turbines with gearless generator systems which are equipped with a direct-drive generator and a fully-rated power electronic converter. The brushless doubly fed induction generator (BDFIG) is a well known drive technology which eliminates the need for brushes and slip rings, increases the lifetime of the machine, and ultimately reduces the maintenance costs (Carlson, Voltolini, Runcos and Kuo-Peng, 2006).

This paper focuses on a 5-MW REpower MM92 wind turbine system (Fig. 5), which is available in both onshore and offshore types. This wind turbine system features a non-integrated drive train with a rotor shaft supported by two bearings, a combined planetary/spur wheel gearbox, and a double-fed asynchronous generator. The three-blade rotor with a diameter of 126 meters is also equipped with an electrical blade angle adjustment and a cast iron rotor hub.

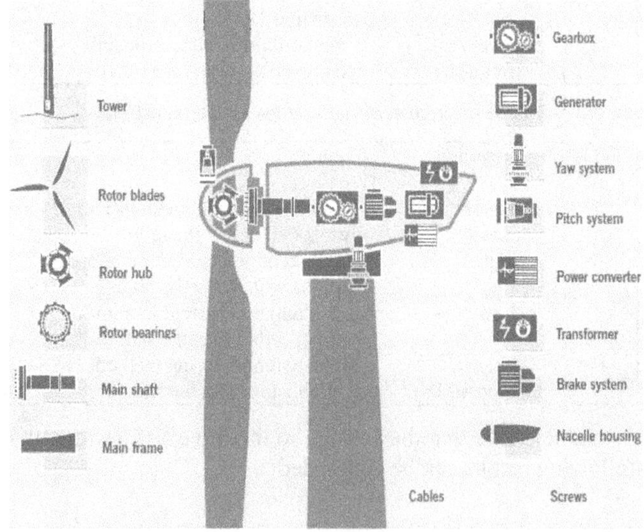

Fig. 5– 5MW REpower MM92 wind turbine system (http://www.REpower.de/)

After recognizing the wind turbine type, we define a general set of the sub-assemblies and main parts. In this study, sixteen sub-assemblies and components with higher failure probabilities and serious consequences have been considered as presented in Table 4. It may be apparent that not all of these components may be available in some types of wind turbine systems.

Table 4– The sixteen sub-assemblies considered in this study (listed in alphabetical order).

ID	Sub-assemblies	Some components
1	Brake system	Brake disk, Spring, Motor
2	Cables	
3	Gearbox	Toothed gear wheels, Pump, Oil heater/cooler, Hoses
4	Generator	Shaft, Bearings, Rotor, Stator, Coil
5	Main frame	
6	Main shaft	Shaft, Bearings, Couplings
7	Nacelle housing	Nacelle
8	Pitch system	Pitch motor, Gears
9	Power converter	Power electronic switch, cable, DC bus
10	Rotor bearings	
11	Rotor blades	Blades
12	Rotor hub	Hub, Air brake
13	Screws	
14	Tower	Tower, Foundation
15	Transformer	Controllers
16	Yaw system	Yaw drive, Yaw motor

After subdivision of the wind turbine system, the potential failure modes of the sub-assemblies are identified using the information gathered from four experts. These experts have experience within the reliability, availability, maintainability and safety (RAMS) of the wind energy industry, ranging from three to six years. The experts used the 'fault tree analysis (FTA)' to describe the complete set of potential system failures (for more see Andrews and Moss, 1993). The FTA is one of the most popular and diagrammatic techniques to analyze the undesired states of a system that uses AND gate (the output occurs only if all inputs occur) and OR gate (the output occurs if any input occurs). Fig. 6 depicts the fault tree diagram for two important sub-assemblies of the wind turbine system: generator and tower.

4. PROPOSED FUZZY-FMEA APPROACH

In this section, a new proposed approach which utilizes the fuzzy IF–THEN rules and grey relation theory is presented. The linguistic terms describing the 'inputs' are *Remote* (R), *Low* (L), *Moderate* (M), *High* (H) and *Very High* (VH), and for 'output' are *Unnecessary* (U), *minor* (mi), *very-low* (vl), *low* (l), *moderate* (mod), *high* (h), *Moderate-high* (Mh), *Very-high* (Vh), *necessary* (n) and *Absolutely-necessary* (A-n).

By using the interpretations of the linguistic terms described in Table 5, the experts were requested to define the membership functions. After receiving the feedback from the experts, the membership function of the linguistic terms defined by triangular fuzzy number (a,b,c) expressing the proposition 'close to b'. Making use of the fuzzy logic toolbox simulator of MATLAB®, the membership functions for the linguistic variables of severity, occurrence, detection, and fuzzy RPN are graphically represented in Fig. 7.

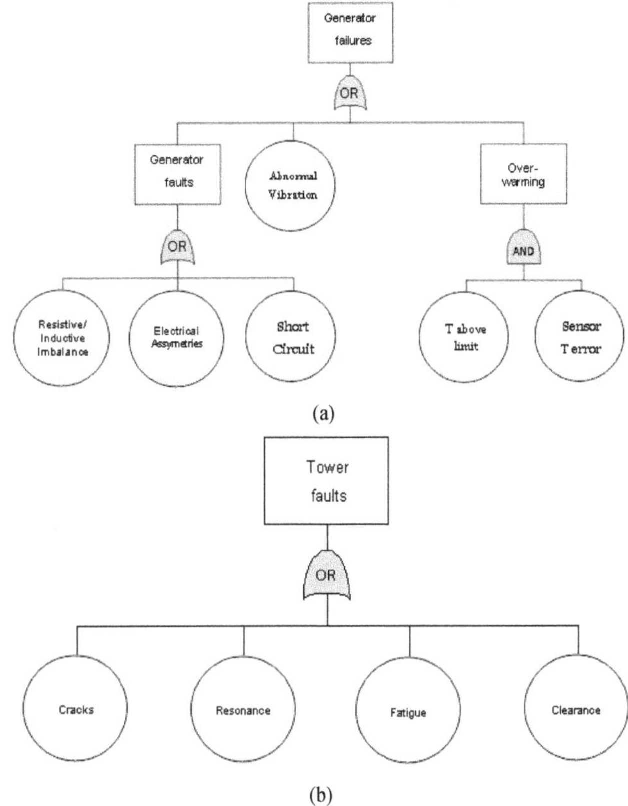

Fig. 6– Fault tree diagram for (a) #4: Generator (b) #14: Tower.

Table 5– Interpretations of the linguistic terms for developing the fuzzy rule system (Guimarães and Lapa, 2007).

Linguistic term	Probability of occurrence	Severity	Detection
Remote	It would be very unlikely for these failures to be observed even once	A failure that has no effect on the system performance, the operator probably will not notice	Defect remains undetected until the system performance degrades to the extent that the task will not be completed
Low	Likely to occur once, but unlikely to occur more frequently	A failure that would cause slight annoyance to the operator, but that cause no deterioration to the system	Defect remains undetected until system performance is severely reduced
Moderate	Likely to occur more than once	A failure that would cause a high degree of operator dissatisfaction or that causes noticeable but slight deterioration in system performance	Defect remains undetected until system performance is affected
High	Near certain to occur at least once	A failure that causes significant deterioration in system performance and/or	Defect remains undetected until inspection or test is carried out

		leads to minor injuries	
Very high	Near certain to occur several times	A failure that would seriously affect the ability to complete The task or cause damage, serious injury or death	Failure remains undetected, such a defect would almost certainly be detected during inspection or test

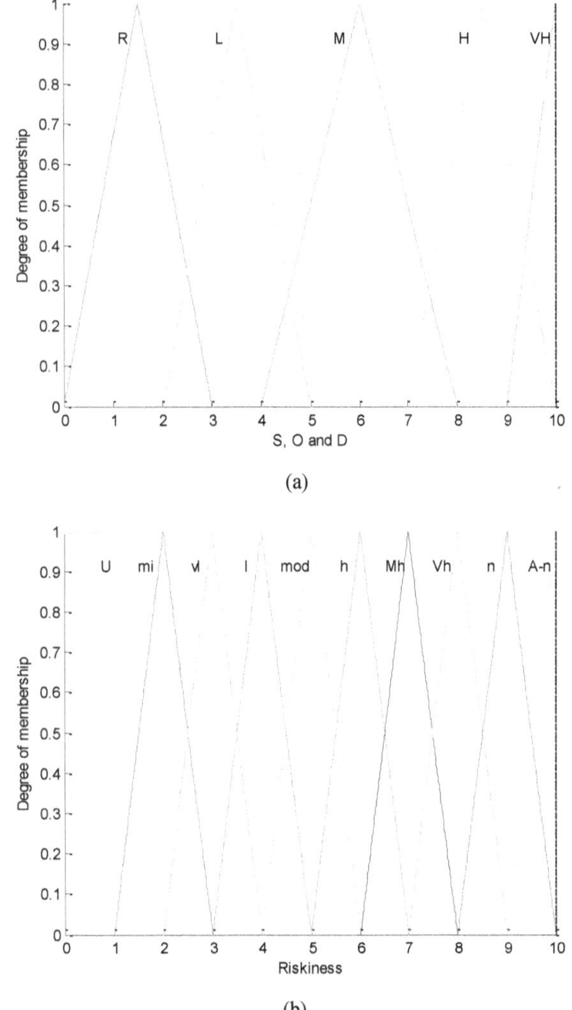

Fig. 7– Membership functions for (a) severity, occurrence, and detection, and (b) riskiness.

4.1. Fuzzy Rule Base

The membership functions derived from the experts are used to generate the fuzzy rule base. A total of $5\times5\times5$ = 125 rules are generated. However, these rules are combined (where possible) and the total number of rules in the fuzzy rule base is reduced to 35 rules. The Rule Viewer of the MATLAB that opens during the simulation can be used to access the 'Membership Function Editor' and the 'Rule Editor'. Through 'Simulator' many results can be evaluated and rules can be removed. For example, consider these three rules:

Rule 1: if Severity is H, Occurrence is M, and Detection is M, then Riskiness is M–h.
Rule 2: if Severity is M, Occurrence is H, and Detection is H, then Riskiness is M–h.
Rule 3: if Severity is H, Occurrence is H, and Detection is M, then Riskiness is M–h.

Rules 1, 2 and 3, can be combined to produce: ''if Severity is H, Occurrence is M, and Detection is M, then Riskiness is M–h'' or any combination of the three linguistic terms assigned to these variables, then Riskiness is M–h.
The results of the fuzzy rule base are then defuzzified using the *centroid* method (see Cheng, 1998) to obtain the crisp value of 'riskiness' for ranking the failure modes. The defuzzified crisp numbers of ten output linguistic terms are given in Table 6.

Table 6–The defuzzified crisp numbers of output linguistic terms

U	mi	vl	l	Mod	h	M-h	V-h	n	A-n
0.81	2.03	3.02	4.01	5.01	6.01	7.01	8.01	9.01	9.67

Fuzzy inference functions used in this application are:
name: 'FMEA_WT'
type: 'Mamdani'
andMethod: 'min'
orMethod: 'max'
defuzzMethod: 'centroid'
impMethod: 'min'
aggMethod: 'max'

4.2. Grey Relational Analysis

In the proposed rule base reduction, the three risk factors S, O and D are assumed to have the same importance. To assign different weights to the three risk factors, grey theory approach is suggested within the FMEA framework. Grey theory was first proposed and developed by Deng (1989) to deal with making decisions characterized by *incomplete* information. Indeed, it provides a measure to analyze relationship between discrete quantitative and qualitative series. The process for carrying out a grey relation analysis in FMEA involves several steps as shown in Fig. 8 (Liu, Liu, Bian, Lin, Dong and Xu, 2011).

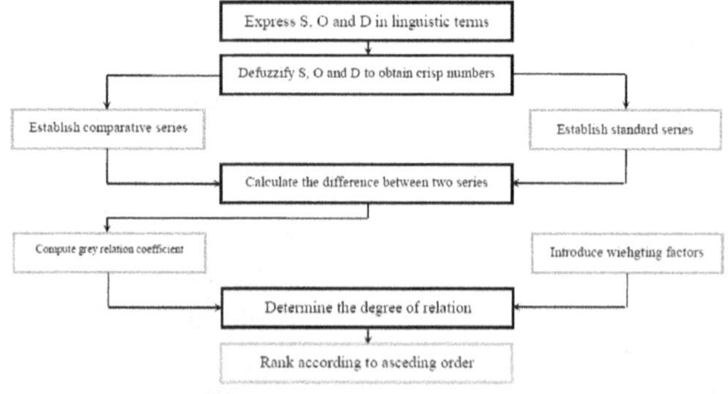

Fig 8–Grey relation analysis in FMEA

(i) Express S, O and D by linguistic terms and the membership functions as shown in Fig. 7(a).

(ii) Defuzzify S, O and D using Chen and Klien's method (1997) for obtaining the crisp number of a fuzzy set as shown in Eq. (1).

As an example, consider the defuzzification of the linguistic term *Low* in Fig. 9. This linguistic term can be defuzzified to

$$K(x) = \frac{\sum_{i=0}^{n}(b_i - c)}{\sum_{i=0}^{n}(b_i - c) - \sum_{i=0}^{n}(a_i - d)}, \quad (1)$$

$$\frac{[5-0]+[3.5-0]}{\{[5-0]+[3.5-0]\}-\{[2-10]+[3.5-10]\}} = 0.370$$

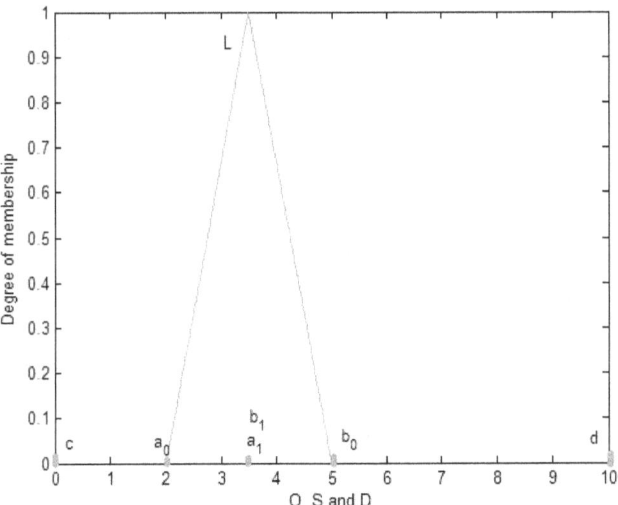

Fig. 9– Defuzzification of the linguistic term *Low*

The defuzzified crisp numbers of five linguistic terms are given in Table 7.

Table 7. The defuzzified crisp numbers of input linguistic terms

R	L	M	H	VH
0.196	0.370	0.583	0.804	0.952

(iii) Establish comparative series, which reflects the various linguistic terms and decision factors of the study. This can be represented in the form of a matrix, X as

$$X = \begin{pmatrix} x_1^1 & x_1^2 & x_1^3 \\ x_2^1 & x_2^2 & x_2^3 \\ \cdot\cdot & \cdot\cdot & \cdot\cdot \\ x_n^1 & x_n^2 & x_n^3 \end{pmatrix}, \quad (2)$$

where n is the number of the failure modes, x_i^1, x_i^2 and x_i^3 are the crisp numbers of three risk factors for i^{th} failure mode. For example, for sub-assembly 6 (main shaft), x_6^1, x_6^2 and x_6^3 are assigned, respectively, 0.370, 0.370 and 0.583.

(iv) Establish standard series, which reflects the ideal or desired level of all the decision factors. This can be represented in a form of a matrix, Y as

$$Y = \begin{pmatrix} y_1^1 & y_1^2 & y_1^3 \\ y_2^1 & y_2^2 & y_2^3 \\ \cdot\cdot & \cdot\cdot & \cdot\cdot \\ y_n^1 & y_n^2 & y_n^3 \end{pmatrix}, \quad (3)$$

where y_i^1, y_i^2 and y_i^3 represent the crisp numbers of the lowest level of three risk factors for i^{th} failure modes. Here, we have $y_i^j = 0.196$, for any $i \in \{1,2,...,n\}$ and $j \in \{1,2,3\}$.

(v) Calculate the difference between the comparative series and standard series. This can be represented in a form of a matrix, D, as

$$D = \begin{pmatrix} d_1^1 & d_1^2 & d_1^3 \\ d_2^1 & d_2^2 & d_2^3 \\ \cdot\cdot & \cdot\cdot & \cdot\cdot \\ d_n^1 & d_n^2 & d_n^3 \end{pmatrix}, \quad (4)$$

where $d_i^j = \left| x_i^j - y_i^j \right|$, for any $i \in \{1,2,...,n\}$ and $j \in \{1,2,3\}$.

(vi) Compute the grey relation coefficient using the following equation (Chang, Wei and Lee, 1999):

$$\gamma_i^j = \frac{D_{\mathbf{min}} + \zeta D_{\mathbf{max}}}{d_i^j + \zeta D_{\mathbf{max}}}, \quad (5)$$

where $D_{\min} \equiv \min_i \min_j d_i^j$, $D_{\max} \equiv \max_i \max_j d_i^j$, and $\zeta \in [0,1]$ is an identifier which only affects the relative value of risk without any change in priority. Here, we have $D_{\min} = 0$, $D_{\max} = 0.756$, and ζ is assumed to be 0.5.

(vii) Introduce the weight vector of three risk factors $(\beta_1, \beta_2, \beta_3)$, where $0 < \beta_1, \beta_2, \beta_3 < 1$, and $\beta_1 + \beta_2 + \beta_3 = 1$. Weight vector of risk factors can be obtained by either directly assigning or indirectly using pair-wise comparisons (Kutlu and Ekmekçioğlu, 2012). Here, we consider the weights vector as in Arabian-Hoseynabadi, Oraee and Tavner (2010), i.e., (0.21, 0.26, 0.53) for (S, O, D).

(viii) Determine the degree of relation using $\Gamma_i = \beta_1 \gamma_i^1 + \beta_2 \gamma_i^2 + \beta_3 \gamma_i^3$ for each failure mode incorporating the weighted variables.

(i) Rank the priority of risk: the stronger the degree of relation, the smaller is the effect of the cause.

5. ANALYSIS OF RESULTS

In this Section, a comparative study is carried out using the traditional and the proposed fuzzy-FMEA methodologies applied to an offshore wind turbine system. The same experts have been surveyed for two methodologies to enable comparisons of the results. Our field failure data has been collected from 10-minute SCADA database, automated fault logs, operation and maintenance reports. Fig. 10 represents the failure rate of the sixteen sub-assemblies of the offshore wind turbine, where the average failure rate of the system (i.e., the expected number of failures per year) is equal to 1.38/year.

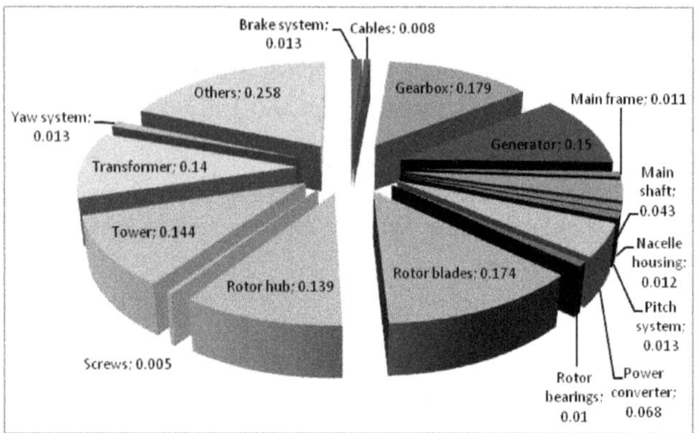

Fig. 10–The failure rates of the sub-assemblies of the offshore wind turbine system (sub-assemblies are listed in alphabetical order).

As shown, the gearbox, rotor blades, generator, tower and the transformer have the highest failure rates.

5.1. Traditional FMEA

On the basis of real data collected from an offshore wind farm database, and the criteria explained in tables 1–3, the traditional FMEA methodology is applied to the offshore wind turbine system. Table 8 gives the RPN values for the sixteen wind turbine sub-assemblies that considered in this study.

Table 8– The RPN values for wind turbine sub-assemblies

ID	Sub-assemblies	S	O	D	RPN	Rank
1	Brake system	2	2	7	28	11
2	Cables	3	2	1	6	14
3	Gearbox	3	5	7	105	2
4	Generator	2	5	7	70	5
5	Main frame	4	2	4	32	10
6	Main shaft	2	3	7	42	8
7	Nacelle housing	3	2	1	6	14
8	Pitch system	4	2	7	56	7
9	Power converter	4	3	7	84	4
10	Rotor bearings	3	2	4	24	12
11	Rotor blades	3	5	7	105	2
12	Rotor hub	2	5	4	40	9
13	Screws	1	2	1	2	16
14	Tower	4	5	7	140	1
15	Transformer	3	5	4	60	6
16	Yaw system	2	2	4	16	13

The results show that the tower (sub-assembly #14) is the most critical and the screws (sub-assembly #13) are the least critical parts in the offshore wind turbine system with the RPN values of 140 and 2, respectively.

From Table 8, the values of S, O and D for both the sub-assemblies of gearbox (sub-assembly #3) and rotor blades (sub-assembly #11) are the same. Hence, the traditional method puts the gearbox and rotor blades as having the same priority in offshore wind turbine systems. However, all the four experts believe that the *hidden* risk implications of these two failure modes are different in practice. One reason for this event is the existing limited number of the severity–occurrence–detection combinations for assigning to the three risk factors. This difference is obvious when the fuzzy rule base method and grey theory is applied.

5.2. Fuzzy FMEA

The results obtained from the proposed fuzzy approach are presented in Table 9.

Table 9– Ranking for proposed approach

ID	Sub-assemblies	Fuzzy rule base	Grey theory	Ranking (Fuzzy rule base)	Ranking (grey theory)
1	Brake system	1.405	0.937	11	12
2	Cables	1.288	0.941	16	16
3	Gearbox	4.873	0.829	3	3
4	Generator	2.032	0.917	6	6
5	Main frame	1.840	0.923	7	7
6	Main shaft	1.492	0.932	9	9
7	Nacelle housing	1.396	0.936	12	11
8	Pitch system	1.798	0.925	8	8
9	Power converter	2.503	0.902	4	4
10	Rotor bearings	1.366	0.939	14	14
11	Rotor blades	7.660	0.739	2	2
12	Rotor hub	1.492	0.935	9	10
13	Screws	1.312	0.940	15	15
14	Tower	8.890	0.700	1	1
15	Transformer	2.077	0.916	5	5
16	Yaw system	1.375	0.938	13	13

Table 9 shows a noticeable similarity between the results obtained from the two 'fuzzy rule base' and 'grey theory' methods. For instance, both approaches are in agreement about the tower (sub-assembly #14) being the most critical, and the cables (sub-assembly #2) being the least critical parts of the offshore wind turbine system. Also, the ranking order of the rotor blades (sub-assembly #11) is obtained higher than the gearbox (sub-assembly #3) in both the methods. The main reason for this event is that the blades are 'stressed' in a harsh maritime environment and extreme weather conditions, and they suffer from different types of external damages (including seasonal affects such as icing and thunderstorms) (Shafiee, Patriksson and Strömberg, 2013).

It should be noted that the ranking order produced by the fuzzy rule base method does not differentiate failure modes that have the same combination of linguistic terms describing the three risk factors. For example, S, O and D for the main shaft (sub-assembly #6) and rotor hub (sub-assembly #12) are assigned, respectively, as 'low/low/moderate' and 'low/moderate/low'. Hence, the defuzzified ranking is obtained the same for these sub-assemblies. This entails that the main shaft and rotor hub should be given the same priority for attention, and it could be misleading. The effects of the weighting coefficient introduced in the grey theory method can be clearly seen in this case. When using the grey theory method, the grey relation ranking is 0.932 and 0.935 for the main shaft and rotor hub, respectively. This entails that the main shaft should be given a higher priority compared to the rotor hub.

5.3. Comparison

In this section, a comparison is made between the ranking orders of the traditional FMEA, fuzzy rule base and grey theory approaches. In Table 10, the results obtained for the offshore wind turbine system from the traditional FMEA using the RPN method is compared with the results obtained from the proposed fuzzy FMEA using the rule base and grey relation methods.

Table 10–Ranking comparisons between the traditional and the fuzzy-FMEA approaches

Rank	Traditional	Fuzzy rule base	Grey theory
1	Tower	Tower	Tower
2	Gearbox / Rotor blades	Rotor blades	Rotor blades
3		Gearbox	Gearbox
4	Power converter	Power converter	Power converter
5	Generator	Transformer	Transformer
6	Transformer	Generator	Generator
7	Pitch system	Main frame	Main frame
8	Main shaft	Pitch system	Pitch system
9	Rotor hub	Main shaft/ Rotor hub	Main shaft
10	Main frame		Rotor hub
11	Brake system	Brake system	Nacelle housing
12	Rotor bearings	Nacelle housing	Brake system
13	Yaw system	Yaw system	Yaw system
14	Nacelle housing / Cables	Rotor bearings	Rotor bearings
15		Screws	Screws
16	Screws	Cables	Cables

As can be seen, the main problem in the traditional FMEA methodology is that it puts two critical sub-assemblies of the gearbox and the rotor blades as having the same priority. The nacelle housing and the cables are also placed at the same ranking level. But, applying the proposed methodology reveals that there is a noticeable difference between their ranking orders.

On the other side, there is some noticeable difference between the ranking orders of some sub-assemblies (such as main frame and nacelle housing) using the traditional and the Fuzzy FMEA methods. This shows that a more accurate ranking can be achieved by the application of the fuzzy rule base and grey theory to FMEA.

6. CONCLUSIONS AND TOPICS FOR FUTURE RESEARCH

The advantages of the proposed fuzzy rule base and grey theory approach for application to FMEA of offshore wind turbine systems can be summarized as follows:

a. The proposed fuzzy-FMEA approach provides an organized framework to combine the qualitative (expert experience) and quantitative (SCADA field data) knowledge for use in an FMEA study;
b. The proposed fuzzy-FMEA approach can be useful when the failure data is unavailable or unreliable;
c. The use of linguistic terms in the analysis enables the experts to express their judgments more realistically and hence improving the applicability of the FMEA technique in offshore wind farms;
d. The relative importance weights of risk factors are taken into consideration in the process of prioritization of failure modes, which makes the proposed FMEA more realistic, more practical and more flexible.

The proposed fuzzy rule base method (without the weighting vector of the risk factors) could be suitable for use in 'risk screening' phase, or during the 'design' stage of a new wind turbine configuration. During the risk-screening phase, only a relative ranking order is needed. This will distinguish the failure modes with a high risk level from those with a low-risk level. The proposed grey theory approach (with the weighting vector of the risk factors) would be suitable for use in 'risk analysis and evaluation' phase, or during the 'operation' stage. At this stage, a more detailed analysis of each failure mode is required to produce a ranking order that would determine the allocation of the limited resources. As the proposed method provides the analyst with the flexibility to decide which factor is more important to the analysis, the outcome of the analysis will provide valuable information for the wind farm managers or the wind turbine manufacturers.

Still, there is a wide scope for future research in improving the traditional FMEA methodology to make it more practical for wind turbine systems. Some of the possible extensions are:

(a) The proposed fuzzy-FMEA approach in this paper has no limitation on the number of risk factors and can be applied to any number of risk factors.
(b) Sometimes, it is observed that the FMEA team members, because of their different expertise and backgrounds have different opinions. The diversity and uncertainty of FMEA team members' assessment information will be considered in our future research.

REFERENCES

Andrawus, J. (2008) *Maintenance optimization for wind turbines*. Ph.D. dissertation, School of Engineering, Robert Gordon University, Aberdeen, UK.

Andrews, J.D. and Moss, T.R. (1993) *Reliability and risk assessment,* Longmans.

Arabian-Hoseynabadi, H., Oraee, H. and Tavner, P.J. (2010), Failure modes and effects analysis (FMEA) for wind turbines. *International Journal of Electrical Power & Energy Systems* 32(7), 817–824.

Bilgili M., Yasar, A. and Simsek, E. (2011) Offshore wind power development in Europe and its comparison with onshore counterpart, *Renewable and Sustainable Energy Reviews* 15, 905–915.

Braglia, M. (2000) MAFMA: multi-attribute failure mode analysis. *International Journal of Quality and Reliability Management* 17(9), 1017–1033.

Carlin, P.W., Laxson, A.S. and Muljadi, E.B. (2003) The history and state of the art of variable-speed wind turbine technology. *Wind Energy* 6(2), 129–159.

Carlson, R., Voltolini, H., Runcos, F. and Kuo-Peng, P. (2006) A performance comparison between brush and brushless doubly fed asynchronous generators for wind power systems, in Proceedings of *the International Conference on Renewable Energies and Power Quality,* Balearic Island, April 5–7, Spain.

Chang, C.-L., Wei, C.-C. and Lee, Y.-H. (1999) Failure mode and effects analysis using fuzzy method and grey theory. *Kybernetes* 28(9), 1072–1080.

Chen, C.B. and Klien, C.M. (1997) A simple approach to ranking a group of aggregated fuzzy utilities. *IEEE Transactions on Systems Man and Cybernetics*, Part B, 27(1), 26–35.

Cheng, C.-H. (1998) A new approach for ranking fuzzy numbers by distance method. *Fuzzy Sets and Systems* 95(3), 307–317.

Chin, K. S., Chan, A. and Yang, J.B. (2008) Development of a fuzzy FMEA based product design system. *International Journal of Advanced Manufacturing Technology* 36, 633–649.

Deng, J. (1989). Introduction to grey system theory. *Journal of Grey Systems* 1(1), 1–24.

Dinmohammadi, F. and Shafiee, M. (2013) An economical FMEA-based risk assessment approach for wind turbine systems. *European Safety, Reliability and Risk Management (ESREL), 30 Sep.–2 Oct., Amsterdam, Netherland.*

European Wind Energy Association, *Wind in power*, 2011 European statistics, published in Feb. 2012.

Gargama, H. and Chaturvedi, S.K (2011) Criticality assessment models for failure mode effects and criticality analysis using fuzzy logic. *IEEE Transactions on Reliability*, 60(1), 102–110.

Gilchrist, W. (1993). Modeling failure mode and effect analysis. *International Journal of Quality & Reliability Management* 10(5), 16–23.

Guimarães, A.C.F. and Lapa, C.M.F. (2007) Fuzzy inference to risk assessment on nuclear engineering systems. *Applied Soft Computing* 7(1), 17–28.

Kahrobaee, S. and Asgarpoor, S. (2011) Risk-based Failure Mode and Effect Analysis for wind turbines (RB-FMEA) North American Power Symposium (NAPS), August 4–6, pp 1–7, Boston, USA.

Keskin, G.-A. and Özkan, C. (2009) An alternative evaluation of FMEA: Fuzzy ART algorithm. Quality and Reliability Engineering International, 25, 647–661.

Kutlu, A.C. and Ekmekçioğlu, M. (2012) Fuzzy failure modes and effects analysis by using fuzzy TOPSIS-based fuzzy AHP. *Expert Systems with Applications* 39, 61–67.

Li, H. and Chen, Z. (2008) Overview of different wind generator systems and their comparisons. *IET Renewable Power Generation* 2(2), 123-138.

Liu, H.-C., Liu, L., Bian, Q.-H., Lin, Q.-L., Dong, N. and Xu, P.-C. (2011) Failure mode and effects analysis using fuzzy evidential reasoning approach and grey theory. *Expert Systems with Applications* 38, 4403–4415.

Márquez, F.P.G, Tobias, A.M., Pérez, J.M.P. and Papaelias, M. (2012) Condition monitoring of wind turbines: Techniques and methods. *Renewable Energy* 46, 169–178.

Muller, S., Deicke, M. and De Doncker, R.W. (2002) Doubly fed induction generator systems for wind turbines. *IEEE Industry Applications Magazine* 8(3), 26–33.

Pillay, A. and Wang, J. (2003) Modified failure mode and effects analysis using approximate reasoning. *Reliability Engineering and System Safety* 79, 69–85.

Shafiee, M., Patriksson, M. and Strömberg, A.-B. (2013) An optimal number-dependent preventive maintenance strategy for offshore wind turbine blades considering logistics. *Advances in Operations Research* (In Print).

Sharma, R.K., Kumar, D. and Kumar, P. (2005) Systematic failure mode effect analysis (FMEA) using fuzzy linguistic modelling. *International Journal of Quality & Reliability Management* 22(9), 986–1004.

Tavner, P.J., Xiang, J. and Spinato, F. (2007) Reliability analysis for wind turbines. *Wind Energy* 10(1), 1–18.

Yang, Z., Bonsall, S., and Wang, J. (2008) Fuzzy rule-based bayesian reasoning approach for prioritization of failures in FMEA. *IEEE Transactions on Reliability*, 57(3), 517–528.

Vibration Analysis and Time Series Prediction for Wind Turbine Gearbox Prognostics

Sajid Hussain[1] and Hossam A. Gabbar[1,2]

[1]*Faculty of Engineering and Applied Science, University of Ontario Institute of Technology, 2000 Simcoe St. North, Oshawa, Ontario, Canada L1H7K4*

sajid.hussain@uoit.ca

[2]*Faculty of Energy Systems and Nuclear Science, University of Ontario Institute of Technology, 2000 Simcoe St. North, Oshawa, Ontario, Canada L1H7K4*

hossam.gaber@uoit.ca

ABSTRACT

Multiple premature failures of a gearbox in a wind turbine pose a high risk of increasing the operational and maintenance costs and decreasing the profit margins. Prognostics and health management (PHM) techniques are widely used to assess the current health condition of the gearbox and project it in future to predict premature failures. This paper proposes such techniques for predicting gearbox health condition index extracted from the vibration signals. The progression of the monitoring index is predicted using two different prediction techniques, adaptive neuro-fuzzy inference system (ANFIS) and nonlinear autoregressive model with exogenous inputs (NARX). The proposed prediction techniques are evaluated through sun-spot data-set and applied on vibration based health related monitoring index calculated through psychoacoustic phenomenon. A comparison is given for their prediction accuracy. The results are helpful in understanding the relationship of machine conditions, the corresponding indicating features, the level of damage/degradation, and their progression.

1. INTRODUCTION

There is a growing interest in renewable energy systems with increased concerns over climate change. Wind energy has an attractive share in renewable energy because it diversifies a resource portfolio and improves overall security of the power system. However, the engineering challenge for the wind industry is to design a reliable wind turbine to harness wind energy and turn it into electricity. Despite all technological advancements in wind turbine design and installation, there is a price to pay in maintaining the wind turbine in harsh operating environments and reduced accessibility. According to two large surveys of European wind turbines, conducted over a span of 13 years, gearbox failure is one of the highest risk events in wind turbines (C. C. James, 2011). Hence, there is a need for efficient condition monitoring system for wind turbine gearbox. Condition monitoring is a good tool to assess the damage early in time in order to plan the maintenance activities in a better way. Condition monitoring can be combined with opportunity maintenance to reduce the turbine's unexpected downtimes.

Typically, faults in wind turbine gearbox arise while in operation. Therefore, it is vital to detect, diagnose and analyze these faults as early as possible. The process should be non-destructive in nature to avoid wind turbine's disassembly. This study presents fault detection, features extraction, and prognostics for wind turbine gearbox based on vibration analysis. Vibration analysis is a non-destructive testing (NDT) technique widely used in the industry and in academia.

1.1. Classification of Vibration Signals

Vibration signals emanating from the rotating gearbox are analyzed to ascertain the current condition of the gearbox. Vibration signals can be classified into stationary and non-stationary, and based on this classification, the nature of their analysis methods differ. For stationary signals, vibration analysis methods are divided into two domains, namely time and frequency. Time-domain methods include statistical, model based, and signal processing based methods. Frequency-domain methods include spectrum and cepstrum based methods. For non-stationary signals, joint time-frequency vibration analysis methods such as short time Fourier transform (STFT) and wavelet analysis (WA) (Hui Li et al., 2011) are used for fault detection in gearboxes. Other methods of non-stationary analysis for gearboxes include Wigner-Ville distribution, Hilbert-Haung transform, and kurtogram analysis (Jerome Antoni, 2007).

1.2. Features Extraction

The features extraction system extracts characteristic signatures from raw vibration signals emanating from gearbox. The extracted features should reflect the changes in the gearbox's health conditions over time. As discussed earlier, the vibration based signal processing analysis is one of the most common non-destructive techniques. Also, with suitable vibration analysis, we can detect many different faults related with gearboxes. We use vibration based features extraction to extract the information that best represents the faulty conditions present in the monitored equipment. Different methods for vibration based features extraction in gearbox fault diagnosis framework have been proposed in research (Halima, E. B. et al, 2008, J. Rafiee et al, 2010). In time-domain, vibration based features such as kurtosis and spectral kurtosis are extensively used (F. Combet et al, 2009). Other studies including statistical-based and transient-based features detection are performed in the past (Hiram Firpi and George Vachtsevanos, 2008; V. Indira et al, 2010). A comprehensive list of time-domain and frequency-domain features for fault detection and diagnosis of gearboxes is discussed in (Yaguo Lei et al, 2010).

1.3. Prognostics

Prognostic plays a very important role in an accurate and reliable decision making. Prognostics can be used effectively in utilization and maintenance of machinery systems. Prognostics uses different machines health related indices including temperature (Jamie Coble et al, 2010), oil-debris analysis (Richard Dupuis, 2010), acoustics, and vibration (Eric Bechhoefer et al, 2010). Among these, vibration based prognostics is quite common. In (B. Samanta and C. Nataraj, 2008), researchers have used different health monitoring indices in gearboxes such as gear-wear, gear-chipped, gear-crack, gear-pitting, and shaft misalignment for projecting the gearbox health information in future. They have used neuro-fuzzy approaches for modeling and prediction of gearbox dynamics. A comprehensive review on prognostics is presented in (Andrew K.S. et al, 2006), where decision making process based on diagnostics and prognostics is discussed. Prognostic is performed by estimating the temporal evolution of the features over time (Wang W., 2007). In vibration based prognostics, vibration signals emanating from sensitive components inside the gearbox are recorded, health related features are extracted, and time series prediction techniques are applied to the features trends for prognostics (L. Gelman et al, 2012). Statistical, evolutionary and soft computing approaches are used to estimate the predictors and the use of neural networks and neuro-fuzy methods are very common.

This paper proposes a novel method to extract machine's health related vibration features based on psychoacoustic phenomenon along with neural networks and neuro-fuzzy approaches for prognostics. In this study, other than vibration features, we also use sunspot data for measuring the performance of the designed predictors. The rest of the paper is organized as follows. Section 2 is the methodology section where we propose the novel feature extraction technique based on psychoacoustics phenomenon followed by wavelet smoothing and prediction approaches. Section 3 simulates the proposed techniques on sunspot data and real-world vibration data emanating from a planetary gearbox inside a wind turbine. Finally, section 4 concludes the paper.

2. THE METHODOLOGY

The process of vibration based features extraction and prognostic is shown in Figure 1. Below we discuss each module one by one.

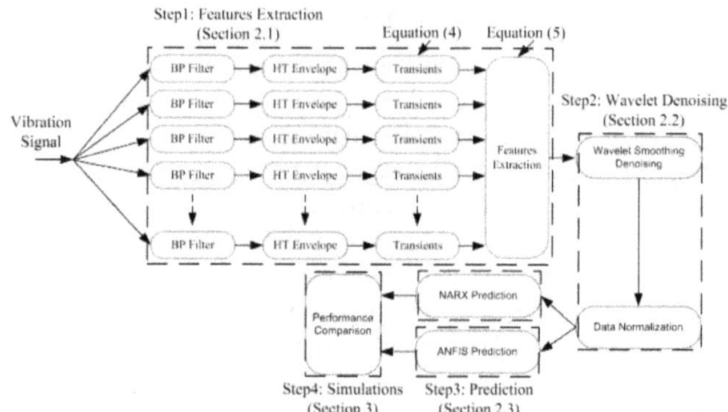

Figure 1. The process of prognostic

2.1. Transient Based Features Extraction

The vibration features extraction algorithm presented in this section works on the principle of transient analysis. Transients are very short and abrupt changes in sound waves due to non-linearity. Non-linearity could be mechanically introduced disturbances in electromechanical systems or unwanted clicks in transmission lines. The transient analysis algorithm calculates a real time estimate of transients caused by non-linearity as perceived by the human ear. The analyzer uses the knowledge about human ear's nature of filtering the signals as presented in (E. Zwicker, and H. Fastl, 2009). The filtering operation ensures that the transients are detected in a way that matches the nature of the cochlea and thereby as perceived by the human ear. Transient analysis gives a much better correlation to the perceived quality of sound than traditional measurements based on frequency analysis.

The vibration features extraction algorithm uses principles of auditory models developed by the auditory physiologists (S. Seneff, 1988). According to psychoacoustic theory, pulses with short rise time or fall time contain a broad spectrum of frequencies. Therefore, it is possible to detect the instantaneous energy in frequency bands in the transient range of the ear. A common method for doing this is to use a filter bank containing a group of band-pass filters covering the frequency interval of interest. The purpose of the band pass filters is to detect the pulses in the frequency band where the pulses have most energy as perceived by the human ear. It will be the filter where the shape of the impulse response of the band-pass filter matches best the shape of the pulses but reverse in time. A gammatone filter bank is used in this study as shown in Figure 2. First designed by Patterson and Holdworth (R. D. Patterson et al, 1992), gammatone filter bank is an array of band-pass filters which simulates the response of the human ear's cochlea. At each point along the cochlea, a psychoacoustic measure of the width of the auditory filter is represented by an equivalent rectangular bandwidth (ERB).

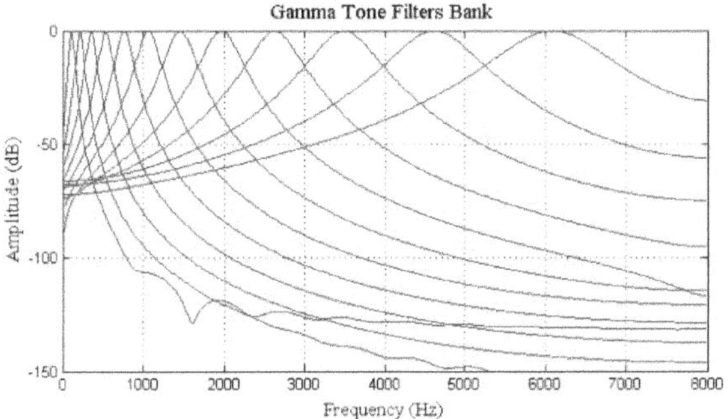

Figure 2. The gammatone filter bank

The ERB is a psychoacoustic measure of a filter's bandwidth in the filter bank. The bandwidth of a filter increases with an increase in its center frequency. The relationship between ERB and center frequency F_c Hz is given by the following equation

$$ERB = 24.7 + 0.108F_c. \tag{1}$$

The impulse response of a band-pass filter is defined by the following relation (R. D. Patterson et al, 1992):

$$h(t) = Rt^{N-1}e^{-2\pi mt}\cos(2\pi F_c t + \phi). \tag{2}$$

Where R is an arbitrary factor that is typically used to normalize the peak magnitude transfer to unity, N is the filter order, m is a parameter that determines the duration of the impulse response and thus the filter's bandwidth, F_c is the filter's center frequency, and ϕ is the phase of the tone. Figure 3 shows an impulse response of a gammatone filter with $F_c = 1000Hz$, $m = 125Hz$ and $N = 4$. To detect the energy in the channels the output signals from the band-pass filter bank are Hilbert transformed. Hilbert transform detects the envelopes of the band-pass filtered signals and extracts the instantaneous energy of the faulty pulses. Hilbert transform can be expressed as

$$\hat{x}(t) = x(t) * \frac{1}{\pi t} = \frac{1}{\pi}\int_{-\infty}^{\infty}\frac{x(\tau)}{t-\tau}d\tau. \tag{3}$$

Hilbert transform creates an artificially complex signal $u(t)$ from $x(t)$. The real part $x(t)$ of the $u(t)$ is the original signal and the imaginary part $\hat{x}(t)$ is the Hilbert transform of the real part. Thus, $u(t)$ is defined as $u(t) = x(t) + j\hat{x}(t)$. The magnitude and phase of $u(t)$ is computed as $A(t) = \sqrt{x(t)^2 + \hat{x}(t)^2}$ and $\theta(t) = \arctan \hat{x}(t)/x(t)$. The magnitude $A(t)$ is the envelope of the signal and is always a positive function.

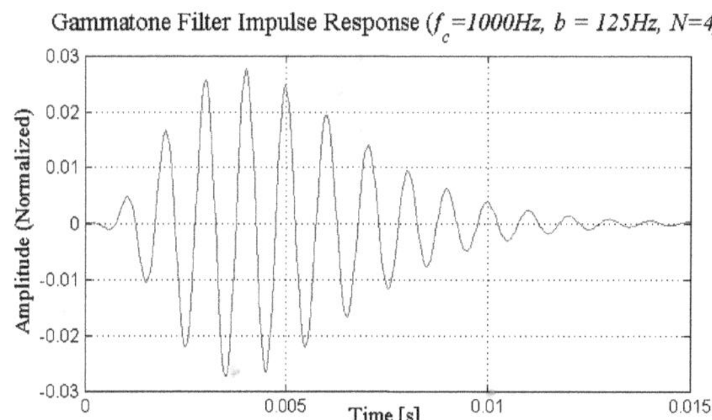

Figure 3. Gammatone filter impulse response

After the pulses are extracted through envelope detection or Hilbert transform, feature extraction block follows. In the feature extraction block, amplitudes and slopes of the pulses are calculated as per Eq. (4) and Figure 4.

$$S_r = \frac{High\ \ Reference\ \ Level - Low\ \ Reference\ \ Level}{Rise\ \ Time}. \tag{4}$$

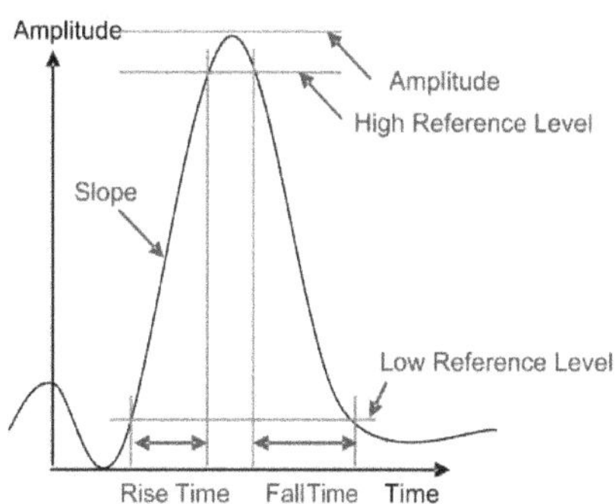

Figure 4. Calculation of amplitude and slope of a pulse

Amplitude (maximum magnitude) is a linear detector for pulses in the full frequency area. Slope is very sensitive for catching nonlinear sounds in the high frequency area. Both metrics can be summarized to a logarithmic index expressing the amount and size of the pulses. The index may be based on either the magnitudes or the slopes of the pulses. It is expected that pulses with a short rise time will be more annoying to the ear, than pulses with longer rise times. Thus, by considering the pulse envelope, it seems reasonable to focus the measure on the slope in the leading edge. Every transient detected is characterized by the 80% amplitude and the maximum slope in the leading edge. In order to obtain an equal number of detected transients in each frequency bands, the output from the transient analyzer is split into blocks of a pre-specified duration. The duration is found as a trade-off between preserving the complexity of the measurement and having an equal number of detected transients in each channel. In each block, only the transient with the maximum slope is gathered. The maximum slope is found by differentiating the signal and finding maximum amplitude of the differentiated signal. The logarithmic index is calculated as follows

$$Vb_S = 10.\log_{10}\left(\frac{\frac{1}{N_B N_C}\sum_{i,channels}\sum_{j=1}^{N_B}\left(S_{max}\right)_{ij}^2}{S_{ref}^2}\right). \quad (5)$$

Where Vb_S is vibration steepness index, i is index for band pass filter and N_C is total number of band pass filters used. Index of sub-blocks in a band pass filtered signal is j and N_B is the total number of sub-blocks. The maximum steepness in band pass filter i and sub-block j is $\left(S_{max}\right)_{ij}$. The argument for squaring the maximum steepness in Eq. (5) is simply to put the large steepness values in favor. $S_{ref} = 1$ if the measured amplitude is acceleration. For amplitude index Vb_A, $\left(S_{max}\right)_{ij}$ in Eq. (5) is replaced by $\left(A_{max}\right)_{ij}$ where A_{max} is the maximum amplitude of the non-differentiated envelope signal. The vibration features extracted are de-noised through wavelet de-noising techniques and normalized before prediction.

2.2. Wavelet Denoising

Features extracted from real signals emanating from complex dynamical systems pose a serious problem of noise. Therefore, it is important to de-noise the extracted features before modeling is performed (Uros Lotric and Andrej Dobnikar, 2005). De-noising the signal is one of the most effective applications of wavelets in signal processing. The wavelet transform-based de-noising methods can produce much higher de-noising quality than conventional methods. Furthermore, the wavelet transform-based methods retain the details of a signal after de-noising (Edmundo G. de Souza e Silva et al, 2010). Wavelets are limited duration, undulatory mathematical functions. The time integral of wavelet functions equals to zero. Figure 5 plots some common wavelets. Similar to Fourier transform, where we use sines and cosines as basis functions, wavelet transform uses wavelets as basis functions. Wavelets are used in many different fields including compression, signal processing, and de-noising (Graps, A, 1995). In Fourier analysis, we approximate a function $f(x)$ by sines and cosines functions with different frequencies and amplitudes. Thus, the approximation equation becomes

$$\hat{f}(x) = a_o + \sum_{k=1}^{\infty}\left(a_i \sin(kx) + b_i \cos(kx)\right). \quad (6)$$

Where, a_o, a_i, and b_i are calculated from Fourier transform as

$$F(\omega) = \int_{-\infty}^{\infty} f(t) e^{-j\omega t} dt. \quad (7)$$

Wavelet analysis is performed in the similar way as Fourier transform but with scaled and translated versions of mother wavelet $\psi(x)$ as basis functions. Mother wavelet can be any one from Figure 5 and the scaling and translation is defined as child wavelets and can be calculated as

$$\psi_{j,k}(x) = \kappa\psi\left(2^j x - k\right). \quad (8)$$

Where κ is a constant, k is wavelet translation and 2^j is scale translation. We can estimate $f(x)$ from the following equation in wavelet analysis

$$\hat{f}(x) = \sum_{\forall j,\forall k} c_{j,k}\psi_{j,k}(x). \quad (9)$$

Where, $c_{j,k}$ are the wavelet coefficients and are obtained through the wavelet transform as

$$c_{j,k} = \int_{-\infty}^{\infty} f(x)\psi_{j,k}(x)dx. \quad (10)$$

Each coefficient $c_{j,k}$ obtained in Eq. (10) is a contribution of the wavelet $\psi_{j,k}(x)$ in the whole approximation for the original signal. If the value of this coefficient $c_{j,k}$ is very small and its contribution to the approximation is considered negligible, we can omit the corresponding child wavelet $\psi_{j,k}(x)$ from the approximation. This procedure is called *thresholding* and it forms the basis for the wavelet de-noising. Wavelet analysis basis functions are finite and limited to one size and this makes wavelet analysis useful technique for detecting local features like discontinuities and spikes in a signal. On the other hand, Fourier analysis basis functions are infinite in nature and an approximation of a specific part of the signal affects the entire signal. Wavelet analysis is joint time-frequency analysis technique as contrast to the Fourier transform that is purely frequency analysis. This feature makes wavelet analysis to detect when in time a particular event took place. Wavelet smoothing or wavelet trend analysis is used to remove high frequency components from the extracted features which can be assumed as noise.

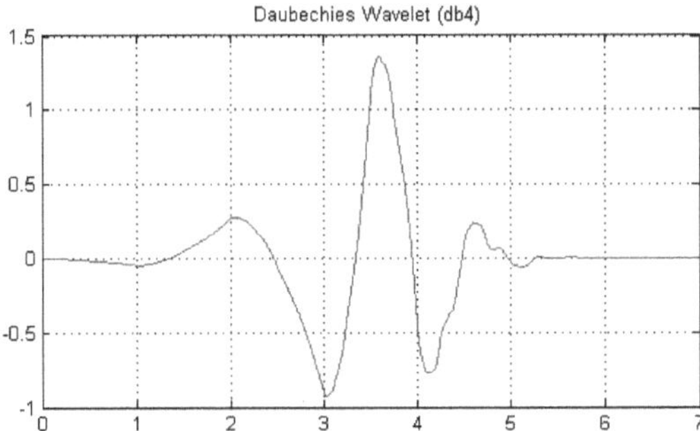

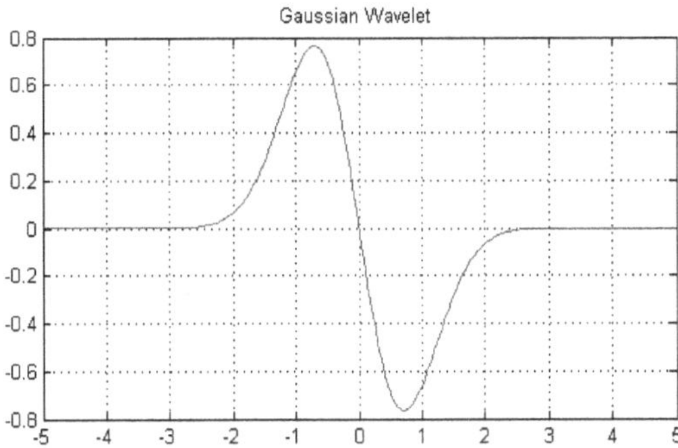

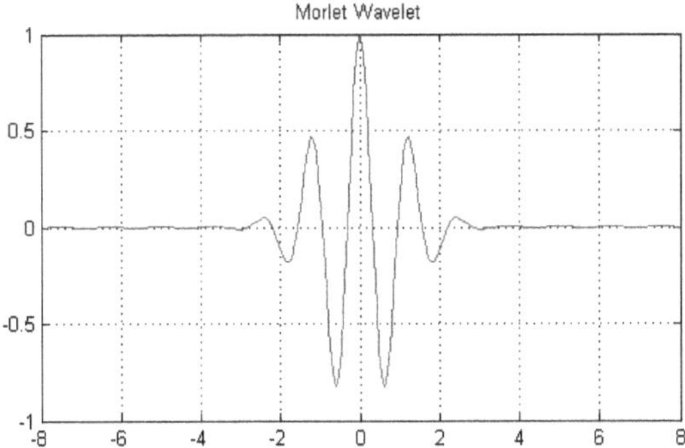

Figure 5. Commonly used wavelets (a) Daubechies (b) Gaussian (c) Morlet

2.3. Time Series Prediction

The prediction of time series $x(t)$ at r time steps ahead, x_{t+r}, is obtained based on its values at present and past time steps $[x_{t-mr}, x_{t-(m-1)r}, x_{t-(m-2)r}, \ldots, x_{t-2r}, x_{t-r}, x_t]$ as $x_{t+r} = \chi(x_{t-mr}, x_{t-(m-1)r}, x_{t-(m-2)r}, \ldots, x_{t-2r}, x_{t-r}, x_t)$. Where χ is a predictor functions and can be approximated through various conventional, statistical and artificially intelligent techniques like Bayesian, support vector regression, adaptive neuro-fuzzy inference system (ANFIS) and neural networks (NN). This paper uses a dynamic neural network called the nonlinear autoregressive model with exogenous inputs (NARX) and ANFIS techniques to approximate the predictor function χ.

2.3.1. The NARX

The NARX is a dynamic neural network, used for modeling nonlinear dynamical systems. The NARX can be represented mathematically as

$$y(n+1) = f\begin{bmatrix} y(n), \ldots, y(n-d_y+1); \\ u(n), u(n-1), \ldots, u(n-d_u+1) \end{bmatrix}. \quad (11)$$

Where, $\mathbf{u}(n)$ and $\mathbf{y}(n)$ are the input and output of the system at time step n, while $d_u \geq 1$ and $d_y \geq 1$, $d_u \leq d_y$, are the input-memory and output-memory orders. Equation (11) can also be written in compact form as $y(n+1) = f[\mathbf{y}(n); \mathbf{u}(n)]$, where $\mathbf{u}(n)$ and $\mathbf{y}(n)$ are the input and output regressor vectors respectively. The nonlinear mapping function $f(.)$ is approximated through a multi-layer perception (MLP) algorithm trained with plain back propagation algorithm. This research deals with nonlinear univariate time series prediction and for this we set $d_y = 0$. This reduces the NARX network to time delay neural network (TDNN) architecture and Eq. (11) reduces to (T. Lin et al, 1997)

$$y(n+1) = f[u(n), u(n-1), \ldots, u(n-d_u+1)]. \quad (12)$$

Figure 6 shows the way the NARX is trained and tested. During the training phase, the feedback loops (dotted lines in Figure 6) are not used. During the testing or prediction phase, if multistep-ahead predictions are required, the output values are fed back to both the input regressor $\mathbf{u}(n)$ and the output regressor $\mathbf{y}(n)$ at the same time. Thus, the resulting predictive model contains two feedback loops, one for the input regressor and another for the output regressor.

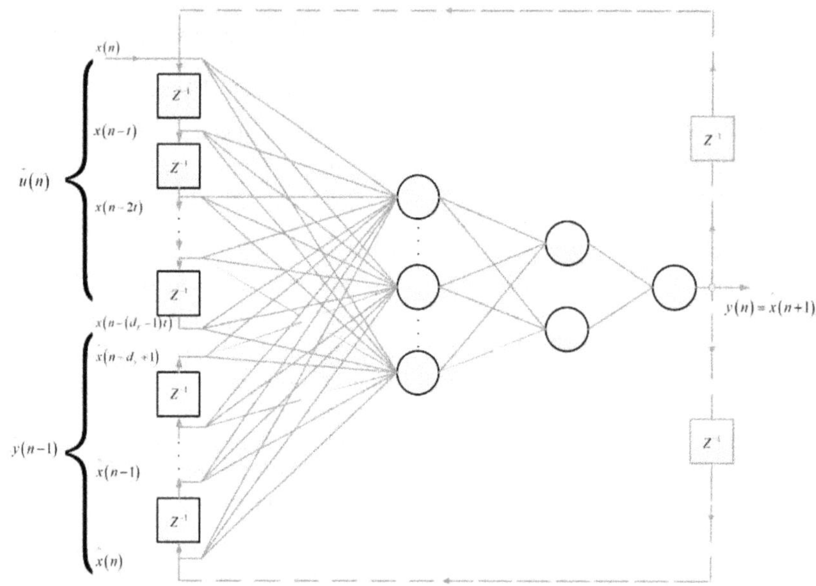

Figure 6. NARX training and testing
(Feedback loops are required only during testing)

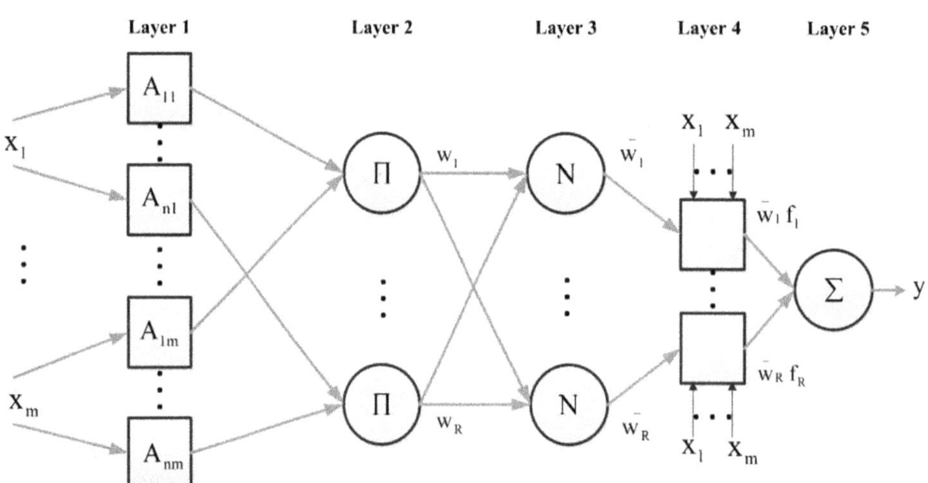

Figure 7. Basic structure of ANFIS

2.3.2. The adaptive neuro-fuzzy inference system

The basic structure of the ANFIS is shown in Figure 7. The ANFIS has m inputs $(x_1, x_2 ..., x_m)$, each with n membership functions (MFs), R rules and one output y. When the ANFIS predicts a time series, the inputs are $(x_{t-mr}, x_{t-(m-1)r}, x_{t-(m-2)r}, ..., x_{t-2r}, x_{t-r}, x_t)$ and the output of the ANFIS is $y = x_{t+r}$. In the above mentioned case, the ANFIS predicts the time series r time steps ahead based on current and the previous m values. We use a Sugeno-fuzzy type inference system with five layers and $m = 4$ inputs. Number of nodes N in layer 1 is the product of number of inputs m and the input MFs n for each input, i.e., $N = m.n$. Number of nodes in layers 2-4 is equal to the number of rules R in the fuzzy rule base. Layer 1 is a fuzzufication layer and it transforms the crisp inputs x_i to linguistic labels A_{ij} . The examples of the linguistic labels are small, medium, large etc., and the transformation occurs with some degree of the MFs as $O_{ij}^1 = \mu_{ij}(x)$. Where, $i = 1,...,m$, $j = 1,...,n$ and μ_{ij} represents the jth membership function for the input x_i . Different types of MFs are used like triangular, trapeziodal, Gaussian etc. Layer 2 of the ANFIS is a product layer, where for each node k , the output represents weighting factor or firing strength of the rule R

associated with k. The output w_k of this layer is $O_k^2 = \prod \mu_{ik}(x_i)$, and it is the product of all its inputs scaled according to the MFs μ_{ik} .Where $i = 1,...,m$ and $k = 1,..., R$. Layer 3 is a normalization layer and the output of each node k in this layer represents the normalized weighting factor $\overline{w}_k$ of the kth rule as $O_k^3 = \frac{w_k}{\sum_k w_k}$. Where $k = 1,..., R$. Layer 4 is a de-fuzzification layer and the output of each node in this layer is a weighted output of the first order Sugeno-type fuzzy if-then rule as $O_k^4 = \overline{w}_k f_k$. Where $f_k = \sum_j p_{kj} x_j + r_k$, $j = 1,...,n$, $k = 1,..., R$, f_k is the output of the kth rule, and the parameters p_{kj} and r_k are called consequent parameters. Layer 5 is the final output layer and it contains only one node inside. The output of the layer 5 is an overall output y of the network as $O^5 = \sum_k \overline{w}_k f_k$. It is also a sum of all the weighted outputs of the rules. We need a training dataset of desired input/output pairs $(x_1, x_2,..., x_m, y)$ to train the ANFIS or model the target system. In training phase, the ANFIS adaptively maps the input features space $(x_1, x_2,..., x_m)$ to the corresponding output y. The mapping in the ANFIS system is done through the membership functions (MFs), the rule base and the related parameters that emulate the training dataset. The training phase of the ANFIS uses hybrid learning method. It uses the gradient descent approach for fine tuning the parameters that define the MFs and applies the least squares method to identify the consequent parameters that define the coefficient of each output equation in Sugeno-type fuzzy rule base. The training process continues till the desired stopping criteria is reached, i.e., number of epochs or error tolerance.

3. SIMULATIONS AND DISCUSSIONS

In this section, the prediction accuracy of both the time series predictors, NARX and ANFIS is compared using standard dataset of sunspot activity for years 1749-2012 (RWC Belgium World Data Center, 2012). The sunspot activity data has non-linear, non-Gaussian and non-stationary characteristics and is suitable to test the performance of the predictors. The entire dataset of sunspot activity was normalized between [0 - 1] and used for training and testing both NARX and ANFIS predictors. To get a reliable prognosis, the data need to be less sensitive to noise. This requirement can be reached using selected signal processing techniques such as wavelet smoothing or denoising discussed in section 2.2. For wavelet denoising, we use Daubechies wavelet (db4) with nine levels. The threshold technique is set to *soft threshold*, and the threshold rule used is *universal*. The *universal* threshold rule is defined as $\sqrt{2 \times \log(L)}$. Where L is the signal length. We also set the rescaling method as single level where the algorithm considers the noise as white and estimates the standard deviation of the noise from the wavelet coefficient at the first level. We use the normalized Akaike Information Criteria (AIC) for assessing the prediction performance. The AIC can be formulated as (Akaike H. 1974).

$$AIC = \ln \sigma^2 + \frac{2u}{P}. \tag{13}$$

Where σ^2 is the variance of the prediction error, u is the number of model parameters to be updated and P is the total number of data points in the predicted dataset. A smaller value of AIC indicates better prediction performance. The sunspot activity data contains 3166 points. For both types of predictors, 70% data is used for training, 15% for testing, 15% for validation, and 500 training epochs. Computation was carried out in MATLAB® environment on a PC with Intel Core i7 with 8GB of RAM. One step ahead prediction is performed and in one step ahead prediction, the target $y(n+1)$ is calculated from previous four values, $x(n)$, $x(n-1)$, $x(n-2)$, and $x(n-3)$ .The input memory we use is $d_u = 4$. Table 1 shows the mean absolute error (MAE) and AIC for NARX and ANFIS for sunspot data. The number of membership functions for ANFIS is 3 with 16 fuzzy rules and 104 parameters. The AIC achieved is -4.457. The neuro-fuzzy system proposed in (Wang WQ 2004) with the same datasets produced AIC of 1.527. The ANFIS method gave better prediction performance as compared to (Wang WQ 2004) for the same datasets. The sunspot training and test datasets were also used for assessing the prediction performance of NARX. For NARX, MAE of testing is 0.0034.

Figure 8(a) shows the sunspot activity data along with wavelet smoothing. NARX and ANFIS prediction results are shown in Figure 8(b) and 8(c) along with error box-plots in Figure 8(d), respectively. In Figure 8(d), the medians of the box-plots are centered at zero for both NARX and ANFIS predictors and the $\pm 2.698\sigma$ lines are at ± 0.01 showing about 99.3% of the error observations within the range of ± 0.01. Where σ is the error's standard deviation. There are some outliers for both NARX and ANFIS cases and NARX depicts better performance as compared to ANFIS in this case. In the case of NARX, the MAE value is approximately same as ANFIS (sunspot data-set) in Table 1.

In the framework of machine condition monitoring, it is very important to know well in advance the expected behavior of a machine system for its proper operation and maintenance scheduling without any major interruptions. Several approaches are adopted to monitor the system behavior and vibration-based approaches are quite popular for condition monitoring.

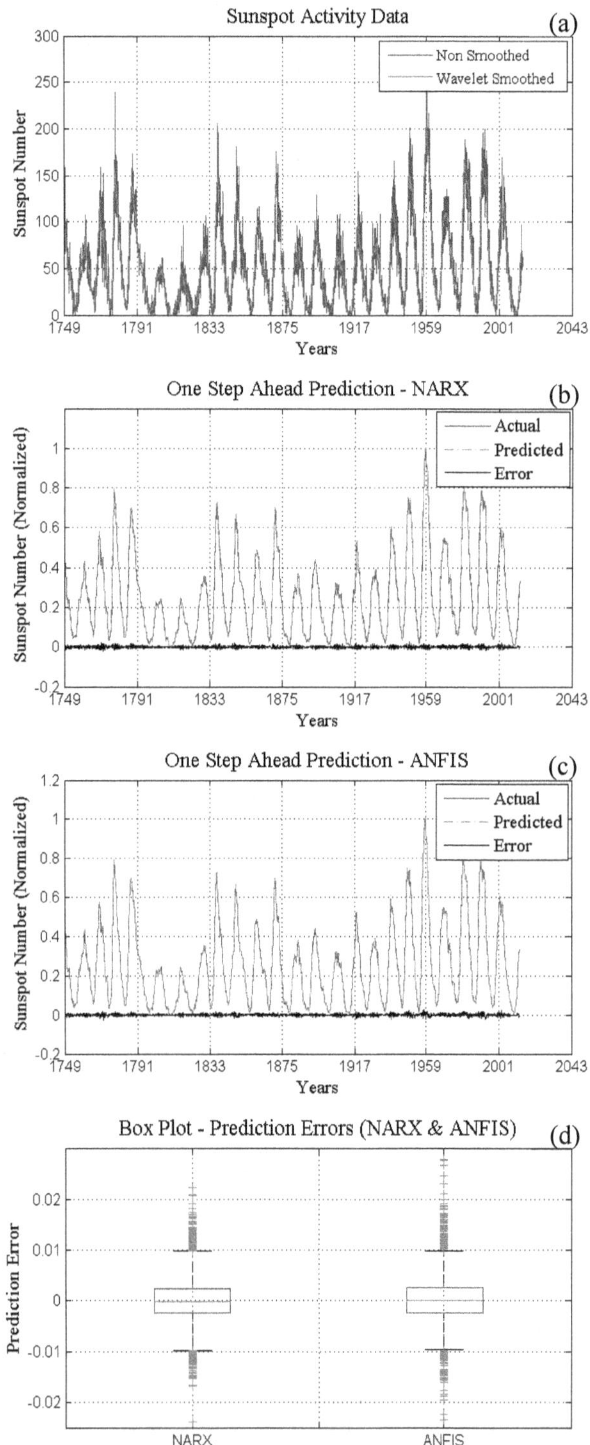

Figure 8. Sunspot data, wavelet smoothing and prediction

The desirable monitoring index should be highly sensitive to the fault-related vibration with low sensitivity to noise. The selection of monitoring index involves different steps of signal processing and feature extraction. These also greatly depend on the type of machine faults. The vibration index we use is proposed in section 2.1. The experimental vibration data emanate from a planetary gearbox inside a wind turbine. The data are provided by the National Renewable Energy Laboratory (NREL), through a consortium called the Gearbox Reliability Collaborative (GRC) (H. Link et. al. 2011). The gearbox under test is one of two units taken from the field and redesigned, rebuilt and instrumented with over 125 sensors. The gearbox first finished its run-in in the NREL dynamometer test facility (DTF) and later was sent to a wind plant close to NREL for field test, where two oil losses occurred.

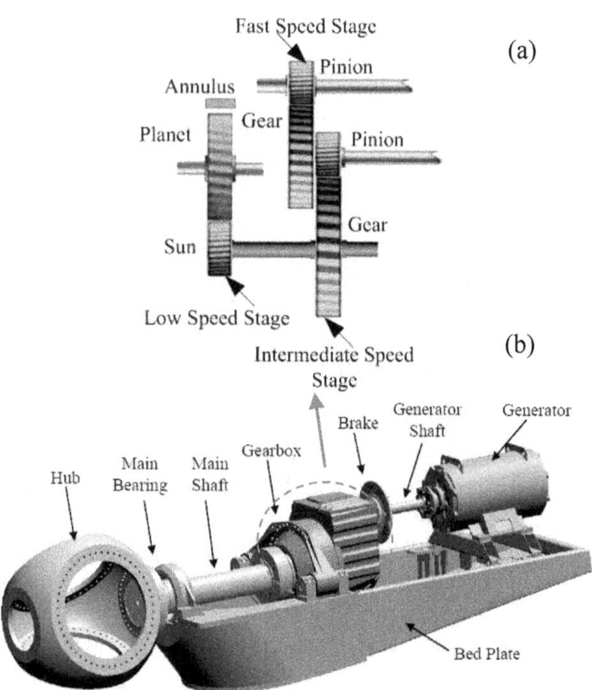

Figure 9. (a) Planetary gearbox (b) GRC Drive train Configuration
(Courtesy of National Renewable Energy Laboratory)

The test turbine in the field is a stall-controlled, three-bladed, upwind turbine with a rated power of 750kW. The turbine generator operates at 1200 RPM and 1800 RPM nominal on two different sets of windings depending on the power. The planetary gearbox has an overall ratio of 1:81.491. It is composed of one low speed (LS) planetary stage and two parallel stages as shown in Figure 9. This study uses data from test case CM_2a with main shaft speed of 14.72 RPM and high speed shaft (HSS) speed of 1200 RPM. The data are collected at a sampling frequency of $F_s = 40\text{KHz}$. Figure 10(a) shows an example of a raw vibration signal, 1sec (40000 samples) in length, collected from the gearbox inside the windmill.

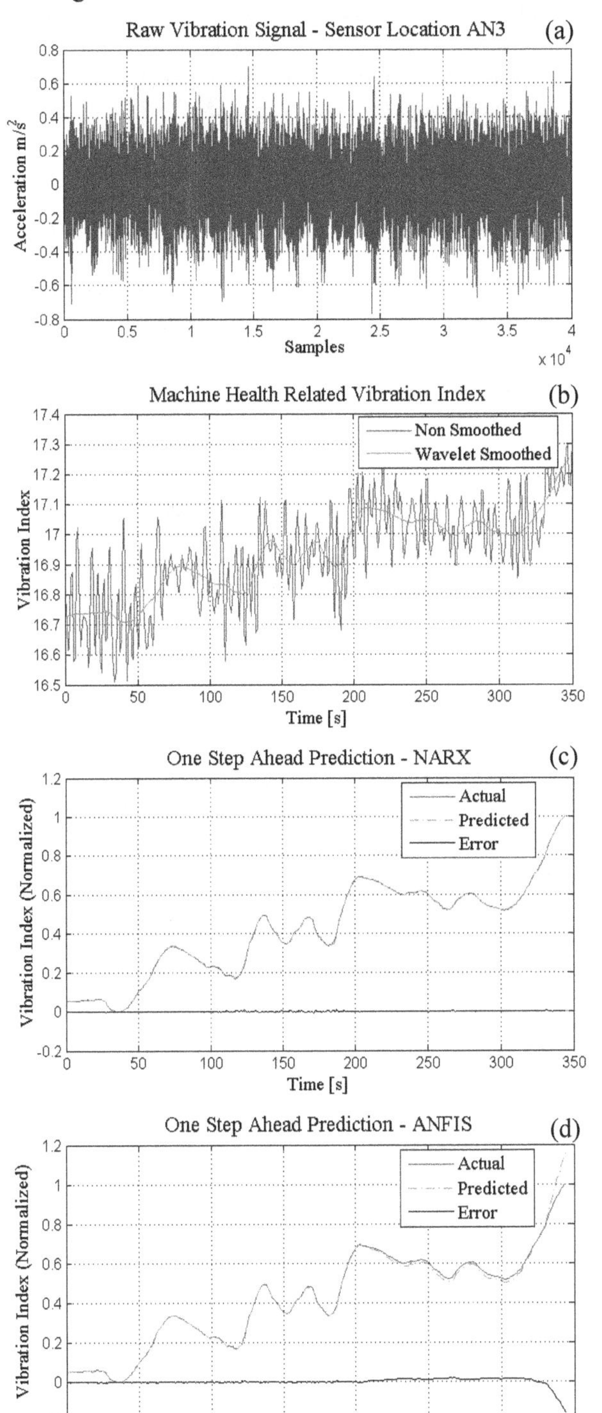

Figure 10. (a) Raw vibration signal (one sample) (b) Vibration index trend (350 samples) (c) one-step prediction NARX (d) one-step prediction ANFIS

A total of 350 such vibration signals are analyzed in this section. Vibration indices are calculated as proposed in section 2.1. For both types of predictors, 70% data is used for training, 15% for testing, 15% for validation, and 500 training epochs. Computation was carried out in MATLAB® environment on a PC with Intel Core i7 with 8GB of RAM. Figure 10(a) shows an example of a raw vibration signal, 1sec (40000 samples) in length, collected from the gearbox inside the windmill. A total of 350 such vibration signals are analyzed in this section. Vibration indices are calculated as proposed in section 2.1. Figure 10(b) shows the vibration index trend with wavelet smoothing. The vibration index is gradually increasing with time. It is because of the oil loss occurred in the field test. The oil loss caused the gearbox to run dry and consequently, a gradual increase in the overall vibration levels and vibration index is observed. Figure 10(c) shows one step prediction for NARX and Figure 10(d) shows one step prediction for ANFIS. In one step ahead prediction, the target $y(n+1)$ is calculated from previous four values, $x(n)$, $x(n-1)$, $x(n-2)$, and $x(n-3)$. The input memory we use is $d_u = 4$.

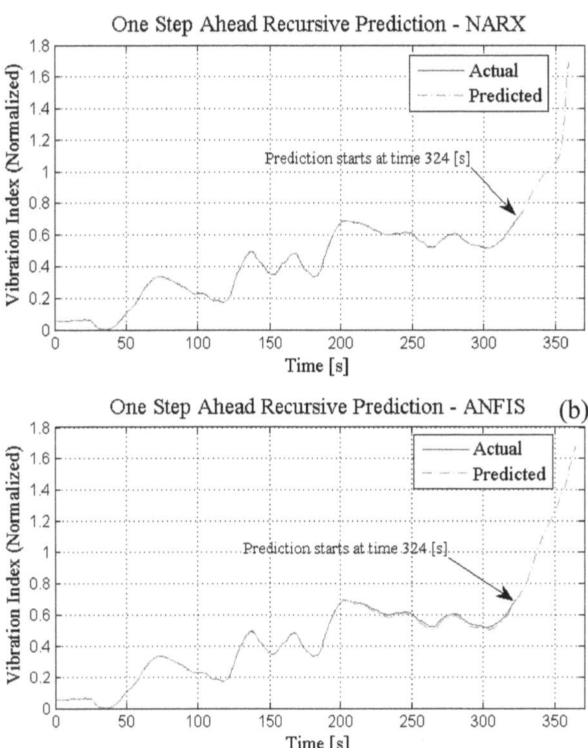

Figure 11. One-step ahead recursive prediction (a) NARX (b) ANFIS

NARX seems to exhibit more promising results as compared to ANFIS in this case. Also, the MAE error value for NARX is less as shown in Table 1 (vibration index data-set). We can also use one step recursive prediction for NARX and ANFIS to predict as many future values as we want as shown in Figures 11(a) and 11(b). For multi-step recursive prediction, we have to loop the output values back as discussed in section 2.3.1 in the NARX case (feedback dotted lines in Figure 6). Similar strategy is adopted in the ANFIS case.

4. CONCLUSION

Two different techniques are used for predicting the wind turbine gearbox health related vibration based index trend. The prediction performance of the predictors, NARX and ANFIS, is illustrated on two data-sets, sunspot activity and vibration index. Both the NARX and ANFIS predictors perform quit well in this study. Results show the effectiveness of the predictors in estimating the variations of the monitoring indices. In this work, one-step-ahead and recursive multi-step ahead prediction is considered for both the NARX and ANFIS cases. Although, the k step ahead prediction is not performed in this research, it can be done easily by predicting the target $y(n+k)$ with the previous values $x(n)$, $x(n-k)$, $x(n-2k)$, $x(n-3k)$ and so on. The potential application of these techniques for the development of on-line prognostic and estimation of remaining useful life (RUL) for machine condition is under consideration for further work.

Data Set	Model	MAE	MSE	AIC
Sunspot	NARX	0.0034	$2.144\text{x}10^{-5}$	-4.943
	ANFIS	0.0035	$2.421\text{x}10^{-5}$	-4.457
Vibration Index	NARX	0.0013	$3.626\text{x}10^{-6}$	-3.467
	ANFIS	0.0086	$4.079\text{x}10^{-4}$	-3.134

Table 1. MAE, MSE ans AIC – NARX and ANFIS

ACKNOWLEDGEMENTS

Authors and thankful to the National Renewable Energy Laboratory (NREL) for providing the wind turbine's gearbox vibration data through a consortium called the Gearbox Reliability Collaborative (GRC).

REFERENCES

Akaike H. A new look at the statistical model identification (1974). *IEEE Trans Automat Control*. vol.17, pp.716–23.

Andrew K.S. Jardine, Daming Lin, and Dragan Banjevic (2006). A review on machinery diagnostics and prognostics implementing condition-based maintenance. *Mechanical Systems and Signal Processing*, vol. 20, no.7, pp. 1483–1510.

B. Samanta and C. Nataraj (2008). Prognostics of machine condition using soft computing. *Robotics and Computer-Integrated Manufacturing*, vol. 24, no. 6, pp. 816-823.

C. C. James., (2011). Condition monitoring techniques for wind turbines. Doctoral dissertation, Durham University, UK. http://etheses.dur.ac.uk/652/.

E. Zwicker, and H. Fastl (2009). *Psychoacoustics - Facts and Models*. Springer 2nd edition.

Edmundo G. de Souza e Silva, Luiz F.L. Legey and Edmundo A. de Souza e Silva (2010). Forecasting oil price trends using wavelets and hidden markov models. *Energy Economics,* vol. 32, pp. 1507–1519.

Eric Bechhoefer, Steve Clark, and David He (2010). A state-space model for vibration based prognostics. *Annual Conference of the Prognostics and Health Management Society,* October 10-16, Portland, Oregon USA.

F. Combet, L. Gelman (2009). Optimal filtering of gear signals for early damage detection based on the spectral kurtosis. *Mechanical Systems and Signal Processing*, vol. 23, no. 3, pp. 652-668.

Graps, A. (1995). An introduction to wavelets. *IEEE Computational Science and Engineering*, vol. 2, no. 2, PP. 50–61.

H. Link, W. LaCava, J. van Dam, B. McNiff, S. Sheng, R. Wallen, M. McDade, S. Lambert, S. Butterfield, and F. Oyague (2011). Gearbox reliability collaborative project report: Findings from Phase 1 and Phase 2 Testing. NREL/TP-5000-51885.

Hui Li, YupingZhang, and Haiqi Zheng (2011). Application of Hermitian wavelet to crack fault detection in gearbox. *Mechanical Systems and Signal Processing*, vol. 25, pp. 1353–1363.

Halima, E. B., Shoukat Choudhury, M. A. A., Shah, S. L., and Zuo, M. J. (2008). Time domain averaging across all scales: a novel method for detection of gearbox faults. *Mechanical Systems and Signal Processing*, vol. 22, no. 2, pp. 261-278.

Hiram Firpi and George Vachtsevanos (2008). Genetically programmed-based artificial feature-extraction applied to fault detection. *Engineering Applications of Artificial Intelligence*, vol. 21, no. 4, pp. 558-568.

J. Rafiee, M.A. Rafiee, P.W. Tse (2010). Application of mother wavelet functions for automatic gear and bearing fault diagnosis. *Expert Systems with Applications*, vol. 37, no. 6, pp. 4568-4579.

Jerome Antoni (2007). Fast computation of the kurtogram for the detection of transient faults. *Journal of Mechanical systems and Signal Processing*, vo. 21, no. 1, pp. 108-124.

L. Gelman, I. Petrunin, I. K. Jennions, and M. Walters (2012). Diagnostics of local tooth damage in gears by the wavelet technology. *International Journal of Prognostic and Health Management*, vol. 3, no. 2.

R. D. Patterson, K. Robinson, J. Holdsworth, D. McKeown, C.Zhang, and M. H. Allerhand (1992). Complex sounds and auditory images. *Auditory physiology and perception, Proc. 9th International Symposium on Hearing*, Pergamon, Oxford, pp. 123-177.

Richard Dupuis (2010). Application of oil debris monitoring for wind turbine gearbox prognostics and health management. *Annual Conference of the Prognostics and Health Management Society*, October 10-16, Portland, Oregon USA.

RWC Belgium World Data Center, Online sunspot data archive, SIDC, [Online]. Available: [Cited: 20th Dec, 2012]: http://sidc.oma.be/index.php3.

S. Seneff (1988). A Joint Synchrony/mean-rate Model of Auditory Speech Processing. *Journal of Phonetics*, vol. 16, no.1, pp. 55-76.

T. Lin, B. G. Horne, P. Tino, and C. L. Giles (1997). A delay damage model selection algorithm for NARX neural networks. *IEEE Transactions on Signal Processing*, vol. 45, no. 11, pp. 2719–2730.

Uros Lotric and Andrej Dobnikar (2005). Predicting time series using neural networks with wavelet-based denoising layers. *Neural Computing and Applications*, vol. 14, pp. 11-17.

V. Indira, R. Vasanthakumari and V. Sugumaran (2010). Minimum sample size determination of vibration signals in machine learning approach to fault diagnosis using power analysis. *Expert Systems with Applications*, vol. 37, no.12, pp. 8650-8658.

Wang W. (2007). An adaptive predictor for dynamic system forecasting. *Mechanical Systems and Signal Processing*, vol. 21, no. 2, pp. 809–23.

Wang WQ (2004), Golnaraghi MF, Ismail F. Prognosis of machine health conditionusing neuro-fuzzy systems. *Mech Syst Signal Process.* Vol. 18, pp. 813–31.

Yaguo Lei, Ming J. Zuo, Zhengjia He, and Yanyang Zi (2010). A multidimensional hybrid intelligent method for gear fault diagnosis. *Expert Systems with Applications*, vol. 37, no. 2, pp. 1419-1430.

BIOGRAPHIES

Sajid Hussain is a PhD student in Electrical and Computer Engineering at University of Ontario Institute of Technology (UOIT) working under supervision of Dr. Hossam A. Gabbar. He did his diploma in Signal Processing from Aalborg University Denmark in 2003 and MSc in Telecommunication from Technical University Denmark in 2006. He worked as a Research Assistant in Computer Systems Engineering at Blekinge Institute of Technology, Sweden for two years. He is the author of more than 18 publications in the area of computer graphics, vibration analysis and risk-based maintenance. He also holds a patent in psychoacoustics for fault detection in electromechanical machines. He has 10 years of industrial experience in computer systems and machines condition monitoring.

Dr. Hossam A. Gabbar is Associate Professor in the Faculty of Energy Systems and Nuclear Science, and cross appointed in the Faculty of Engineering and Applied Science, University of Ontario Institute of Technology (UOIT). He obtained his Ph.D. degree (Safety Engineering) from Okayama University (Japan), while his undergrad degree (B.Sc.) is in the area of automatic control from Alexandria University, Egypt. He is specialized in safety and control engineering where he worked in process control and safety in research and industrial projects in Japan and Canada. Since 2004, he was tenured Associate Professor in the Division of Industrial Innovation Sciences at Okayama University, Japan. And from 2001, he joined Tokyo Institute of Technology and Japan Chemical Innovative Institute (JCII), where he participated in national projects related to advanced distributed control and safety design and operation synthesis for green energy and production systems. He developed new methods for automated control recipe synthesis and verification, safety design, and quantitative and qualitative fault simulation.

He is a Senior Member of IEEE, the founder of SMC Chapter - Hiroshima Section, the founder and chair of the technical committee on Intelligent Green Production Systems (IGPS), and Editor-in-chief of International Journal of Process Systems Engineering (IJPSE), president of RAMS Society, and editorial board of the technical committee on System of Systems and Soft Computing (IEEE SMCS). He is invited speaker in several Universities and international events, and PC / chair / co-chair of several international conferences. Dr. Gabbar is the author of more than 110 publications, including books, book chapters, patent, and papers in the area of safety and control engineering for green energy and production systems. His recent work is in the area of risk-based safety and control design for energy conservation and supply management, and smart grid modeling and planning with distributed generation.

Analysis of Acoustic Emission Data for Bearings subject to Unbalance

Seyed A. Niknam [1], Tomcy Thomas [1], J. Wesley Hines [2], and Rapinder Sawhney [1]

[1] Department of Industrial and Systems Engineering, University of Tennessee, Knoxville, TN 37996, USA
sniknam@utk.edu
tomcy@utk.edu
sawhney@utk.edu

[2] Department of Nuclear Engineering, University of Tennessee, Knoxville, TN 37996, USA
jhines2@utk.edu

ABSTRACT

Acoustic Emission (AE) is an effective nondestructive method for investigating the behavior of materials under stress. In recent decades, AE applications in structural health monitoring have been extended to other areas such as rotating machineries and cutting tools. This research investigates the application of acoustic emission data for unbalance analysis and detection in rotary systems. The AE parameter of interest in this study is a discrete variable that covers the significance of count, duration and amplitude of AE signals. A statistical model based on Zero-Inflated Poisson (ZIP) regression is proposed to handle over-dispersion and excess zeros of the counting data. The ZIP model indicates that faulty bearings can generate more transient wave in the AE waveform. Control charts can easily detect the faulty bearing using the parameters of the ZIP model. Categorical data analysis based on generalized linear models (GLM) is also presented. The results demonstrate the significance of the couple unbalance.

1. INTRODUCTION

Acoustic emission (AE) is defined as transient elastic waves generated due to localized physical changes in a solid material under mechanical or thermal stresses (Tan et al., 2007). AE is also referred to as the practical non-destructive technology to investigate the behavior of stressed materials using the transient elastic waves. The major advantage of this technology is its sensitivity to capture surface and subsurface micro-damage. It has been proven that in some cases AE can ensure superiority over vibration-based monitoring systems in early fault detection (Tan et al., 2007, Alghamd & Mba, 2006, Tandon & Mata, 1999, Eftekharnejad & Mba, 2009). In case of rolling-element bearings, AE detects the fault earlier compared to other technologies (Mba & Rao, 2006, Yoshioka & Fujiwara 1982). Significant changes in vibration signatures can be observed when the remaining operational life of a bearing is very short. Hence, AE offers good potential for prognostic capabilities. Additionally, insensitivity to structural resonance and mechanical background noise gives AE an additional advantage over typical vibration-based monitoring systems. However, AE signals may suffer severe attenuation and reflections due to sensor positioning and machine complexity. AE is a well-established diagnostic method for static structures. In recent decades, AE applications in structural health monitoring have been extended to other areas such as rotating machineries and cutting tools (Niknam & Liao, 2011). The readers are referred to ISO 22096:2007 for the general principles of AE application.

AE hit is defined as the process of detecting and measuring an AE signal on a channel (PAC, 2007). The fundamental features of the AE hit include amplitude, duration, count, and rise time. These parameters can be used to provide additional signal features such as root mean square (RMS), AE cumulative event count, counts to peak, rise time slope, crest factor and Kurtosis (Alghamd & Mba, 2006, Tandon & Mata, 1999, K. Miyachika et al. 1995, He et al. 2010, Bansal et al. 1990, Choudhury & Tandon, 2000). Figure 1 illustrates the diagram of AE hit feature extraction (PAC, 2007).

AE count is defined as the number of times the signal crosses the threshold in an AE hit. In the literature, this parameter is also known as ring down count or threshold crossing count. In effect, AE counts imply the existence of a transient wave in the AE waveform. One major drawback of this parameter is its dependence on the threshold level. It is to be noted that the average counts is a measure of AE intensity i.e. the size of the emission signals detected.

Furthermore, AE counts divided by duration gives the average frequency of the signal. The next section provides more details of the application of AE count for fault detection. The parameter of interest in this study is PAC-energy (PAC is the registered trademark of Physical Acoustics Corporation). PAC-energy is a 2-byte parameter derived from the integral of the rectified voltage signal over the duration of AE hit (PAC 2007). The unit of PAC-energy is micro-volt-seconds per count and the range is 0 to 65535 in each AE hit. Therefore, this parameter is a discrete random variable that covers the significance of count, duration and peak amplitude. Thus, the statistical models for count data can be applied for the PAC-energy.

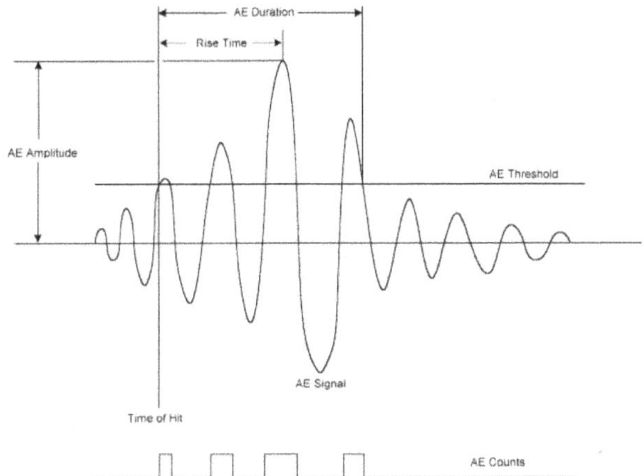

Figure 1. Diagram of AE hit feature extraction (PAC 2007)

This research attempts to provide effective diagnostic and prognostic algorithms for wind turbine drive trains. In practice, rotor unbalance is a major concern in wind turbine reliabilities. Rotor unbalance in wind turbines is caused by manufacturing defects, accumulated damage to the rotor blades, and non-uniform accumulation of ice, dust, and moisture (Lu et al. 2009, Hyers et al., 2006). Rotor unbalance has an important influence on the pitch mechanism, the yaw brakes and the variations in the rotor speed which can even affect the output power waveform. Furthermore, rotor unbalance imposes additional vibrations of the nacelle, tower and the drivetrain components (Lu et al. 2009, Hyers et al., Hameed et al., 2009). It may also shorten the life of the rotor blades. In particular, this research investigates the application of AE data in analyzing unbalanced rotary systems. Categorical data analysis based on generalized linear models (GLM) is presented in this study. The objective of this analysis is to provide a meaningful interpretation of the test variables' effect on AE features. In addition, a statistical model based on Zero-Inflated Poisson (ZIP) regression is proposed to handle over-dispersion and excess zeros of the counting data. The ZIP model considers the data set as a mixture of processes that generates only zeros or non-negative integers.

The remainder of the paper is organized as follows: Section 2 provides the background for unbalanced systems, a review of using AE count in fault detection, and modeling of count data. Section 3 presents a detailed description of the experimental setup and the test procedure. Section 4 provides the analytical results and discussions. Finally, section 5 concludes this paper.

2. BACKGROUND

2.1. AE count

Traditionally, AE count has been widely used in the literature as condition indicator. Tandon and Mata (1999) realized that defects in gear systems would result in broader distribution of AE counts and peak amplitude. They claimed that AE counts showed better results than other AE parameters in gear defect detection. Miyachika et al. (1995) postulates that AE cumulative event count increases with crack growth in the case-hardened gears. Tandon and Nakra (1990) considered AE counts for condition monitoring of radially loaded ball bearing. They observed a direct relationship between AE count and speed in case of outer race defect. The results showed that AE count is a practical indicator for defects less than 250μm in diameter. However, Morhain and Mba (2003) declared the AE count is able to detect large defects up to 15 mm in lengths and 1 mm in width. This study also emphasized the sensitivity of AE counts to the level and grade of lubricant. AE parameter analysis for tool condition monitoring has received attentions mainly for real time applications. AE count rate was introduced as a reliable parameter for monitoring tool wear during turning, although AE signals highly depend on process parameters (Li 2002, Sharma et al. 2007). Carpinteri et al. (2007) applied AE technology to monitor concrete and masonry buildings. The cumulative number of counts was found to be informative for analyzing the evolution of cracks and determining the released strain energy. Carpinteri et al. (2007) observed that the maximum counting of AE corresponds to the maximum velocity of crack propagation. Considering the proven usefulness of AE count for various AE applications, it seems an appropriate modeling and analysis of AE count or its derivatives would lead to fruitful diagnostic. As mentioned earlier, PAC-energy is a function of AE counts and it will be used for the statistical modeling in this study.

2.2. Modeling count data

In the literature, linear regression models have been primarily utilized to correlate the AE signatures and the physical features of interest. Traditional regression (non-Bayesian) methods have been widely used to model count data in both natural and social sciences. In effect, the response variable in such models is a nonnegative integer. The most common regression-based count data model is Poisson GLM which is an extension of ordinary least squares regression and agrees with distributions from the exponential family. The Poisson GLM has the form

$$y \sim Poisson(\lambda)$$

The log link is specified by

$$\log\lambda = \beta_0 + \sum_i \beta_i x_i \quad (1)$$

The logistic model where p is the probability of success is given by

$$y \sim Binomial\ (p,n)$$

The logit link is given by

$$\text{logit } p = \log\frac{p}{1-p} = \beta_0 + \sum_i \beta_i x_i \quad (2)$$

The major drawback of Poisson regression is the restrictive assumption of equality between the variance and the mean i.e. equidispersion (Liu & Cela 2008, Lambert 1992, Guikema & Coffelt 2008). On the other hand, overdispersion occurs when the variance is greater than the mean. The overdispersion can be handled through negative binomial GLM. Generalized linear mixed models (GLMM) add an error term to GLM which is not part of our discussion.

In practice, depending on the preset reference threshold value, there may be a high frequency of zero counts in the AE signals, which indicates excess zeroes. In this case, a simple Poisson regression would not satisfactorily fit the data. Zero-inflated Poisson (ZIP) model, introduced by Lambert (1992), fittingly handles overdispersion and excess zeros. Principally, ZIP model considers the data set as a mixture of a process that generates only zeros and a process that generates counts from a Poisson or a negative binomial model. In other words, ZIP models calculate the probability (p) of having observations of 0. Therefore, 1-p would be the probability of having non-negative integers. Therefore the count response can be written as

$$Y_i \sim 0 \quad \text{with probability } p_i$$
$$Y_i \sim Poisson(\lambda_i) \quad \text{with probability } 1 - p_i$$

Thus,

$$Y_i = 0 \quad \text{with probability } p_i + (1-p_i)e^{-\lambda_i}$$
$$Y_i = k \quad \text{with probability } (1-p_i)e^{-\lambda_i}(\lambda_i)^k/k!$$

Consider the following regression models for

$$\lambda = [\lambda_1, \lambda_2, \dots, \lambda_n]' \text{ and } \mathrm{p} = [p_1, p_2, \dots, p_n]':$$
$$\ln(\lambda) = \mathbb{B}\beta, \qquad logit(\mathrm{p}) = \ln\left(\frac{\mathrm{p}}{1-\mathrm{p}}\right) = \mathbb{G}\gamma$$

where $\mathbb{B}$ and $\mathbb{G}$ are covariate matrices (design matrices), β and γ are regression coefficients.

Accordingly, the log-likelihood function is

$$L(\beta,\gamma|y) = \sum_{y_i=0}^{n} \ln\left(e^{\mathbb{G}_i\gamma} + \exp\left(-\left(e^{\mathbb{B}_i\beta}\right)\right)\right)$$
$$-\sum_{i=1}^{n}\ln(1+e^{\mathbb{G}_i\gamma}) + \sum_{y_i>0}^{n}(y_i\,\mathbb{B}_i\beta - e^{\mathbb{B}_i\beta}) - \ln(y_i!) \quad (3)$$

By using $\ln(\lambda) = \mathbb{B}\beta$ and $logit(\mathrm{p}) = \ln\left(\frac{\mathrm{p}}{1-\mathrm{p}}\right) = \mathbb{G}\gamma$ we conclude

$$\lambda_i = \exp(\mathbb{B}_i\beta), \text{ and } p_i = \frac{\exp(\mathbb{G}_i\gamma)}{1+\exp(\mathbb{G}_i\gamma)} \quad (4)$$

Hence, using the delta method, the variances of $\hat{p}$ and $\hat{\lambda}$ are

$$cov(\hat{\lambda}_i, \hat{p}_i) = [[\frac{\partial\lambda_i}{\partial\beta}]'[\frac{\partial p_i}{\partial\gamma}]']cov(\hat{\gamma}, \hat{\beta})[[\frac{\partial\lambda_i}{\partial\beta}]'[\frac{\partial p_i}{\partial\gamma}]']'$$

and $\hat{\sigma}_{0(\hat{p}_i)} = \sqrt{var(\hat{p}_i)}$, $\hat{\sigma}_{0(\hat{\lambda}_i)} = \sqrt{var(\hat{\lambda}_i)}$

2.3. Unbalanced rotary systems

The lack of balance causes excessive vibration in rotary systems. Vibrations impose centrifugal force and oscillatory force. For wind turbines, unbalance occurs primarily due to uneven material deposit on the rotor about its rotating centerline. High vibration amplitude at the rotating speed is the primary indicator of this fault. Other sources of unbalance include imperfect manufacturing and operational changes. Bearings are a significant contributor to unbalance in rotary system because of internal clearance and run-out. In effect, the existence of heavy spot in rotors leads to shaft bending and cyclical forces on bearing. Both centrifugal and oscillatory forces affect the bearing life. The magnitude of centrifugal force can be obtained by

$$F_c = m\,r\,\omega^2 \quad (5)$$

where m represents mass, ω represents the angular speed and r is the radius from center of rotation.

There are various types of unbalance based upon the relation between the center of gravity (CG) and the heavy spot. In static unbalance, the CG and the heavy spot are in the same plane. Couple unbalance occurs due to existence of two equal heavy spots which are 180 degrees apart. In this case, the shaft axis intersects the principal mass at the CG. Couple unbalance imposes radial force on bearings. Dynamic unbalance is the most common type of unbalance and it represents a combination of static and couple unbalance. In essence, balancing treats the cause rather than symptom. Balancing may involve the followings: correction of mass distribution, creating centrifugal force, changing orientation of parts and adding/removing mass from non-rotating part (Wowk 1995). These actions would provide mass symmetry, change the center of gravity, and affect bending moments. It is important to note that cyclical forces on bearings measured through vibration analysis are utilized for balancing. However, in this research, the magnitudes of the balancing masses were found graphically. Details pertaining to the unbalance disks are presented in the next section.

3. EXPERIMENTAL SET-UP AND PROCEDURE

A multi-purpose test rig was designed and developed to simulate the drive train of wind turbines. Figure 2 shows the test rig. A WindMax 2 kW wind turbine generator was used in this setup. The test rig's motor provided a rotational speed in the range of 10- 1760 round/min. The control drive along with a LabVIEW-based program allowed the motor to provide variable speed according to a pre-defined speed pattern. Other main components of the test rig include a single phase planetary gearbox, support bearings, and two test bearings of different size.

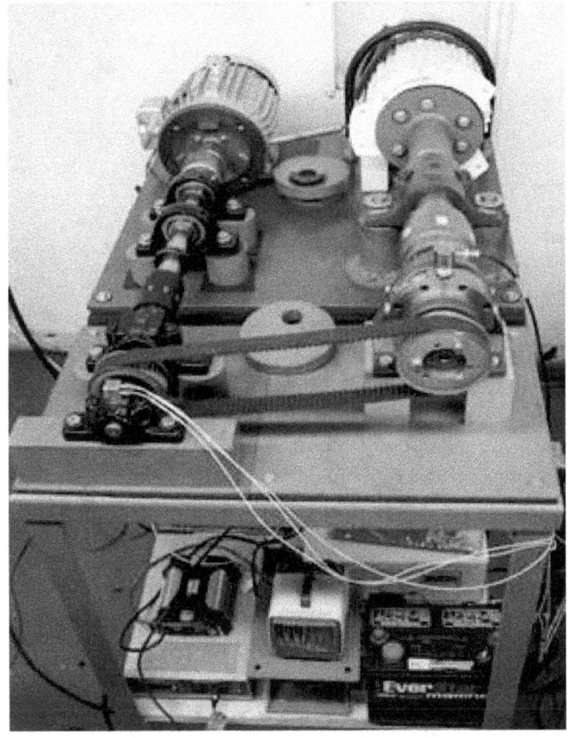

Figure 2. Test Rig

The test bearing used was a SKF self-aligning ball bearing type 1205 ETN9 (Figure 4). The characteristics of the test bearing are as follows: internal (bore) diameter 25 mm, external diameter 52 mm, 26 rollers, and diameter of roller 7.4 mm. To make a faulty bearing, a groove with the width of 0.5 mm and the maximum height of 1.41 mm was made by using electrical discharge machine on the inner race of a bearing (Figure 4). Prior to the main tests, two bearings were run for 20 hours at the speed of 1700 rpm i.e. over 2 million cycles. These bearings will be called "used bearings" in the analysis. It is important to note that all the bearings were used in dry condition. One reason for this is to ensure that the used bearings will not remain as good as new bearing after 2 million cycles.

To impose the unbalance force two uniform disks with the diameter of 6 inch and the thickness of 1 inch were used. To be able to attach the disks to the shaft and also change the angle of the disks, a shaft collar with a thickness of 0.5 inch was welded to each disk (Figure 3). An off-center hole with the diameter of 1.375 inch was then made for each disk as shown in Figure 5.

The mass of the disk with and without the hole were approximately 8 pounds and 7.6 pounds respectively. The new center of gravity moved off-center about 0.078 inch. The disks could be placed between the test bearings as shown in Figure 2. Consequently, several different sets of experiments could be performed by changing the number and orientation of the disks. In this study, the following loading configurations were used: no disk, one disk, two disks with an angle of zero, two disks with an angle of 120 and two disks with an angle of 180 namely couple unbalance.

Figure 3. Sensor placement and an unbalance disk

Figure 4. SKF 1205 ETN9 with a seeded fault

In this study, three major variables were considered: (1) bearing type, (2) shaft speed and (3) unbalance. Table 1 shows different categories of the variables. As shown in the table, three speed levels , three bearing types, and 5 loading patterns were designed for the experiments.

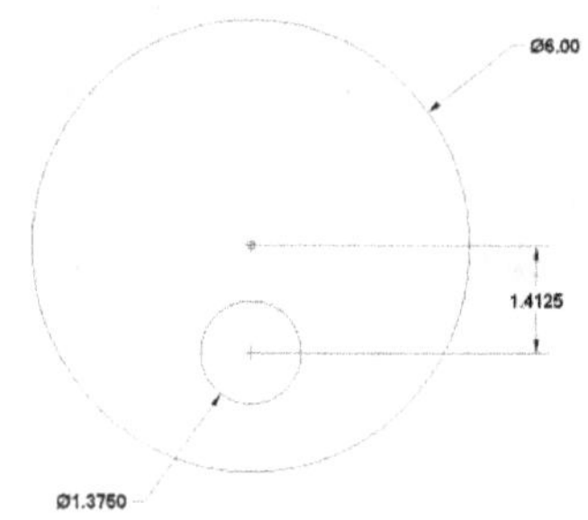

Figure 5. Disk with off-center hole

	Settings	Category
Unbalance	No Disk	1
	1 Disk	2
	2 Disk w 0 °	3
	2 Disk w 120 °	4
	4 Disk w 180°	5
Bearing	New	1
	Used	2
	Faulty	3
Shaft Speed (rpm)	150	1
	300	2
	450	3

Table 1. Test Data Description

A 2-channel PCI-2 based AE system was utilized for data acquisition. The PAC R15I-AST acoustic sensor was used. This sensor has an operating frequency range of 80-200 kHz. The sensor housing contains a filter and an integral preamplifier of 40 dB. With an improved 18 bit analogue to digital conversion scheme, PCI-2 board provides a sampling rate of up to 40 MHz and a dynamic range of more than 85 dB. This data acquisition system is able to record up to 65535 counts per hit. The data acquisition was performed with a sampling rate of 5 mega points per seconds. The upper and lower limit of filter was set to 10 KHz and 1 MHz respectively. The sensors were placed on the bearing housing as shown in Figure 3. The sensor which was placed on the top provided more data than the sensor which was on the side of the housing. In essence, one can expect less data from the side sensor due to the interfaces that exist between the source and the side sensor. In figures 6 and 7 the left graph shows the energy collected from the top sensor and the graph on the right shows the energy collected from the side sensor. As shown in these figures, data from the side sensor is not really helpful. In Figure 6 the seeded fault was exactly below the top sensor. Experimenting various position of the seeded fault, it was realized that placing the fault exactly below the top sensor provides more energy.

In Figure 7 the seeded fault was placed 90 degrees apart from the top sensor and 180 degrees apart from the side sensor. For our analysis, only the data collected from the top sensor was utilized. Moreover, the movement of the bearing with inside the housing was negligible.

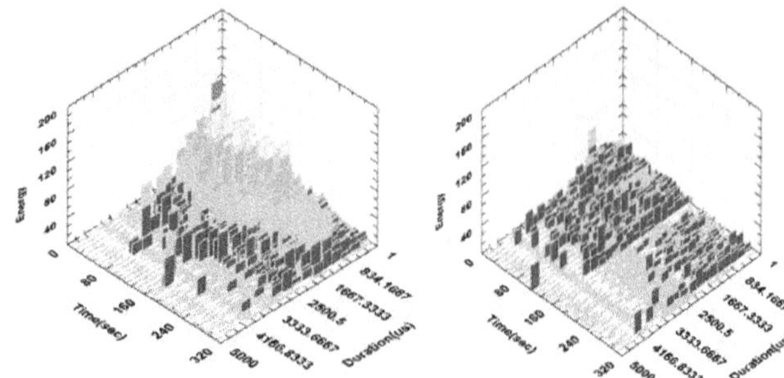

Figure 6. PAC-energy from the sensors while the fault was under the top sensor

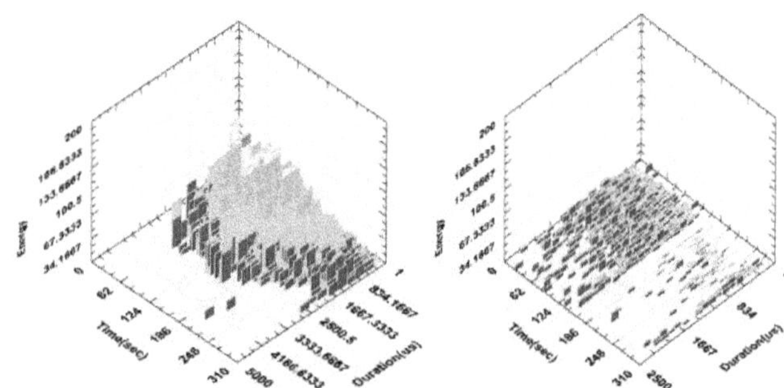

Figure 7. PAC-energy from the sensors while the fault was 90 degrees apart from the top sensor

The results of variable speed tests indicate the sensitivity of AE count to the shaft speed as shown in Figure 8 where the speed was gradually changed from 0 to 300 rpm and dropped back to 0.

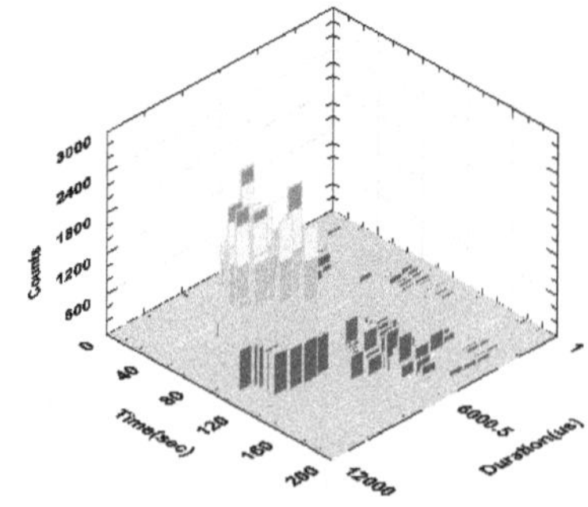

Figure 8. Dependence of AE parameters to the shaft speed

4. RESULTS AND DISCUSSION

4.1. Categorical data analysis

The data set for the categorical data analysis consisted of data from six bearings: one faulty bearing, two used bearing and three new bearings. The cumulative energy (CE) was selected as the response variable. The objective of this categorical analysis is to reveal the effects of explanatory variables (i.e. bearing, speed and unbalance) on the response variable. Due to time limit, it was not possible to perform all the loading patterns for the speed level of 150 and 450 rpm. Although, this can be a drawback for the categorical analysis, the results are satisfactory. The GENMOD procedure in SAS was used to conduct the statistical tests. This procedure fits a GLM to the data by maximum likelihood estimation. Through an iterative fitting process, the parameters of the model are numerically estimated in this procedure. Based on the asymptotic normality of maximum likelihood estimators, standard errors, and p-values of the estimated parameters are computed. The parameter with the smallest p-value will be the most significant parameter in that category. The GENMOD procedure provides a number of probability distributions and link functions such as log-link function.

4.1.1 Analysis of AE data with various speed levels

The categorical data analysis (CDA) shows that speed has the most significant effect among the explanatory variables. The bearing type has the least significant effect with the p-value of 0.6295 (Table 2). Figure 6 supports the SAS results, that speed would positively affect the number of count and the energy of the signal. The interaction between explanatory variables (speed, unbalance and bearing) is added to the model. This 3-way interaction is significant (p-value = 0.1041). This model shows that speed and unbalance have significant effects on the CE level (Table 3).

Parameter	Estimate	Standard Error	Pr > Chi Sq.
intercept	0.1278	0.3054	0.6757
bearing	-0.0475	0.0985	0.6295
speed	0.0011	0.0007	0.1254
unbalance	0.0336	0.0493	0.4962

Table 2. CDA– Model 1

Parameter	Estimate	Standard Error	Pr > Chi Sq.
intercept	-0.6358	0.5584	0.2549
bearing	0.1680	0.1617	0.2987
speed	0.0022	0.0010	0.0229
unbalance	0.1717	0.0971	0.0769
Interaction	-0.0002	0.0002	0.1041

Table 3. CDA– Model 1 with 3-way interaction

A new model was built based on the interactions between every two explanatory variables (Table 4). The most significant interaction is the bearing-unbalance. In this case, unbalance type is the most significant variable. The effect of categories of each variable on the AE generation was then investigated. One category was selected as the baseline and the significance of the other categories were analyzed. For speed, 450 rpm was selected as the baseline. As expected, the results indicate that this speed level is more significant than the other two levels. The bearing type 3 (i.e. faulty) was the baseline to analyze bearing types. As shown in the Table 5, type 1 (new bearing) and 2 (used bearing) are not significant compared to the faulty bearing. For the unbalance, type 5 was the baseline. As shown in the Table 6, type 1 and 2 are not significant at all compared to the couple unbalance i.e. type 5. Interestingly, the SAS output results indicate that couple unbalance is the most significant type of unbalance in comparison with type 3 and 4. This result is very similar to the graphical balancing calculation that shows couple unbalance needs the highest amount of balancing mass.

Parameter	Estimate	Standard Error	Pr > Chi Sq.
intercept	-1.0091	0.9006	0.2625
bearing	0.0924	0.3834	0.8095
speed	0.0013	0.0025	0.6070
unbalance	0.6373	0.2128	0.0027
bearing*speed	0.0013	0.0010	0.1604
bearing*unbal.	-0.1836	0.0679	0.0069
speed*unbal.	-0.0008	0.0005	0.1094

Table 4. CDA – Model 1 with 2-way interactions

Parameter	Type	Estimate	Standard Error	Pr > Chi Sq.
intercept		0.1088	0.3034	0.7199
bearing	1	0.0101	0.199	0.9593
bearing	2	-0.2117	0.2195	0.3347
speed		0.0011	0.0007	0.1256
unbalance		0.0338	0.0493	0.4928

Table 5. CDA - Bearing 3 is the baseline

Parameter	Type	Estimate	Standard Error	Pr > Chi Sq.
intercept		0.2736	0.4105	0.505
bearing		-0.0421	0.0986	0.6696
speed		0.0009	0.0009	0.3432
unbalance	1	0.0808	0.214	0.7058
unbalance	2	-0.2115	0.2814	0.4521
unbalance	3	-0.4037	0.2917	0.1664
unbalance	4	0.3487	0.251	0.1648

Table 6: CDA – Unbalance type 5 is the baseline

4.1.2 Analysis of AE data without the effect of speed

Since it was realized that speed has the most significant effect on CE, the next set of statistical tests were conducted with the speed level of 300 rpm. Having more data was the reason for selecting this speed level. The results of this data analysis indicated that unbalance is more significant in the absence of speed effect (Table 7). Similar to the case with various speeds, adding the interactions of these variables lead to the results that all the inputs are significant.

Parameter	Estimate	Standard Error	Pr > Chi Sq
intercept	0.4184	0.3326	0.2084
bearing	-0.1906	0.1393	0.1712
unbalance	0.1362	0.0706	0.0538

Table 7: Categorical data analysis – without the effect of speed

4.2. ZIP model

The independent variables in this study were shaft speed and unbalance. The amount of mass required for balancing was used as the unbalance values in the ZIP model. The graphical balancing techniques were utilized to calculate the amount of the balancing mass. The procedure of balancing considers both the balance of forces and couples. It is to be noted that couple unbalance (type 5) needs the highest amount of mass for balancing.

The COUNTREG procedure in SAS was used to develop the ZIP models. This procedure performs nonlinear optimization. Two iterative minimization method were applied; (1) the quasi-Newton method and (2) the Newton-Raphson method. The ZIP models provide the probability (p) of obtaining zeros and the parameter (λ) for the Poisson model. The probability of obtaining zero for the faulty bearing is almost zero for all the tests. This implies the existence of strong burst signal in all the AE hits. To provide a fair comparison, Figure 9 shows the λ for different unbalance types at the speed level of 300 rpm.

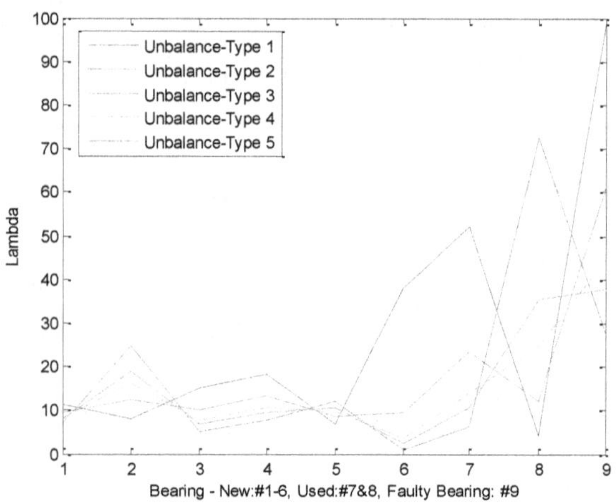

Figure 9. The Lambda from the ZIP model for different unbalance type at 300 rpm

In this graph, bearing 9 represents the faulty bearing; bearing 7 and 8 represent the used bearings. It is clear that the highest values of λ belong to either used or the faulty bearings. It implies that these bearing generate stronger burst signals in comparison with all the six new bearings used in this study. It is interesting to note that the highest λ (= 429) was obtained for the couple unbalance of the faulty bearing.

In order to provide a platform for diagnostics, cumulative sum (CUSUM) chart was utilized (He et al., 2011, Montgomery2001, Leger et al., 1998). This chart detects the deviation of the process mean through cumulative sums of the shift between sample averages from a target value. CUSUM charts are sensitive to small and moderate changes in the process mean. Such changes, e.g. one-sigma shift in the mean, are hardly detectable by Shewhart-charts. The run chart of CUSUM displays the successive differences between the sample average and the target i.e. process mean. The two-sided tabular CUSUM chart is defined by

$$S^{\pm}(n)\begin{cases} 0, & if\ n = 0 \\ \max\{0, S^{\pm}(n-1) \pm (x_n - \hat{\mu}_0) - k\hat{\sigma}_x\}, & if\ n = 1,2,... \end{cases}$$

where $x_n = \lambda_n$ (or p_n), $S^{+}(n)$ and $S^{-}(n)$ give the cumulation on high side and low side respectively and k is the threshold for cumulation which is also called allowable slack. This parameter is the minimum difference between the target and sample average. We have H = $h\sigma_x$ = $\tan(\theta)\, d\, \sigma_x$ as the decision interval and h is called the decision parameter as shown below.

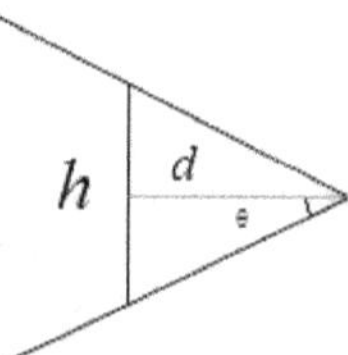

If $S^{+}(n)$ or $S^{-}(n)$ exceeds H, the process is out- of-control. Figures 10 and 11 show the two-sided CUSUM charts of λ and p for the case when only the new bearings were used. Here, as none of the plotted points cross the arms of the V-mask we conclude that the process is in control.

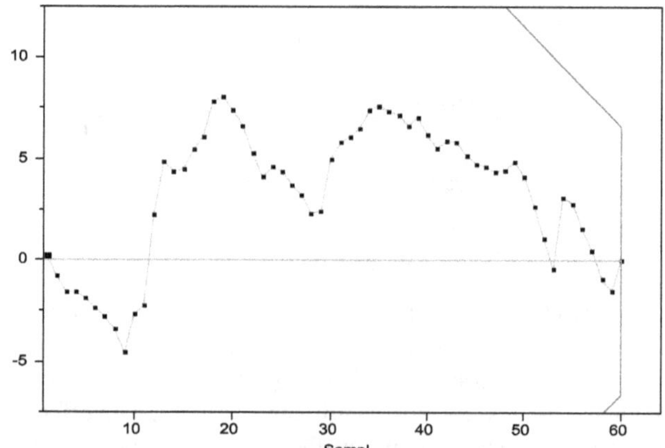

Figure 10. CUSUM charts of λ for new bearings

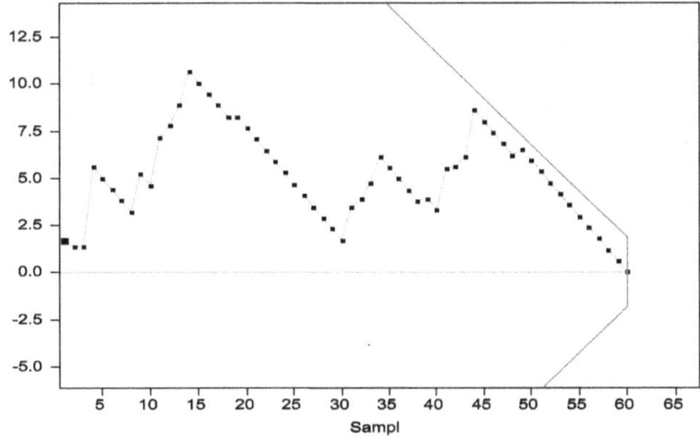

Figure 11. CUSUM charts of p for new bearings

Figures 12 and 13 show the CUSUM charts of λ and p for the case when used bearings were also taken into account. The chart for λ clearly shows that the bearings experienced certain deviation and they are not as good as new bearings.

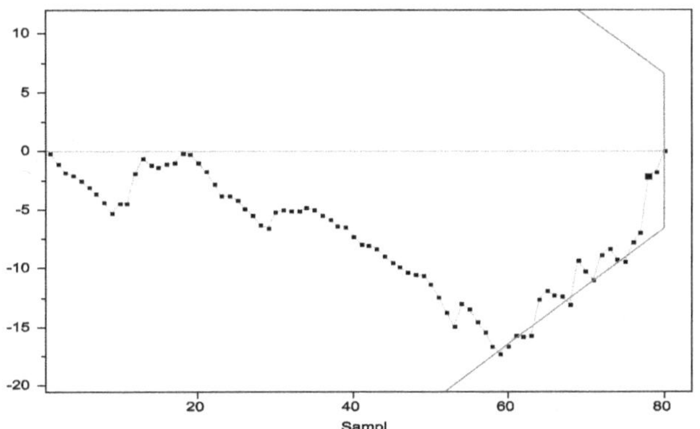

Figure 12. CUSUM charts of λ for new and used bearings

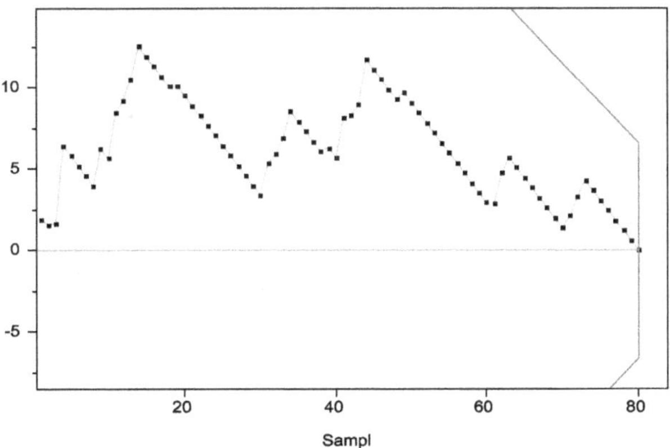

Figure 13. CUSUM charts of p for new and used bearings

Figures 14 and 15 show the CUSUM charts of λ and p for the case when all types of bearings were taken into account. These charts clearly depict the out of control condition.

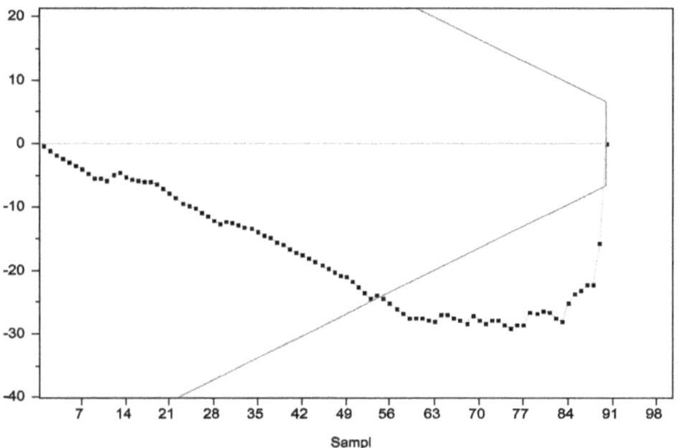

Figure 14. CUSUM charts of λ for all bearings

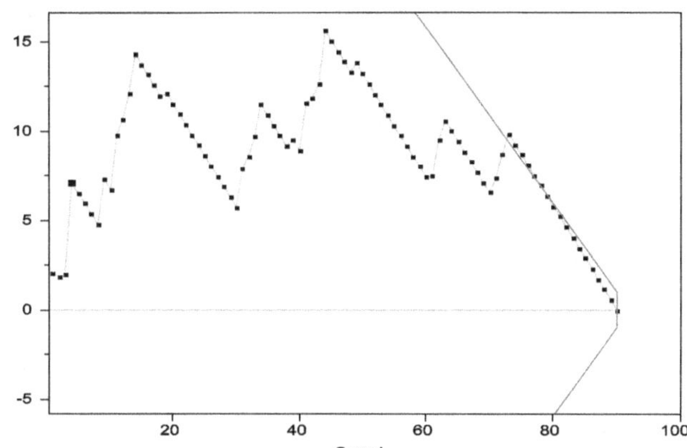

Figure 15. CUSUM charts of p for all bearings

5. CONCLUSION

This paper investigates the usefulness of AE data for unbalance analysis in rotary systems. The parameter of interest in this study is the PAC-energy which covers the significance of count, duration and peak amplitude in AE signals. The categorical data analysis based on generalized linear models is presented. The results of this analysis confirm the visual observations of the AE signals during the data acquisition process. It was discovered that speed has the most significant effect on PAC-energy. The results of the categorical analysis indicated that loading patterns with two disks were more significant. More importantly, the results specified the significance of couple unbalance. In addition, it was also understood that bearing types influence the level of cumulative energy. It was realized that the faulty bearing is the most significant in producing more PAC-energy. Interestingly, the effect of used bearing was more significant than new bearings in this respect.

A statistical model based on Zero-Inflated Poisson regression is presented to handle over-dispersion and excess zeros of the counting data. Combined with CUSUM chart, the ZIP model can provide a platform to diagnose the faults in bearings. It was observed that CUSUM charts of both λ and p can be utilized for fault detection and also to track the health of bearings. It was also observed that while unbalance influences the amplitude of signal, speed directly affects the probability of getting more zero counts.

For more effective diagnosis, the pattern of control charts can be recognized through artificial intelligence methods. Moreover, distribution-free tabular CUSUM chart can be investigated. It is noteworthy to point that AE can be utilized to detect and quantify unbalance in the rotary systems. However, it is essential to consider the sources of uncertainty such as the instrumentation errors, rotor system nonlinearities, stiffness (rotor, bearing support and foundation), and the nature of bearings.

REFERENCES

Tan, C., Irving, P., Mba, D. (2007). A comparative experimental study on the diagnostic and prognostic capabilities of acoustics emission, vibration and spectrometric oil analysis for spur gears. *Mechanical Systems and Signal Processing,* vol. 21, pp. 208-233.

Alghamd, A., & Mba, D. (2006). A comparative experimental study on the use of acoustic emission and vibration analysis for bearing defect identification and estimation of defect size. *Mechanical Systems and Signal Processin,* vol. 20, pp. 1537-1571.

Tandon, N., & Mata, S. (1999). Detection of Defects in Gears by Acoustic Emission Measurements. *Journal of Acoustic Emission,* vol. 17, pp. 23-27.

Eftekharnejad, B., & Mba, M. (2009). Seeded fault detection on helical gears with acoustic emission. *Applied Acoustics,* vol. 70, pp. 547-555.

Mba, D., & Rao, R. B. K. N. (2006). Development of Acoustic Emission Technology for Condition Monitoring and Diagnosis of Rotating Machines; Bearings, Pumps, Gearboxes, Engines and Rotating Structures. *The Shock and Vibration Digest*, vol. 38, pp. 3-16.

Yoshioka, T., & Fujiwara, T., (1982). New acoustic emission source locating system for the study of rolling contact fatigue. *Wear*, vol. 81, pp. 183-186.

Niknam, S. A. & Liao, H. (2011). Diagnostic and Prognostic Modeling of High-Speed Milling Machine Cutters. *The 7th International Conference on Mathematical Methods in Reliability - Theory, Methods, Applications*, Beijing, China.

PAC (Physical Acoustic Corporation) (2007). *PCI-2 Based AE System*. Princeton Jct, NJ

Miyachika, K., Oda, S., Koide, T. (1995). Acoustic Emission of Bending Fatigue Process of Spur Gear Teeth. *Journal of Acoustic Emission*, vol. 13, pp. 47-53.

He, D., Li, R., Bechhoefer, E. (2010). Split Torque Type Gearbox Fault Detection using Acoustic Emission and Vibration Sensors. *International Conference on Networking Sensing and Control.*

Bansal, V., Gupta, B. C., Prakash, A., Eshwar, V. A. (1990). Quality inspection of rolling element bearing using acousic emission technique. *Acoustic Emission,* vol. 9, pp. 142-146.

Choudhury, A. & Tandon, N. (2000). Application of acoustic emission technique for the detection of defects in rolling element bearings. *Tribology International* , vol. 33, pp. 39-45.

Lu, B., Y. Li, Wu, X., Yang, Z. (2009). A review of recent advances in wind turbine condition monitoring and fault diagnosis. *Power Electronics and Machines in Wind Applications*, Milwaukee, WI.

Hyers, R. W., McGowan, J. G., Sullivan, K.L., Manwell, J.F., Syrett, B.C. (2006). Condition monitoring and prognosis of utility scale wind turbines. *Energy Materials*, vol. 1, pp. 187-203.

Hameed, Z., Hong, Y.S., Cho, Y.M., Ahn, S.H., Song, C.K. (2009). "Condition monitoring and fault detection of wind turbines and related algorithms: A review. *Renewable and Sustainable Energy Reviews*, vol. 13, pp. 1-39.

Tandon, N., & Nakra, B. C. (1990). Defect Detection of Rolling Element Bearings by Acoustic Emission Method. *Journal of Acoustic Emission*, vol. 9, pp. 25-28.

Morhain, A., & Mba, D. (2003). Bearing defect diagnosis and acoustic emission. *Engineering Tribology*, vol. 217, pp. 257-272.

X. Li (2002). A brief review- acoustic emission method for tool wear monitoring during turning. *International Journal of Machine Tools & Manufacture*, vol. 42, pp. 157-165.

Sharma, V.S., Sharma, S. K., Sharma, A. K. (2007). An approach for condition monitoring of a turning tool. *Journal of Engineering Manufacture*, vol. 22, pp. 635-646.

Carpinteri, A., Lacidogna, G., Pugno, N. (2007). Structural damage diagnosis and life-time assessment by acoustic emission monitoring. *Engineering Fracture Mechanics*, vol. 74, pp. 273-289.

Liu, W.S., & Cela, J. (2008). Count Data Models in SAS. *SAS Global Forum.*

Lambert, D. (1992). Zero-Inflated Poissoin Regression, with an application to defects in manufacturing. *Technometrics*, vol. 34, pp. 1-14.

Guikema, S. D. & J. P. Coffelt (2008). Modeling count data in risk analysis and reliability engineering. In K. B. Misra (ed.). *Handbook of Performability Engineering,*. London, Springer.

Wowk V. (1995). *Machinery Vibration: Balancing*. New York, McGraw-Hill.

He, S., Huang, W., Woodall, W.H . (2011). CUSUM Charts for Monitoring a Zero-inflated Poisson Process. *Quality and Reliability Engineering International*, vol. 28, pp. 181-192.

Montgomery, D. C. (2001). Introduction to Statistical Quality Control, 4th ed. *John Wiley & Sons*, New York.

Leger, R.P., W.J. Garland, and W.F.S. Poehlman, (1998). Fault detection and diagnosis using statistical control charts and artificial neural networks. *Artificial Intelligence in Engineering*, vol. 12, pp. 35-47.

BIOGRAPHIES

Seyed A. Niknam received the B.Sc. degree in mechanical engineering in 2003, the M.Sc. degree in advanced manufacturing systems from Brunel University in 2006, and the M.Sc. degree in reliability and maintainability engineering from the University of Tennessee, Knoxville in 2011. Currently, he is a PhD candidate in the department of Industrial and Systems Engineering at the University of Tennessee, Knoxville. He has been working as the instructor of the core courses in the reliability and maintainability engineering program since January 2012. His research interests include reliability of repairable systems, diagnostics and prognostics.

Tomcy Thomas received his B.E. (Electronics) from Marathwada University, India in 1991. He received his MBA (Operations Management) from IGNO University, India in 2000. He received his MS (Engineering Management) from the University of Tennessee in 2010. He is currently a Ph.D. candidate in the Department of Industrial and Systems Engineering at the University of Tennessee, Knoxville. His research interests include energy management, reliability engineering and supply chain optimization.

Dr. J. Wesley Hines is currently the Nuclear Engineering Department Head at the University of Tennessee. He has served as Interim Associate Dean for Research in the College of Engineering and more recently as Interim Vice Chancellor for Research for the Campus. He received the BS degree in Electrical Engineering from Ohio University in 1985, and then served as a nuclear qualified submarine officer in the US Navy. He later received both an MBA and an MS in Nuclear Engineering from The Ohio State University in 1992, and a Ph.D. in Nuclear Engineering from The Ohio State University in 1994. He has been with the University for 18 years. Dr. Hines teaches and conducts research in artificial intelligence and advanced statistical techniques applied to process diagnostics, condition based maintenance, and prognostics; and has made notable accomplishments in the invention and development of reliability enhancing condition monitoring technologies. These technologies have allowed operators to understand the condition of vital assets and in several cases to predict their remaining useful life with quantifiable confidence bounds. He has authored over 275 papers and has three patents that have been implemented in commercial products.

Dr. Rapinder Sawhney has been a faculty member with the Department of Industrial and Systems Engineering (ISE) for almost two decades. He was named as head of ISE and Weston Fulton Professor in 2010. He is also a faculty member of the Center for Interdisciplinary Research and Graduate Education (CIRE), a joint effort between UT and ORNL focused on renewable energy. Sawhney received his Ph.D in engineering science and mechanics from the University of Tennessee. He is a member of the Institute for Industrial Engineers and the American Society of Quality and is also a committee member for the Boeing Welliver Fellowship Committee and the Institute of Industrial Engineers Lean Division.

Classification and Detection of Wind Turbine Pitch Faults Through SCADA Data Analysis

Jamie L. Godwin[1] and Peter Matthews[2]

[1,2]*School of Engineering and Computing Sciences, Durham University, Durham, DH1 3LE, United Kingdom*

j.l.godwin@durham.ac.uk
p.c.matthews@durham.ac.uk

ABSTRACT

The development of electrical control system faults can lead to increased mechanical component degradation, severe reduction of asset performance, and a direct increase in annual maintenance costs. This paper presents a highly accurate data driven classification system for the diagnosis of electrical control system faults, in particular, wind turbine pitch faults. Early diagnosis of these faults can enable operators to move from traditional corrective or time based maintenance policy towards a predictive maintenance strategy, whilst simultaneously mitigating risks and requiring no further capital expenditure. Our approach provides transparent, human-readable rules for maintenance operators which have been validated by an independent domain expert. Data from 8 wind turbines was collected every 10 minutes over a period of 28 months with 10 attributes utilised to diagnose pitch faults. Three fault classes are identified: "no pitch fault", "potential pitch fault" and "pitch fault established". Of the turbines, 4 are used to train the system with a further 4 for validation. Repeated random sub-sampling of the majority fault class was used to reduce computational overheads whilst retaining information content and balancing the training and validation sets. A classification accuracy of 85.50% was achieved with 14 human readable rules generated via the RIPPER inductive rule learner. Of these rules, 11 were described as "useful and intuitive" by an independent domain-expert. An expert system was developed utilising the model along with domain knowledge, resulting in a pitch fault diagnostic accuracy of 87.05% along with a 42.12% reduction in pitch fault alarms.

1. INTRODUCTION

Maintenance costs for wind energy represent between 20-25% of total asset cost, of which, up to 75% is due to unscheduled maintenance (WWEA, 2012). This deters future investment, increases the cost of wind energy and as such, reduces the long term economic viability of wind energy. As corrective maintenance can be up to 40 times more expensive than a proactive strategy (Hatch, 2004) there is the potential for significant cost savings on wind turbine operations and maintenance (O&M) costs. For this reason, maintenance is moving from a "fail and fix" reactive approach to maintenance, to a "predict and prevent" strategy for maintenance (Levrat, et al 2008). Maintenance savings of 20-25% can be achieved using condition based maintenance (CBM) (Djurdjanovic, et al 2003), this is echoed by Wu & Clements-Croome (2005) who have shown the potential for proactive maintenance actions to be performed at 10 times to 40 times less than respective corrective maintenance actions. However, uptake across all domains of prognostic technologies for the prediction of future failure modes has been slower than anticipated. It is believed that within the UK, CBM and prognostic technologies have only reached 10-20% penetration into industry (Moore & Starr, 2006). This is believed to be due to many factors, such as: the lack of transparency of some expert systems, the capital outlay required for data collection and analysis, the uncertainty and inaccuracy present within some techniques, staff training costs and no proven track record in similar domains. Whilst strategies such as reliability centred maintenance (RCM) can help optimise available maintenance resources, they are static in nature in that they do not take into account the current level of asset degradation or external conditions. This means that whilst cost savings can be made through RCM (Niu, et al 2010), severe degradation is likely to go unnoticed for extended periods, causing secondary damage to auxiliary systems, reducing component efficiency and as a result, reduce overall return on investment for stakeholders. Due to as few as 20% of assets failing within the manufacturers prescribed times (Eti, et al. 2006), there is a need to move away from a static analysis towards a more dynamic, real-time approach to maintenance. Currently, maintenance is often seen by senior management as a cost minimisation exercise, rather than an attempt to maximise benefit (Marais

& Saleh, 2009). This is due to the ease of quantifying the cost of maintenance, but not the benefit provided. This attitude towards maintenance means that most efforts to reduce annual maintenance expenditure result in a direct loss of availability or reduction in the quality of service provided (Gomez-Fernandez & Crespo-Marquez, 2009). Typically, condition based monitoring is performed using high frequency data – acoustic emissions and vibration data – collected for the remote diagnosis and prognosis of the gearbox, generator and main bearing (Crabtree, 2010). However, being able to establish and track the development of a fault over longer lengths of time through utilising low frequency data is interesting as it provides feedback into the maintenance planning and scheduling process, enabling the optimisation of available resources, thereby reducing annual maintenance costs. In this paper we present a new methodology for the development of a transparent expert system for the detection of wind turbine pitch faults utilising a data-intensive machine learning approach. This approach describes a classifier to determine the current condition of the pitch system on a wind turbine through analysis of low frequency SCADA data, and if a fault is observed within the pitch system, an expert system recommends the correct action to take depending upon its severity. Severe pitch faults requiring potential maintenance actions can then be presented to the maintenance operator whilst filtering out unnecessary information and reducing the cognitive load which is placed upon them. As the data utilised for this methodology is from a pre-existing SCADA system, no further sensors are required and no additional capital expenditure is incurred. This mitigates many of the risks associated with moving to a proactive maintenance strategy.

2. WIND TURBINE PITCH FAULTS

Wind turbine pitch faults are deviation of the blade pitch angle from a predefined optimum for a given wind speed and are the most common fault mode to occur. As can be seen in Table 1, pitch faults account for over one third of all faults which are present within the SCADA system which are then presented to the maintenance operator. It is not uncommon for over 2,000 SCADA pitch fault alarms to occur over a year. However, less than 5% of these directly correlate to a maintenance action within the maintenance log; wasting available maintenance resources with undue inspection and analysis. As such, there is a need to develop a data-driven expert system to allow the encapsulation of the behaviours both during and immediately preceding a pitch fault so that maintenance operators can further understand the extent of the fault, the causation of the fault and the maintenance action required. Accurate identification of pitch faults is of particular interest to maintenance operators and decision makers, as these faults are often the result of the electrical control system, and not due to severe physical degradation of the pitch motors controlling the wind turbine blades. As such, when a pitch fault is identified, the potential exists to remotely reset the turbine pitch system. This enables the turbine to return to normal operating conditions, without the need for excessive downtime for the required inspection. As such the energy generation can be increased, with the potential risk of increased degradation on auxiliary components reduced. Should a mechanical fault be observed, this will then be diagnosed by the system presented in Section 4, enabling the effective scheduling and planning of maintenance activities.

3. RELATED WORK

Over recent years, interest in improving the efficiency of all aspects of the wind turbine life cycle has become of paramount importance to ensure a continued transition to a low carbon economy and ending the reliance on fossil fuels. As up to 25% of total cost is manifested as maintenance for a wind turbine, effective maintenance through condition based maintenance and proactive maintenance is essential to increasing global investment in wind energy, reducing energy prices to consumers and ensuring continued reliable operation as transitions are made to the smart grid (Massoud Amin & Wollenberg, 2005). Prognosis of the wind turbine enables 5 key benefits to be provided to the operator as stated in Hameed et al (2009). They are:

1. The avoidance of premature failures - reducing secondary damage to components and also reducing catastrophic failures.
2. A reduction in maintenance costs - by reducing catastrophic failures and optimising inspection intervals.
3. The capability of remote diagnosis – essential due to the remote nature of offshore turbines.
4. An increase in generation capacity - prognosis enables maintenance to be performed at low wind speed to ensure maximal utilisation.
5. Optimised future designs - large quantities of data can be analysed to ensure new generation turbines are more reliable.

Typically, condition monitoring on a wind turbine focuses on the high value components; the gearbox, generator and main bearing (Crabtree, 2010). Strong prognostic capability is prevalent within the literature. For example, the work done by Lin & Zuo (2003) and Rafiee et al (2010) use wavelet filters to provide condition based maintenance on these components. Also, Wang & Makis (2009) utilise statistical methods (such as autoregressive models) to achieve similar aims. However, these techniques require the installation of various additional sensors to each turbine to be monitored, which can be costly to the operator. For a full review of high frequency techniques, please see the work of Jardine et al (2006) and Hameed et al (2009). Techniques utilising low frequency data, such as SCADA data, do exist.

Sub-system	Turbine 1	Turbine 2
Pitch	4035	4130
Weather	2775	2866
Inverter	1438	1751
Gearbox	504	374
Yaw	316	385
Communications	285	827
Total	9353	10333

Table 1. SCADA alarms aggregated by subsystem over a 28 month period for 2 typical turbines.

Work done by Kim et al (2011) has shown the electrical system of a wind turbine is the most prone to establishing a fault condition. It has been shown that low frequency SCADA data can be used in conjunction with both PCA and self-organising feature maps for fault classification. However, diagnosis to determine the turbine sub-assembly at fault is not performed. As such, whilst maintenance managers may know a turbine requires inspection, further manual analysis will be required to determine the cause of the fault. Chen (2011) utilises an artificial neural network for the automatic analysis of SCADA alarm data. This is utilised as a filter to determine which SCADA alarms are novel and warrant further analysis. Work done by Kusiak & Li (2011) has shown that a variety of data mining approaches (neural networks, ensembles of neural networks, the boosting tree algorithm, support vector machines and classification and regression trees) can be used to diagnose and prognose irregular wind turbine states. However, even when utilising many different data driven approaches, a low prognostic horizon (less than an hour) is achieved, and accuracy of the classification of fault instances ranges from 40% to 71%.

4. METHODOLOGY

SCADA data from 8 wind turbines was collected over a period of 28 months and sampled every 10 minutes, across 190 channels. All of these wind turbines had pitch faults noted in their histories as assessed by their maintenance log book. There had been 243 recorded pitch faults across the 28 months for the 8 turbines, ranging from 6 – 60 pitch faults per turbine ($M = 30.38$, $SD = 16.16$). In total, 999,944 records were retrieved. This data was combined with SCADA alarm system data and maintenance log data to give a holistic overview of the condition of the turbine and so that pitch fault events could be analysed. Due to the inherent nature of the data acquisition, erroneous and missing values are common; these are manifested as implausible values, missing data and duplicate data. This is ascribed to malfunction of the sensors, mechanical systems, data collection systems and also imperfections within the SCADA system itself (Sainz, et al 2009). Due to these problems, the data must be cleansed before processing can take place. Both missing and duplicate values were removed; missing values cannot accurately describe the current state of the wind turbine, and duplicate values provide no additional information whilst simultaneously increasing computational overhead. Once this is complete, attribute selection is performed. Based upon the work of Chen et al (2011) and also Kusiak & Verma (2011), 8 attributes were selected for their consistently strong performance for wind turbine pitch fault diagnosis. Chen et al (2011) presents an artificial neural network (ANN) approach to pitch fault diagnosis, however, the diagnosis accuracy ($M = 42.07\%$; $SD = 17.49\%$) is relatively poor and black box nature of the approach is difficult to interpret by both domain experts and maintenance operators. Whilst the work of Kusiak & Verma (2011) provides improved accuracy for the prediction of wind turbine pitch faults ($M = 76.70\%$; $SD = 5.62\%$), the genetic algorithm used provides human readable rules which are not necessarily transparent (that is, easy to interpret by operators). As such, the attributes chosen for the model based upon the work in the literature (Chen et al., 2011 and Kusiak & Verma, 2011) were:

- Average wind speed
- Maximum wind speed
- Blade 1 pitch motor torque maximum
- Blade 2 pitch motor torque maximum
- Average pitch motor torque
- Blade 1 pitch angle average
- Blade 2 pitch angle average
- SCADA pitch fault alarm status

In conjunction with these attributes, 2 additional derived parameters were utilised based upon the work of Chen et al (2011). These are:

- The absolute difference in torque across pitch motors
- The absolute difference in blade angle position

These attributes were chosen as they fully encapsulate the current operating characteristics of the wind turbine pitch fault system. The feathering control strategy for variable pitch wind turbines is described in detail by Bianchi et al. (2006). For a given wind speed, each blade should be set to a pre-determined pitch based upon the strategy employed by the individual turbine. The pitch of all the wind turbine blades should be identical, and as such, deviations in either pitch or torque across the blades can be used to identify the presence of a pitch fault. The wind speed and SCADA alarms status provide additional context to the classifier to aid in the classification accuracy.

Following this, the data was classified into three distinct groups; "No pitch fault", "Potential pitch fault" and "Pitch fault established". These represent the development of a fault over time within the wind turbine. By classifying the data in this way we can identify both the wind turbines which urgently require maintenance and also the turbines with a reduced remaining useful life (RUL). Maintenance logs were used to determine when pitch faults had been severe enough to warrant a maintenance action.

The SCADA data from the 48 hours preceding this maintenance action was used to describe the "Pitch fault established" class. The SCADA data prior to this where the SCADA-alarm for the pitch fault was active was used to describe the "Potential pitch fault" class. Finally, all other data was used to describe the "no pitch fault" class. Annual maintenance costs can then be reduced utilising this classification; either by scheduling further turbines into existing maintenance actions, or by pre-emptively scheduling those which require maintenance before they become inaccessible to external factors. Repeated random sampling with 20 samples was utilised to remove the majority class bias inherent within the data. As "No pitch fault" was the dominant class and the turbine remains in this state for a prolonged period, a data-driven classifier would be stronger if it encapsulates this class well and ignores the pitch faults. However, as the aim of the system is the quality of the rules which describe the behaviour of the pitch faults, it is essential that this bias is removed so that the minority fault classes are encapsulated and characterised effectively. Within our data, the imbalance was typically between 125 to 380 instances per fault instance. Whilst other minority oversampling techniques could have been used such as SMOTE, MSMOTE and FSMOTE (Garcia, et al 2012) no significant increase in rule accuracy was attained over using traditional repeated random sampling within our dataset. As such, the majority class was under sampled, and the minority class oversampled until the data was balanced. After the data had been pre-processed, the RIPPER propositional rule learning algorithm (Cohen & Singer, 1999) was used to generate order independent, distinct encapsulations of explicit knowledge from the dataset. This technique was chosen due to its transparent, human-readable nature; ensuring trust was placed in the derived rules. An example of rules generated by the RIPPER algorithm can be seen in the appendix. Although other techniques such as artificial neural networks can achieve high quality classifications, their "black box" nature makes them difficult to extract meaningful rules from. Similarly, although techniques such as clustering and instance-based classification seem intuitive, the high-dimensionality of the dataset and high levels of noise present means that decision regions are non-convex in nature and neither a high level of accuracy nor good quality of rules can be extracted from the system. Decision tree algorithms could have been utilised, however, each rule generated cannot be understood independently from the system, and as such, can be difficult to extract and encapsulate as a single unit of knowledge.

4.1. Ripper Algorithm

The RIPPER algorithm (Cohen, 1995), is an extension to the IREP algorithm proposed by Fürnkranz and Widmer (1994), utilizing reduced error pruning (REP) used in decision tree algorithms. However, where the rule induction from decision trees is done in a breadth-first manner (as per C4.5), rule induction is performed in a depth-first manner.

There are two main stages within the ripper algorithm as described by Cohen (1995). Firstly, the data is split into "growing" and "pruning" dataset, with two thirds typically used for growing. This is done by random partitioning of the data. After this, rules are grown. This is done by adding conditions to a rule (greedily) until it is 100% accurate (that is, it covers no negative instance in the growing dataset). This is done by maximizing Foil's information gain criterion (Quinlan, 1990):

$$FOIL(L,R) = t \cdot (\log_2 \frac{p_1}{p_1+n_1} - \log_2 \frac{p_0}{p_0+n_0}) \quad (1)$$

Where L is the condition to be added to R, t is the number of positive instances covered by $R+L$, p_1 and p_0 are the number of positive instances covered by R and $R+L$ (respectively), and n_1 and n_0 are the number of negative instances covered by R and $R+L$ (respectively). This favours rules which have high accuracy and cover many positive instances.

Once the rule has been grown, it is pruned immediately. This is done within RIPPER by considering the removal of the final sequence of conditions from the rule that maximise rule value:

$$v^*(Rule, PrunePos, PruneNeg) = \frac{p-n}{P+n} \quad (2)$$

Where p is the number of examples in *PrunePos* covered by *Rule* and n is the number of examples in *PruneNeg* covered by *Rule*. This is done until no deletion increases the value of v^* (Cohen, 1999).

Once the rules have been generated, optimisation is performed. In this stage, for each rule which has been grown and pruned, two variants are produced; the replacement and the revision. The replacement is generated by growing and pruning a rule where the pruning stage is guided to maximize the accuracy of the entire rule base. The revision is generated by greedily adding conditions to the rule. The rule with the minimum descriptive length (of the original, revision or replacement rule) is then chosen for the final rule base. For completeness, the full pseudo-code for the RIPPER algorithm (Cohen, 1995) is presented in Figure 1 (Alpaydin, 2004).

```
Ripper(Pos,Neg,k)
    RuleSet ← LearnRuleSet(Pos,Neg)
    For k times
        RuleSet ← OptimizeRuleSet(RuleSet,Pos,Neg)
LearnRuleSet(Pos,Neg)
    RuleSet ← ∅
    DL ← DescLen(RuleSet,Pos,Neg)
    Repeat
        Rule ← LearnRule(Pos,Neg)
        Add Rule to RuleSet
        DL' ← DescLen(RuleSet,Pos,Neg)
        If DL'>DL+64
          PruneRuleSet(RuleSet,Pos,Neg)
          Return RuleSet
        If DL'<DL DL ← DL'
          Delete instances covered by Rule from Pos and Neg
    Until Pos = ∅
    Return RuleSet
PruneRuleSet(RuleSet,Pos,Neg)
    For each Rule ∈ RuleSet in reverse order
        DL ← DescLen(RuleSet,Pos,Neg)
        DL' ← DescLen(RuleSet-Rule,Pos,Neg)
        IF DL'<DL Delete Rule from RuleSet
    Return RuleSet
OptimizeRuleSet(RuleSet,Pos,Neg)
    For each Rule ∈ RuleSet
        DL0 ← DescLen(RuleSet,Pos,Neg)
        DL1 ← DescLen(RuleSet-Rule+
                  ReplaceRule(RuleSet,Pos,Neg),Pos,Neg)
        DL2 ← DescLen(RuleSet-Rule+
                  ReviseRule(RuleSet,Rule,Pos,Neg),Pos,Neg)
        If DL1=min(DL0,DL1,DL2)
                  Delete Rule from RuleSet and
                         add ReplaceRule(RuleSet,Pos,Neg)
        Else If DL2=min(DL0,DL1,DL2)
                  Delete Rule from RuleSet and
                         add ReviseRule(RuleSet,Rule,Pos,Neg)
    Return RuleSet
```

Figure 1. Pseudo-code of the RIPPER algorithm (Alpaydin, 2004).

4.2. Training and Validation Turbine Selection

Of the 8 wind turbines, 4 were used for training with the remaining turbines used for validation. In order to ensure the robustness of the methodology against training turbine selection, all combinations of turbines for both training and validation were considered. In total, 70 combinations of varying training and validation turbines were created. These models created a Pareto surface compromising the trade-off between the number of rules and rule accuracy which were then presented to an independent domain expert. This allows for both a quantitative and qualitative analysis of these rules so that the causation and diagnosis of pitch faults could more effectively be understood. This enables operators to understand the underlying physical properties of pitch faults so that they can be trained or assisted to identify pitch faults before further damage occurs, which may lead to the turbine being shut down for corrective maintenance which is often expensive.

5. RESULTS

The RIPPER propositional rule learner was trained on 70 models so that the robustness of the methodology could be ensured. Pruning of the rule set was enabled to reduce the quantity of rules to prevent potential cognitive overload, and was utilized in conjunction with four optimization iterations with three fold partitioning of the data.

5.1. Robustness to data scarcity

As can be seen in Table 2, the quantity of data available for training influences the accuracy of the system developed and also the size of the rule base. In addition to the analysis described in section 4 (on the full dataset), analysis was also carried out on 4, 8, 12, 16, 20 and 24 months of available data to determine the influence of the quantity of data on both classification accuracy and size of the rule base.

Each analysis in Table 2 was performed on the full set of 70 models generated by choosing each combination of the 8 training and testing turbines. As such, in total 490 models were developed and assessed to analyse the robustness of the system to the quantity of training data which was available.

With regards to model accuracy, a Pearson product-moment correlation was used to assess the relationship between mean classification accuracy attained and the quantity of data used. Preliminary analyses showed this relationship to be linear with both variables normally distributed, as assessed by Shaprio-Wilk test ($p > .05$), and there were no outliers. There was a strong positive association between classification accuracy and the quantity of data, $r(7) = .91, p < .01$. This is also the case for maximum classification accuracy; $r(7) = .91$, $p < .01$, and minimum accuracy attained, $r(7) = .92, p < .01$.

This shows that there is a strong positive correlation between the quantity of data available for training and the accuracy of the RIPPER algorithm. As such, it was determined that as much data as is available should be utilised when performing rule extraction. It should be noted that the lower bound of classification accuracy increased by 17.62% from utilising 4 months of data to using the entire data set (28 months), whereas the upper bound increased by 1.68% over the same period. The mean accuracy increase was 3.48% over this period; however, the standard deviation of accuracies was reduced by 1.81% in this period. This indicates less sensitivity to the selection of wind turbines used for testing as more data to be available. This was to be expected.

With regards to the size of the rule base, another Pearson product-moment correlation was used to assess the

	4 Months	8 Months	12 Months	16 Months	20 Months	24 Months	Full Dataset
Mean Accuracy	77.29%	77.92%	78.53%	78.37%	78.89%	78.91%	80.77%
Max Accuracy	85.73%	86.18%	86.94%	86.53%	87.25%	87.73%	87.41%
Min Accuracy	51.41%	59.11%	56.49%	65.74%	63.75%	66.39%	69.03%
Accuracy (SD)	6.49%	5.72%	5.86%	5.15%	5.12%	5.32%	4.68%
Mean Rule Base	7.57 rules	9.10 rules	10.94 rules	12.87 rules	13.57 rules	14.77 rules	16 rules
Max Rule Base	15 rules	16 rules	23 rules	24 rules	32 rules	34 rules	38 rules
Min Rule Base	3 rules	4 rules	4 rules	4 rules	5 rules	6 rules	6 rules
Rule Base (SD)	2.42 rules	2.90 rules	4.10 rules	4.27 rules	4.76 rules	4.78 rules	5.77 rules

Table 2. Robustness to data scarcity with descriptive statistics for classification accuracy and rule base size.

relationship between mean rule base size and the quantity of data used. Preliminary analyses showed this relationship to be linear with both variables normally distributed, as assessed by Shaprio-Wilk test ($p > .05$), and there were no outliers. There was a very strong positive association between the size of the rule and the quantity of data, $r(7) = .99, p < .01$. This was also the case for the maximum size of the rule base, $r(7) = .97, p < .01$, and also the case for the minimum size of the rule base, $r(7) = .97, p < .01$.

This correlation is to be expected based upon the behaviour of the RIPPER algorithm. However, due to this, a trade off does exist. Increasing the quantity of data available to the propositional rule learner would increase the quality of the classifier produced, but would also increase the quantity of rules generated for analysis. This is detailed below.

5.2. Model selection

Due to the higher mean accuracy and larger rule base variance attained by models utilising the full 28 months of data, this was chosen for further analysis. The accuracy of the classification for the full data models was in the range of 69.03% - 87.41% (M = 80.77%; SD = 4.68%), with the number of rules generated by each model being in the range of 6 – 38 (M = 16; SD = 5.77). After removal of the models which were dominated by those with stronger classification accuracy but the same number of rules, 21 models were eligible to be presented to an independent domain expert and for critical analysis of the rules generated allowing for further understanding of wind turbine pitch fault behaviour.

The 21 models developed had classification accuracy in the range of 69.99% - 87.41% (M = 82.70%; SD = 4.26%). Similarly, the quantity of rules generated were in the range of 6 – 38 (M = 16.5, SD = 7.65). A Pearson product-moment correlation was used to assess the relationship between classification accuracy and the number of rules generated by the model. Preliminary analyses showed the relationship to be linear with both variables normally distributed, as assessed by Shaprio-Wilk test ($p > .05$), and there were no outliers. There was no association between classification accuracy and the number of rules present, $r(21) = .056, p > .05$. This can clearly be seen in Figure 2. As such, it is beneficial to maintenance operators and decision makers that a smaller set of rules are analysed and understood. This enables a holistic understanding of the underlying behaviour and development of wind turbine pitch faults whilst reducing cognitive load and whilst providing comparable classification accuracy to the models with a larger rule base.

Within the model selected, 14 rules were generated leading to an overall classification accuracy of 85.50%. For completeness, the knowledge base determined by the RIPPER algorithm has been included in the appendix. It can be noted that although a high classification accuracy has been attained in this model, it is still difficult to differentiate between no pitch fault existing and a pitch fault being present, with expert analysis required to certify classifications.

As can be seen in Table 3, the Matthews Correlation Coefficient (MCC) (Matthews, 1975) for all classes is strong, showing high correlation between the learnt rules and the validation data. A substantial level of agreement was found between the developed model and the validation data (Cohen's k = 0.78; p < .05). After deriving the classification, the 14 rules were presented to an independent domain expert so that qualitative and quantitative analysis could be performed. Due to the min-max normalization process during pre-processing, values had to be converted back to ensure they were human readable. Once this had been done, a full analysis was performed.

Class	TP Rate	FP Rate	Precision	Recall	F-Measure	MCC	ROC	PRC
No Pitch Fault	0.81	0.12	0.77	0.81	0.79	0.68	0.91	0.74
Potential Pitch Fault	1.00	0.00	1.00	1.00	1.00	1.00	1.00	1.00
Pitch Fault Established	0.75	0.01	0.80	0.75	0.78	0.67	0.91	0.79
Weighted Average	0.85	0.07	0.86	0.86	0.78	0.78	0.85	0.85

Table 3. Descriptive statistics of the developed model.

6. EVALUATION AND EXPLOITATION OF GENERATED RULES

Due to the size of the knowledge base, it was practical to have the domain expert evaluate each rule individually. This is done as the expert can provide a context sensitive ground truth to the analysis, along with experience of situations and conditions which may not have been present within the training data. As domain experts have subjective opinions with regards to what constitutes interesting, novel and important, it is difficult to quantify these characteristics.

However, various artefacts are present within the rule-base which is expected given the nature of the classification. To assess the quality of the rules, a 56- item questionnaire was presented to an independent domain expert who has over 6 years wind turbine diagnostic and prognostic experience within academia. This questionnaire contained a 5-point Likert response scale ranging from 1 (Not intuitive, useful, clear or interesting) to 5 (Highly intuitive, useful, clear or interesting). There were 4 questions presented per rule generated from the model, assessing whether or not the rule was intuitive, useful, clear and interesting

The results of this analysis can be seen in Table 4. As can be seen in Table 4, an average response of 2.89 was recorded; indicating that the rules are typically not particularly intuitive, clear, useful or interesting. This was unexpected. Rules were often regarded as just as useful (M = 2.79) as intuitive (M = 2.71). This is likely due to the nature of the complex nature of the underlying pitch faults. By having the independent domain expert drive the discussion it was found that of the 14 rules, 11 of the rules were deemed "interesting" and warranted further analysis.

After performing this analysis, the independent domain expert was then presented with a further 13 rules, taken from the work of Kusiak & Verma (2011). To remove potential bias, the expert was not informed of the origin of either set of rules. A 52-item questionnaire was used containing a 5-point Likert scale from 1 (Not intuitive, useful, clear or interesting) to 5 (Highly intuitive, useful, clear or interesting). This was to provide an objective analysis of the intuitiveness, usefulness, clearness and interestingness.

Initially, the expert could not understand the rules due to their format and abstract nature, however, after some time, analysis could be performed. The comparative analysis showed that whilst the rules were found to be less intuitive (M = 1.53; SD = 0.63) and clear (M = 1.46; SD = 0.49), they were still regarded as somewhat useful (M = 2.23; SD = 0.79) and interesting (M = 2.07; SD = 0.61). When questioned regarding this, the expert responded that as long as the rules were accurate and accountable, they could be disseminated at a later date. As such, it was determined that an expert system should be developed to aid maintenance operators with enquiries and to handle the large quantities of data present within the system.

6.1. Rule sensitivity to wind turbine location

As different geographical locations have inherently distinct operating conditions, it is expected that the accuracy of the expert system would be reduced when applying the rules to similar wind turbines in a different location. As such, a new expert system would have to be developed for each wind farm as described by the methodology described in Section 4. In order to assess the impact of the geographical location on the accuracy of the rules, data was collected from an additional turbine in the same manner as the previous turbines and was located at a different wind farm within the same country. The wind turbine also had 28 months of SCADA data available, with 3 pitch faults recorded in the historical maintenance log. This wind farm was subject to different external conditions due to being located in a different region.

Validation of the selected model on this additional wind turbine yielded a classification accuracy of 68.68%; somewhat lower than the 85.50% accuracy of the original model. The model was able to identify 2 of the 3 pitch faults which had been recorded in the maintenance log; giving a diagnostic accuracy of 66.67%.

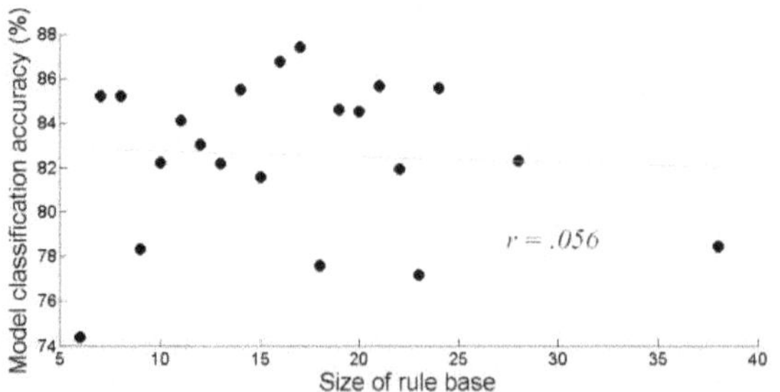

Figure 2. Dominant model classification accuracy plotted against the number of rules generated in each model. No strong correlation existed ($r(21) = 0.06, p > .05$).

7. EXPERT SYSTEM DEVELOPMENT

Due to the strong classification gained from the model, an expert system was developed to aid maintenance managers and decision makers so that available resources could be optimized. Due to the often inaccessible nature of offshore wind turbines, predicting failures can significantly reduce operations and maintenance (OM) costs, thereby increasing the competitive nature of wind energy. The model developed in Section 4 was combined with domain knowledge (meta-data) elicited from the independent domain expert to reduce the high dimensionality of the SCADA data and provide filtering. This was done so that the maintenance operator did not have to analyse 190 channels of data coming from over 40 wind turbines per farm, every 10 minutes. In order to assist the operator or decision maker in their role, the expert system must aid their ability to perform analysis and make decisions based upon relevant information. These decisions become more difficult due to various stressors which exist in the working environment. Kontogiannis & Kossiavelou (1999) identify these stressors as:

- Environmental stressors:
  - Noise
  - Temperature
  - Vibrations
- Task complexity stressors:
  - Time Pressure
  - Workload
  - Uncertainty
  - Threat/High error consequences
  - Negative feedback
- Group and organisational stressors:
  - Occupational stress
  - Shift/continuous work
  - Lack of team cohesion
  - Communication problems

Due to the nature of the domain, the expert system aims to reduce task complexity stressors. Specifically, reducing time pressure by providing automated analysis and reducing workload by reducing the initial quantity of information presented to the operator per wind turbine. The expert stated that typically, SCADA-alarms for pitch fault are noisy, and only when constant irregularities are noticed over an extended period, is maintenance considered on the turbine.

This is typically due to imperfections within the SCADA system itself causing duplicate, missing and implausible values to be recorded (Sainz, et al 2009). Also, as SCADA data quickly accumulates to create large and unmanageable volumes of data, attempts to deduce the current state of a wind turbine can be severely hindered (Zaher, et al 2009), it is therefore essential that this data can be adequately filtered in an automated manner.

Question	*N*	*M*	*SD*
Intuitive	*14*	*2.71*	*1.09*
Useful	*14*	*2.79*	*0.93*
Clear	*14*	*3.00*	*1.00*
Interesting	*14*	*3.07*	*0.96*

Table 4. Independent domain expert evaluation.

As such, based upon the expert-knowledge, a threshold was set that should either the "Potential pitch fault" or "Pitch fault established" classification be active for over 90 minutes, an alert would be sent to the maintenance operator. 90 minutes was deemed by the expert to be the minimum length of time an alarm was active before action would be taken and was used as a filter to reduce the noise of the SCADA system. Lower values would increase the noise within the expert system whereas higher values may miss the potential development of pitch faults. This, therefore, reduces the quantity of SCADA alarms presented to the maintenance operator, whilst still presenting those which warranted further investigation.

This reduces the potential cognitive overload of the maintenance operator, allowing for their analysis to be focused on the wind turbines which are current exhibiting potential pitch fault state. This optimises the available maintenance resources by reducing the time spent analysing large quantities of false-positive alarms provided from SCADA system. With regards to the imperfections within the SCADA system, a threshold was also set for missing and implausible values. As missing data cannot fully encapsulate the current operating condition of the wind turbine, it would be difficult to establish if the fault was caused by either a mechanical fault on the turbine, an electrical fault on the turbine or an electrical fault on the SCADA system. As such, 90 minutes of continuous operation in this state provides an alert to the maintenance operator, as above. A similar strategy is employed for implausible data, with expert defined maximum and minimum values for each attribute. Should a single attribute fall outside of this pre-defined range for a full 90 minutes, the operator is also alerted to this.

It should also be noted that one of the alarms on a separate turbine (outside of the training and test data) was active for over 100 days continually. Clearly this is undesirable and hinders the efforts of maintenance managers and decision makers to correctly diagnose and both plan and schedule maintenance.

As such, the ability to correctly filter and classify SCADA-data so that the false-positive instances such as this do not occur is essential. In the best case, these false-positive instances are simply a minor hindrance and require further manual analysis by the maintenance operator to determine if a turbine warrants inspection. In the worst case, they provide a basis for maintenance actions which may not be required. In an offshore situation, these un-necessary maintenance actions can be expensive due to the equipment

Turbine	Pitch Fault Alarm Time	Number of Pitch Alarms	Number of Maintenance Jobs	Expert System Alarms	Expert System Time Active
01	15.46 days	193	25	97	10.06
02	17.68 days	222	25	106	12.72
03	12.04 days	27	26	75	8.45
04	19.64 days	215	9	138	9.84

Table 5. Comparison of Expert System against SCADA-Alarm system

and skills required, and as such, can potentially account for a large portion of maintenance expenditure.

8. EVALUATION OF EXPERT SYSTEM

In order to assess the validity of the expert system developed, historical SCADA-data from 4 wind turbines was used to determine the number of maintenance alerts issued in comparison to the on-board SCADA-alarm system. The validation turbines were independent of those used within the training model, and were located in the same geographical location as the turbines used for the model development and training.

As can be seen in Table 5, in each of the 4 wind turbines analysed, a reduction in the number of alarms generated was observed compared to the turbines integrated SCADA alarm system. This was between 35.80% - 52.26% (*M* = 44.69%; *SD* = 6.62%), effectively reducing the workload of the maintenance operator when analysing data to diagnose potential pitch faults. Similarly, this was the case for active alarm time; the reduction was between 28.06% - 49.90% (*M* = 35.68%; *SD* = 8.60%). This, again, reduces the quantity of information the maintenance operator has to manage. It is worth noting that although 85 pitch maintenance actions were undertaken over the 28 month period in which this historical data was analysed, 11 of these maintenance actions were not detected by the expert system. This is mainly due to malfunction of the sensors, mechanical systems, and the data collection systems; Of the 11 instances, 7 occurred when data acquisition failed for an extended period. Due to the design of the expert system, missing data does not fully encapsulate the correct turbine condition, and as such, the accuracy is significantly reduced. It is believed that the remaining 4 cases are partly due to time-based preventive maintenance which may not have had sufficient basis for action based upon the observed SCADA-data

9. CONCLUSIONS

In this paper we have presented a robust, accurate expert system for the classification and detection of wind turbine pitch faults, as validated by the 85.50% classification accuracy achieved. Transparent, human readable rules were extracted, analysed and verified by an independent domain expert enabling trust in the expert system one of the key barriers to wide scale adoption of CBM technology. These rules were found to be more intuitive than other rules within the literature, and provided the basis for an expert system to aid maintenance operators and decision makers. The number of SCADA alarms was reduced by an average of 44.68%, with a mean reduction of active alarm time by 35.68%. The developed expert system reduced the potential cognitive load on maintenance operators and decision makers by significantly reducing the number of alarms presented to them. This frees maintenance resources, enabling a reduction in annual maintenance costs whilst retaining an equal quality of service. Additionally, no further capital expenditure was necessary due to using pre-existing technological capability. A diagnostic accuracy of 87.05% is achieved in the system, although it is believed that this could be further increased should more reliable sensor technology become available. Our methodology provided a robust strategy to classify SCADA data as having no pitch fault, an established pitch fault or a potential pitch fault. This provides a means to both condition based maintenance and proactive maintenance strategies. By performing remote diagnosis through the expert system, the opportunity for remote maintenance arises due to the nature of the electrical system. In some cases, resetting the control system remedies the existing electrical fault, increasing availability whilst reducing unnecessary maintenance inspections and mitigating the associated costs. By understanding the severity of the fault through the expert system classification, maintenance managers can make informed decisions regarding the most appropriate course of action.

Future work will look to utilise statistical techniques to reduce the quantity of data required for accurate classification. Whilst 4 months of data provided an average classification of 77.29%, had no historical pitch fault data been available, the expert system would not have been able to encapsulate the pitch fault behaviour and would not be fit for purpose. Thus, the expert system would not be effective. As such, the use of suspension histories to classify normal operating behaviour through utilising robust statistical methods would be more appropriate in these circumstances. This would remove the need for fault data present within the training data, providing a strategy for the prognosis and diagnosis of new wind turbines.

APPENDIX

For completeness, the 14 rules learnt by the RIPPER algorithm are presented here. This represents the knowledge base of the expert system.

1. If Alarm is Not Active, and Difference Between Blade Angles is ≤ 18.32 degrees Then Pitch Fault Established.
2. If Alarm is Not Active, and Difference Between Blade Angles is ≤ 18.56 degrees, and Wind Speed ≥ 7.11 m/s Then Pitch Fault Established.
3. If Blade 1 Pitch Motor Torque Maximum ≥ 14.81 kN but ≤ 30.13 kN, and Blade 2 Angle ≥ -12.52 degrees, and Wind Speed ≥ 7.69 m/s, and Then Pitch Fault Established.
4. If Blade 1 Pitch Motor Torque Maximum ≥ 15.59 kN but ≤ 24.35 kN, and Blade 1 Angle ≥ 95.52 degrees, and Wind Speed ≥ 6.73 m/s, and Difference Between Pitch Motor Torques ≤ 41.0 kN Then Pitch Fault Established.
5. If Blade 2 Angle ≤ -0.28 degrees, and Blade 1 Angle ≥ 0.52 degrees, and Wind Speed ≥ 6.44m/s, and Average Pitch Motor Torque ≤ 9.67 kN Then Pitch Fault Established.
6. If Blade 2 Angle ≤ -0.28 degrees, and Blade 1 Angle ≥ -19.74 degrees, and Average Pitch Motor Torque ≤ 10.22 kN, and Wind Speed ≥ 7.42 m/s Then Pitch Fault Established.
7. If Blade 2 Angle ≤ -0.35 degrees, and Blade 1 Angle ≥ -17.13degrees, and Wind Speed ≥ 6.11 m/s, and Difference Between Pitch Motor Torques ≤ 1.08 kN, and Average Pitch Motor Torque ≤ 11.58 kN Then Pitch Fault Established.
8. If Blade 2 Angle ≤ -2.85 degrees, and Blade 1 Angle ≥ -17.13 degrees, and Wind Speed ≥ 7.34 m/s, and Average Pitch Motor Torque ≤ 13.44 kN, Then Pitch Fault Established.
9. If Blade 2 Angle ≤ -2.85 degrees, and Blade 1 Angle ≥ -17.14 degrees, and Wind Speed ≥ 6.19 m/s, and Blade 2 Pitch Motor Torque Maximum ≤ 21.91 kN, and Wind Speed ≥ 6.81 m/s Then Pitch Fault Established.
10. If Blade 2 Angle ≤ -2.98 degrees, and Blade 1 Angle ≥ -17.23 degrees, and Difference Between Pitch Motor Torques ≥ 2.35 kN, and Average Pitch Motor Torque ≤ 10.53 kN, and Wind Speed ≥ 6.56m/s, and Blade 2 Pitch Motor Torque Maximum ≤ 25.02 kN, and Difference Between Blade Angles is ≥ 18.7 degrees Then Pitch Fault Established.
11. If Blade 2 Angle ≤ -3.02 degrees, and Blade 1 Angle ≥ -23.33 degrees, and Wind Speed ≥ 8.25 m/s Then Pitch Fault Established.
12. If Blade 1 Pitch Motor Torque Maximum ≥ 22.58 kN, and Blade 1 Angle ≥ -17.24 degrees, and Wind Speed ≥ 5.80 m/s, and Average Pitch Motor Torque ≤ 10.22 kN, and Blade 2 Angle ≥ -19.08 degrees Then Pitch Fault Established.
13. If Blade 2 Angle ≥ 4.50 degrees Then Potential Pitch Fault Exists.
14. Otherwise, No Pitch Fault Is Present.

ACKNOWLEDGMENT

This research is funded through an EPSRC Industrial CASE award in collaboration with 5G Technology Ltd.

REFERENCES

Alpaydin, E. (2004). *Introduction to Machine Learning*. Cambridge, MA: The MIT Press.

Bianchi, F., De Battista, H., & Mantz, R.J. (2006). *WT Control Systems – Principles, Modelling and Gain Scheduling Design*. London: Springer.

Chen B., Qiu Y., Feng Y., Tavner P., & Song W. (2011). Wind turbine scada alarm pattern recognition. *Renewable Power Generation* (1-6). 6-8 September, Edinburgh, UK.

Cohen, W. W. (1995). Fast effective rule induction. *International conference on Machine Learning* (115-123). 9-12 July, California, USA.

Cohen W. & Singer Y., (1999). A simple, fast, and effective rule learner," *National. Conference on Artificial Intelligence*. (335 – 342). 18-22 July, Florida, USA.

Crabtree C.J. (2010). *Survey of commercially available condition monitoring systems for wind turbines.* Durham University, 2010.

Djurdjanovic D., Lee J., & Ni J., (2003). Watchdog agent an infotronics-based prognostics approach for product performance degradation assessment and prediction. *Advanced Engineering Informatics*. 17(3), pp. 109–125.

Eti M., Ogaji S., & Probert S. (2006). Reducing the cost of preventive maintenance (pm) through adopting a proactive reliability-focused culture. *Applied Energy* 83(11) pp. 1235-1248.

Garcia V., Sanchez J., Martin-Felez R., & Mollineda R., (2012). Surrounding neighborhood-based smote for learning from imbalanced data sets. *Progress in Artificial Intelligence*, 1(4) pp. 1–16.

Gomez Fernandez J. & Crespo Marquez A. (2009). Framework for implementation of maintenance management in distribution network service providers. *Reliability Engineering & System Safety* 94(10) pp.1639-1649.

Hameed Z., Hong Y., Cho Y., Ahn S., & Song C. (2009). Condition monitoring and fault detection of wind turbines and related algorithms: A review. *Renewable and Sustainable energy reviews*, 13(1) pp. 1–39.

Hatch C. (2004). Improved wind turbine condition monitoring using acceleration enveloping. *GE Energy Journal of Electrical Systems*, 3(1) pp. 26-38.

Jardine A., Lin D., & Banjevic D. (2006). A review on machinery diagnostics and prognostics implementing condition-based maintenance. *Mechanical Systems and Signal Processing*, 20(7) pp. 1483–1510.

Kim K., Parthasarathy G., Uluyol O., Foslien W., Sheng S., & Fleming P. (2011). Use of SCADA Data for Failure Detection in Wind Turbines. *Energy Sustainability and Fuel Cell Science.* ESFuelCell2011-54243. 7-10 August, Washington DC, USA.

Kontogiannis, T., & Kossiavelou, Z. (1999). Stress and team performance: principles and challenges for intelligent decision aids, *Safety Science*. 33(3) pp. 103-128.

Kusiak A & Li W. (2011).The prediction and diagnosis of wind turbine faults.*Renewable Energy*.36(1)pp.16-23.

Kusiak A. & Verma A. (2011). A data-driven approach for monitoring blade pitch faults in wind turbines. *Sustainable Energy, IEEE Trans. on*, 2(1) pp. 87–96.

Levrat E., Iung B., & Marquez A. (2008). E-maintenance: review and conceptual framework. *Production Planning and Control*, 19(4) pp. 408–429.

Lin J. & Zuo M. (2003). Gearbox fault diagnosis using adaptive wavelet filter. *Mechanical Systems and Signal Processing*, 17(6) pp. 1259–1269.

Marais K. & Saleh J. (2009). Beyond its cost, the “value” of maintenance: An analytical framework for capturing its net present value. *Reliability Engineering & System Safety*, 94(2) pp. 644 - 657.

Massoud Amin S. & Wollenberg B. (2005). Toward a smart grid: power delivery for the 21st century. *Power and Energy Magazine, IEEE*, 3(5), pp. 34–41.

Matthews B. (1975). Comparison of the predicted and observed secondary structure of t4 phage lysozyme. *Biochemica et Biophysica Acta*, 405(2) pp. 442-451.

Moore W. & Starr A. (2006). An intelligent maintenance system for continuous cost-based prioritisation of maintenance activities. *Computers in Industry*, 57(6) pp. 595–606.

Niu G., Yang B., & Pecht M., (2010). Development of an optimized condition-based maintenance system by data fusion and reliability centered maintenance. *Reliability Engineering & System Safety*, 95(7) pp. 786–796.

Quinlan, J. R. (1990). Learning logical definitions from relations. *Machine Learning*, 5(3), pp. 239-266.

Rafiee J., Rafiee M., & Tse P. (2010). Application of mother wavelet functions for automatic gear and bearing fault diagnosis. *Expert Systems with Applications*, 37(6) pp. 4568–4579.

Sainz E,, Llombart A., & Guerrero J. (2009). “Robust filtering for the characterization of wind turbines: Improving its operation and maintenance. *Energy Conversion and Management*, 50(9) pp. 2136–2147.

Wang X. & Makis V. (2009). Autoregressive model-based gear shaft fault diagnosis using the kolmogorov–smirnov test. *Journal of Sound and Vibration*, 327(3), pp.413–423.

Wu S. & Clements-Croome D. (2005). Preventive maintenance models with random maintenance quality. *Reliability Engineering & System Safety*. 90(1) pp. 99-105.

WWEA (2012). “Quarterly bulletin,” *World Wind Energy Association Bulletin*, 3(1), October, pp. 1 – 40.

Zaher, S. McArthur, D. Infield, & Y. Patel. (2009). Online wind turbine fault detection through automated scada data analysis. *Wind Energy*, 12(6) pp.574–593.

BIOGRAPHIES

Jamie L. Godwin received the BSc(Hons) degree in computer science from the University Of Durham, England in 2010. At present, he is working towards the degree of Doctor of Philosophy within the department of Engineering and Computing sciences, also at the University of Durham. He has researched areas such as SCADA data analysis and robust multivariate prognostic techniques. He has provided consultancy to various public and private entities, including 5G technologies, Northumbrian Water, various electricity distributors and North Yorkshire Police. His current doctoral research focuses on metrics for maintenance effectiveness, SCADA data analysis and robust multivariate statistical measures for prognosis.

Peter Matthews is a Lecturer in Design Informatics at the School of Engineering and Computing Sciences and holds degrees in Mathematics (1994), Computer Science (1995) and Engineering Design (2002), which were all taken at Cambridge University. He is the author or co-author of numerous books, technical papers and EU patents. His current research interests are centred around industrial data analysis involving collecting and analysing data obtained either from production process monitoring (e.g., SCADA logs from various production machinery) or service life data (e.g. maintenance logs) and utilising Monte Carlo simulations, Evolutionary Algorithms, Bayesian Belief Networks, and interval probabilities (p-boxes) to form future design and operation decisions.

Detection of Damage in Operating Wind Turbines by Signature Distances

Dylan D. Chase[1], Kourosh Danai[2], Mathew A. Lackner[3], and James F. Manwell[4]

[1,2,3,4] *Dept. of Mechanical and Industrial Engineering, Univ. of Massachusetts Amherst, Amherst, Massachusetts, 01003, USA*
ddchase@student.umass.edu
danai@ecs.umass.edu
manwell@ecs.umass.edu
lackner@ecs.umass.edu

ABSTRACT

Wind turbines operate in the atmospheric boundary layer and are subject to complex random loading. This precludes using a deterministic response of healthy turbines as the baseline for identifying the effect of damage on the measured response of operating turbines. In the absence of such a deterministic response, the stochastic dynamic response of the tower to a shutdown maneuver is found to be affected distinctively by damage in contrast to wind. Such a dynamic response, however, cannot be established for the blades. As an alternative, the estimate of blade damage is sought through its effect on the third or fourth modal frequency, each found to be mostly unaffected by wind. To discern the effect of damage from the wind effect on these responses, a unified method of damage detection is introduced that accommodates different responses. In this method, the dynamic responses are transformed to surfaces via continuous wavelet transforms to accentuate the effect of wind or damage on the dynamic response. Regions of significant deviations between these surfaces are then isolated in their corresponding planes to capture the change signatures. The image distances between these change signatures are shown to produce consistent estimates of damage for both the tower and the blades in presence of varying wind field profiles.

1. INTRODUCTION

Condition monitoring of wind turbines has become increasingly more important as progressively larger turbines are situated in remote locations that are exorbitantly costly and time-consuming to inspect. This exorbitant cost of access and routine inspection also precludes the use of nondestructive methods based on ultrasound or acoustics, which require overhaul and disassembly of the turbines and their blades. Therefore, the most promising recourse is the development of an automated structural health monitoring (SHM) system that relies on remote sensory data to continually assess the condition of the turbine tower and its blades.

Among the variables influenced by damage, structural vibration is particularly easy to measure by remote sensing as the basis of an automated SHM system. Vibration-based damage detection can be inverse or direct (Farrar & Doebling, 1997; Carden & Fanning, 2004; Santos, Maia, Soares, & Soares, 2008). Inverse methods update the structural model periodically to duplicate the measured response. They then estimate the physical properties of the structure from the updated model to asses the health of the structure (e.g., reduction of stiffness due to the onset of a crack or loosening of a connection) (Friswell, 2007). Inverse models, therefore, require not only an accurate model of the structure but also a complete knowledge of the input conditions that produce the measured vibration. This, unfortunately is never true for in-operation wind turbines because of unknowable wind conditions at various locations of the blade (i.e., wind profile). In contrast to inverse methods, direct methods of damage detection focus directly on identifying the effect of damage on the dynamic response of the structure (Danai, Civjan, & Styckiewicz, 2011) or its modal properties (Doebling, Farrar, Prime, & Shevitz, 1996). Given that dynamic response histories are never the same for operating wind turbines due to variable wind conditions, a challenge in health monitoring of wind turbines is to identify the dynamic time history or modal properties that are distinctly affected by damage as opposed to wind.

We have identified a shutdown maneuver of the turbine, consisting of pitching the blades to feather and braking, to produce a dynamic response that is distinctly affected by tower damage. However, a dynamic response representative of blade damage could not be identified. The blade response during the shutdown maneuver is considerably more sensitive to the wind profile because of the continuous pitching of the blades during the maneuver and the cyclic and stochastic loads on the blades that depend on the wind profile. This makes the blade time history response more sensitive to the wind profile and to the operating and initial conditions at the time of the maneuver. As a recourse, blade damage estimation is performed via modal analysis. Several studies have reported the effect of structural damage on the higher modal frequencies and the corresponding mode shapes (Farrar & Doebling, 1997). By following their lead, we have found in our studies that blade damage, as represented by localized reduction of stiffness coefficients, reduces the third and fourth natural frequencies of the blade independent of the wind conditions. Our studies also indicate that these frequency shifts (off-sets) are proportional to the level of damage and that their estimates are sensitive to the location of vibration measurement as well as the damage location.

Even though the damage effects are visually distinct from the wind effect, their isolation across different measurement locations and wind conditions is not robustly possible in either the time or frequency domain. The contribution of this paper, therefore, is to introduce a method of damage detection that differentiates the effect of damage from the wind on vibration time histories or their spectra, alike. The differentiation between the damage and wind effects in this method is facilitated by transforming the corresponding data series to surfaces in the time-scale or frequency-scale domain via continuous wavelet transforms (CWTs). The difference between the pairs of surfaces, comprising the dynamic response and its baseline, is characterized by isolating regions of the plane that represent localized significant deviations between them. The image distances between these isolated regions are then found to be significantly larger for damages than wind effects. They are, therefore, used as estimates of the damage for different measurement locations and wind conditions.

2. WIND TURBINE MODEL

Damage is characterized in this study by a reduction of tower or blade stiffness, which would occur from fatigue, cracking, loosening of connections or delamination. Both the tower and blade damage are estimated from the acceleration of the tower and blade. The wind turbine is modeled using the aero-elastic design code FAST (Jonkman & Buhl Jr, 2005) developed by the National Wind Technology Center (NWTC). The NWTC program Modes is used to calculate the blade and tower mode shapes (Buhl, n.d.) and TurbSim is used to simulate full-field, turbulent, stochastic wind files (Kelly & Jonkman, n.d.). For this study the NREL 5-MW reference wind turbine (Jonkman, Butterfield, Musial, & Scott, 2009) is simulated to replicate the effect of damage in large scale wind turbines.

In FAST distributed properties are specified at discrete locations along the length of the tower and the blades. The properties of the blade include: (1) flap-wise stiffness, (2) edge-wise stiffness, (3) torsional stiffness, and (4) mass per unit length, and those of the tower are the fore-aft and side-side stiffness coefficients. The stiffness values are also inputs to the program Modes, which computes the mode shapes of the tower and the blades. The NREL 5-MW reference offshore wind turbine consists of a tower that is 87.6 m high, with its distributed properties specified at 11 nodes, and blades that are 63 m in length, with their distributed properties specified at 49 span-wise nodes. The location of the distributed properties are shown in Fig. 1. Tower damage is simulated at six locations by reduction of the corresponding parameters p1 through p6. Blade damage is simulated at three locations: (1) close to the root, (2) mid-span, and (3) tip of the blade. The geometry of the blade is much more complicated than that of the tower and therefore many more stiffness values need to be specified for each blade. Since the reduced stiffness value of an individual node has no discernable effect on the blade vibration, damage at a location is represented by the reduced stiffness of six nodes associated with that location. Furthermore, damage level is simulated by the proportional reduction of the corresponding stiffness values. For instance, 30% damage at the blade mid-span is simulated by a 30% reduction in the flap-wise stiffness values of its 6 mid-span nodes, denoted by parameter p2 in Fig. 1. The other parameters p1 and p3, associated with the other blade locations, are also shown in Fig. 1.

FAST balances forces at a set of locations along the length of the tower or the blade. For the NREL 5 MW wind turbine model, FAST balances the forces at 20 nodes along the length of the tower and 17 analysis nodes along the length of the blade. FAST can be configured to output data at any of these analysis nodes, but for a given simulation, data can only be output at 9 blade and 9 tower analysis nodes. Therefore, 9 analysis nodes were chosen on the tower and blade each, as also shown in Fig. 1. At each of the sensor locations, fore-aft acceleration is recorded for the tower and flap-wise acceleration for the blade.

The primary input to the wind turbine is the stochastic wind input profile. In order to consider the effect of different wind profiles, five different wind data sets (referred to as Wind 1 to Wind 5) were generated by TurbSim to represent wind speeds in a square grid of 31 by 31 elements. At each grid location, the wind speed and direction are represented by a vector. The x-direction wind speeds of the five wind profiles at the hub location are shown in Fig. 2. Each of the wind profiles uses the Risø smooth terrain spectral model, has an average wind speed of 12 m/s and a power law exponent of 0.143, to account for wind shear. The turbulence intensity of the five wind profiles are: 0.1063, 0.8617, 0.10088, 0.11187 and 0.11418. In addition to wind profiles, other inputs of interest include the dynamics associated with yawing the wind turbine and pitching the wind turbine blades. Pitching of the wind turbine blade occurs in normal operation when the wind

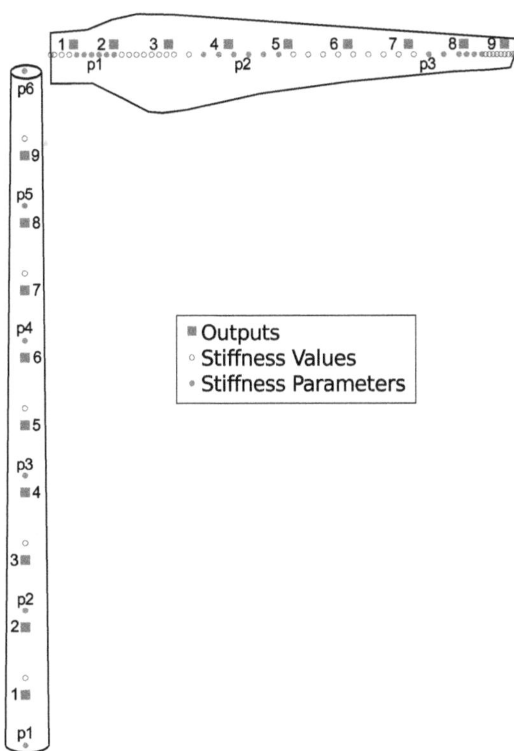

Figure 1: Schematics of the tower and a blade showing the locations of the distributed parameters, the parameters associate with each damage, and the measurement locations. The symbol, ○, corresponds to the locations where blade flap-wise stiffness values are defined. The symbol, •, corresponds to the flap-wise stiffness values associated with damage. The stiffness values of six node groups associated with the damage at that location are labeled as p1 - p3. The symbol, ■, corresponds to the nine locations where tower fore-aft or blade flap-wise acceleration is recorded.

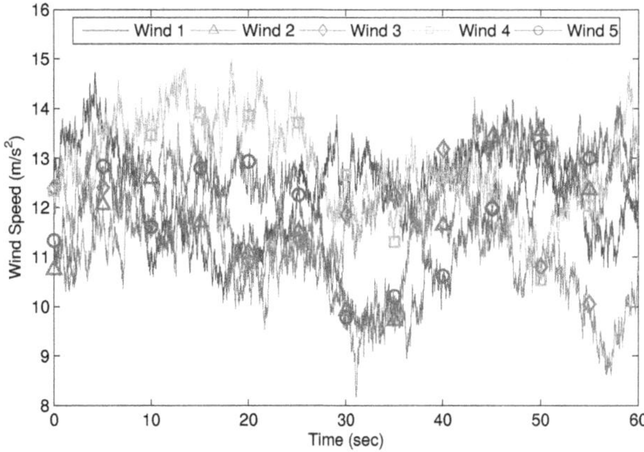

Figure 2: Hub height horizontal, perpendicular to the rotor, wind speeds of the five wind input profiles during a 60 sec time window

speed is above the rated wind speed of the wind turbine and also during shutdown when the blade is pitched out of the wind.

3. DYNAMIC RESPONSE OVERVIEW

A requisite of direct damage detection is the presence of a deterministic dynamic response that would be representative of the damage. This deterministic response recorded for the healthy system is then used as the baseline for identifying the damage from the dynamic responses that are periodically acquired for evaluating the health of the system. Ordinarily, such a deterministic dynamic response would comprise the response to a uniform excitation of the system (e.g., harmonic excitation of the structure by an eccentric mass shaker) (Danai et al., 2011). However, such a deterministic dynamic response cannot be established for wind turbines, due to the stochastic and ever present effect of wind on the vibration of in-operation wind turbines. A first task of this research, therefore, is identifying a dynamic response that despite its stochastic nature would distinctly represent the effect of damage in contrast to wind. To this end, we have explored the response of the turbine to a shutdown maneuver, consisting of pitching the blade to feather and applying the brake, to provide uniform initial conditions for the free response of the turbine. The acceleration of the tower and blade under different conditions (healthy and a faulty condition with two different wind profiles) before and after the application of this shutdown maneuver is shown in Fig. 3. For improved clarity, these acceleration time series are also shown in Fig. 4 for a smaller time window directly after the application of the shutdown maneuver in the 500-508 s time window. Vibration time series of the tower during the shutdown maneuver, shown in the top plot of Fig. 4, indicate a phase shift due to damage independent of the wind. However, the effect of damage on the blade acceleration, shown in the bottom plot of Fig. 4, is not distinguishable from the effect of wind. This is attributed to the more complex loading conditions of the blade due to the time-varying distributed load from the wind and its cyclic gravitational and wind shear load that depends on the azimuth angle of the rotor.

In the absence of an acceleration time history to distinctly reflect the effect of blade damage in spite of the wind effect, the modal properties of the blade acceleration are studied. For this, the power spectral density of the blade acceleration under steady operating conditions are studied. The motivation for this study is to capture the effect of damage on the higher frequencies independent of the wind, which is expected to affect the lower frequency acceleration (Avendano-Valencia & Fassois, 2012). For illustration purposes, the power spectral density of the first two modes of output 7 on the blade are shown for two healthy cases (Wind 1 and Wind 2) and a damaged case (Wind 1, 50% damage) in Fig. 5. The results indicate that the effect of the damage on the natural frequen-

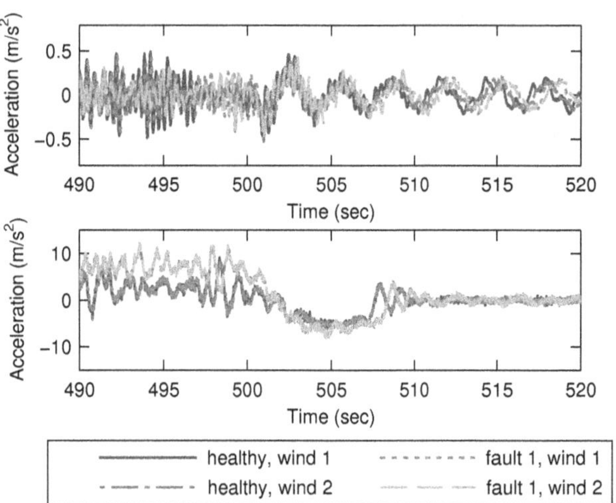

Figure 3: Acceleration of the tower at location 7 (top) and the blade at location 7 (bottom) before and after the shutdown maneuver at 500 s

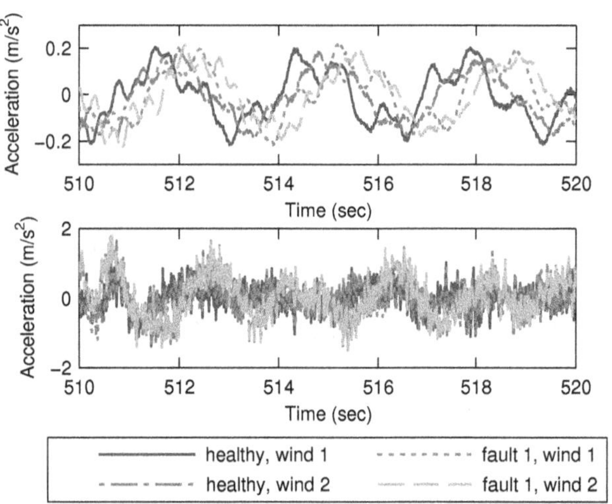

Figure 4: Acceleration of the tower at location 7 (top) and the blade at location 7 (bottom) after the brake maneuver is performed

cies of the first two modes is difficult to distinguish from the effect of wind. This motivated a study of the higher modes of acceleration in search of a more pronounced effect of the damage, independent of the wind conditions.

To evaluate the effects of wind conditions and damage on the third and fourth modes, acceleration data were generated with FAST using the third and fourth flap-wise mode shapes of the blade. The power spectral density counterparts of Fig. 5 for the third and fourth modes are shown in Fig. 6. The results in Fig. 6 indicate a more pronounced shift of the third and fourth natural frequencies caused by damage beyond any shifts by wind conditions. This observed shift in the third and fourth

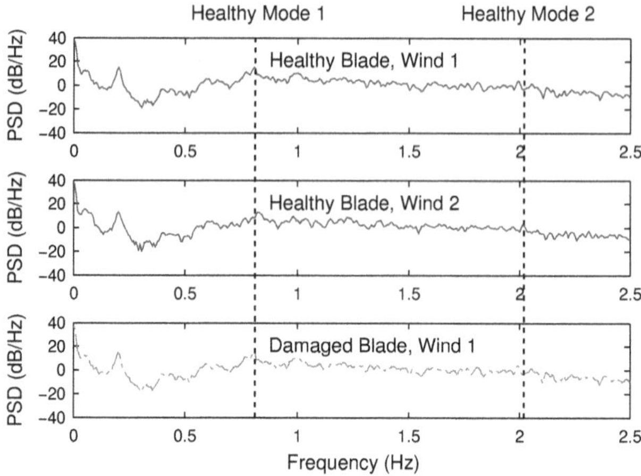

Figure 5: Power spectral density of blade acceleration at output 7. The top plot corresponds to the healthy blade excited by wind profile 1 (i.e., the baseline). The middle plot corresponds to the healthy blade excited by wind profile 2, and the bottom plot is associated with a damaged blade excited by wind profile 1. The damage consists of a 50% reduction in stiffness at the p1 nodes.

mode natural frequencies motivates the use of blade acceleration power spectral density as the dynamic response representative of blade damage. It should be noted here that tower damage could potentially be identified as well from the spectral density of its higher frequency vibration. However, such an approach is unnecessary because it is more straightforward to detect tower damage from the dynamic time response history which precludes the complexities of spectral analysis.

4. DAMAGE DETECTION METHOD

The objective of the proposed damage detection method is to provide clear and irrevocable indication of tower and blade damage of various levels by the majority of the measured accelerations with different wind speeds and profiles. To this end, it uses continuous wavelet transforms (CWTs) to represent and enhance various shape attributes of dynamic responses in order to identify the responses that are different in shape due to damage and wind. Continuous wavelet transforms have the noted feature of representing the shape attributes of transformed signals in the time-scale domain as well as the capacity to accentuate their differences (Danai et al., 2011). Therefore, they have been extensively used in structural damage detection for accentuating the effect of damage on mode shapes (Chang & Chen, 2004; Ovanesova & Suarez, 2004; Loutridis, Douka, Hadjileontiadis, & Trochidis, 2005; Rucka & Wilde, 2006). The damage detection strategy used here, instead, applies the CWTs directly to the tower dynamic time responses or blade frequency spectra to identify regions, referred to as "change signatures," in the time-scale domain or frequency-scale do-

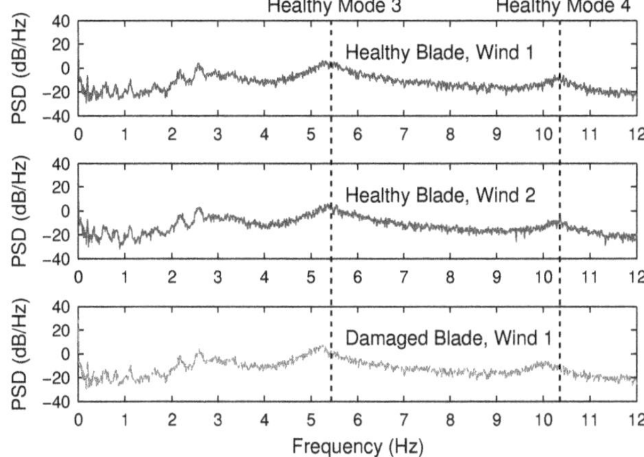

Figure 6: Power spectral density of blade acceleration at output 7 using flap-wise bending modes 3 and 4 in the FAST Model. The top plot corresponds to the healthy blade excited by wind profile 1 (i.e., the baseline). The middle plot corresponds to the healthy blade excited by wind profile 2, and the bottom plot is associated with a damaged blade excited by wind profile 1. The damage consists of a 30% reduction in stiffness at the p1 nodes.

main wherein the change in the corresponding surfaces, due to wind or damage, exceeds a dominance factor. This signal detection strategy has been shown to be effective in characterizing localized differences among the dynamic time histories of a structural model in order to establish the pattern of faults in a nine-storey building (Danai et al., 2011). In this paper, the change signatures are not independently sufficient for damage detection, since signal change may have been caused by the randomness of wind, instead. Therefore, to differentiate the effect of damage from the wind, the change signatures are evaluated further by image distances so as to identify the effect of damage by their larger image distances they produce.

4.1. Transformation to the Time-Scale Domain

Briefly, a wavelet transform (WT) is obtained by the convolution of a wavelet function $\psi_s(t)$ with the signal $f(t)$ (Mallat, 1998), as

$$W\{f\}(t,s) = f * \psi_s(t) = \int_{-\infty}^{\infty} f(\tau)\frac{1}{\sqrt{s}}\psi^*\frac{(\tau - t)}{s}d\tau \quad (1)$$

where $W\{f\}$ denotes the WT of the time function f, $*$ denotes convolution, ψ^* is the complex conjugate of ψ, $\psi_s(t) = \frac{1}{\sqrt{s}}\psi(\frac{t}{s})$ represents the wavelet function, and t and s denote the time (translation) and scale (dilation or constriction) parameters, respectively. The wavelet function can be manipulated in two ways: (i) it can be moved sideways (translated) to coincide with different segments of the signal, and (ii) it can be widened (dilated) or narrowed (constricted) to align with a larger or smaller segment of the signal at its current location (current time). The wavelet coefficients $W\{f\}$ that result from the convolution integral of Eq. (1) at each time (t) and scale (s) denote the cross-correlation between the wavelet function $\psi_s(t)$ and time function $f(t)$, with the wavelet function positioned at time t and dilated at scale s. Numerically, the computation of WTs is facilitated for dyadic time data. Here, we have chosen to use 128 data points; i.e., $t_k = t_1 \ldots t_{128}$ of each dynamic response to obtain the WTs for 72 scales; i.e., $s_l = s_1 \ldots s_{72}$. These choices result in a time-scale plane of 128×72 pixels, each pixel (t_k, s_l) having unity time and scale dimensions. For illustration purposes, the wavelet transform of the tower acceleration time signal at location 7 is shown in Fig. 7.

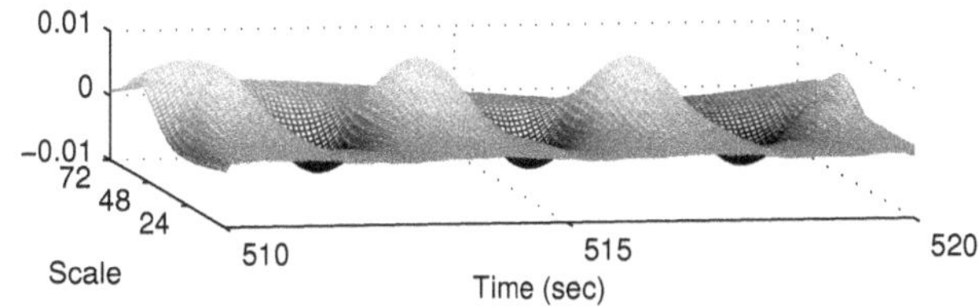

Figure 7: Gauss WT of tower acceleration at location 7 after the shutdown maneuver is performed

4.2. Representation of Shape Attributes

Representation of shape attributes of time signals by CWTs stems from their multiscale differential feature (Mallat, 1998). Consider $\psi(t)$ to be the nth order derivative of the smoothing function $\beta(t)$; i.e.,

$$\psi(t) = (-1)^n \frac{d^n(\beta(t))}{dt^n} \quad (2)$$

then this wavelet transform is a multiscale differential operator of the smoothed function $f * \beta_s(t)$ in the time-scale domain (Mallat & Hwang, 1992); i.e.,

$$W\{f\}(t,s) = s^n \frac{d^n}{dt^n}\left(f * \beta_s(t)\right) \quad (3)$$

Using this feature, one can utilize the CWT to represent the first derivative of a time signal for its slope, or its second derivative to represent the rate of slope change. For instance, one may consider the smoothing function $\beta(t)$ to be the Gaussian function. In this case, the Gauss wavelet is the first derivative of the Gaussian function. This results in a wavelet transform that is the first derivative of the signal $f(t)$ smoothed by the Gaussian function, and orthogonal to this smoothed signal. Similarly, the Sombrero wavelet is the second derivative of the Gaussian function, and produces a wavelet transform that is the second derivative of this smoothed signal in the time-scale domain. Therefore, the Gauss WT represents the Gaussian smoothed slope of the signal $f(t)$ and the Sombrero WT denotes its rate of slope change.

4.3. Delineation of Localized Dissimilarities

For a view of the delineation capacity of CWTs, which is of significance to this research, let us consider the WT of a time signal $f(t)$ at a particular coordinate (t_1, s_1):

$$W\{f\}(t_1, s_1) = \int_{-\infty}^{\infty} f(\tau)\frac{1}{s_1}\psi\left(\frac{t_1 - \tau}{s_1}\right) d\tau \quad (4)$$

The wavelet coefficient, $W\{f\}(t_1, s_1)$, which represents the cross-correlation of $f(t)$ with $\psi_{s_1}(t_1)$, depends upon the magnitude of $f(t)$ as well as the conformity of $f(t)$ with the shape of the dilated $\psi_{s_1}(t_1)$. Therefore, the wavelet coefficients can accentuate minute differences between time signals at the lower scales by capturing the conformity of a narrow $\psi_s(t)$ with a small segment of the time signal.

To illustrate the enhanced delineation provided by CWTs, let us consider the highly correlated pairs of acceleration signals ($\rho = 0.9830$) shown in the top plot of Fig. 8. The two signals are different due to a damage. Although the two signals have near identical shapes, as represented by their correlation coefficient, they have distinct local differences that can be accentuated by their wavelet coefficients. The points of deviation between the slopes of the two signals are accentuated by the differential Gauss wavelet coefficients in the bottom plot of Fig. 8. The peaks and valleys in the plot are reflections of the differences in the slopes of the two signals in the top plot of the figure. Whereas such local dissimilarities are masked by a lumped measure such as the correlation coefficient, the pixels associated with these peaks and valleys mark the regions of slope difference between the two signals. Therefore, the analysis can be focused on where the difference is prominent. As a result of the enhanced delineation described above, regions of significant deviation between two signals; i.e., the signal and its baseline, can be identified to characterize the effect of wind or damage on the signal.

4.4. Signature Extraction

For signal change detection, the change (due to wind or damage) in a dynamic response (tower acceleration history or blade power spectral density) is identified by comparing the wavelet coefficients of the response with those of their baseline. Change identification is performed by isolating regions of the corresponding plane (comprised of the sample points $t_k, k = 1, \ldots, N$ and scales: $s_l, l = 1, \ldots, M$) wherein a dynamic response deviates from the undamaged response by a *dominance factor*, η_d. The union of the isolated regions associated with a dynamic response is called the 'change signature,' formally defined for the jth sensor location as the union of all pixels $(t_k^j, s_l^j) \in \delta_j$ in the plane wherein the nonzero (relative to the threshold h) normalized wavelet coefficient of the output y_j exceeds the normalized wavelet coefficient of its baseline y_j^n by a dominance factor η_d, expressed mathematically as

$$\left|\overline{W\{y_j\}}(t_k, s_l)\right| > h > \eta_d \left|\overline{W\{y_j^n\}}(t_k, s_l)\right| \implies (t_k, s_l) = (t_k^j, s_l^j) \in \delta_j \quad (5)$$

where

$$\overline{W\{y_j\}} = \frac{W\{y_j\}}{\max_{(t,s)} |W\{y_j\}|} \quad (6)$$

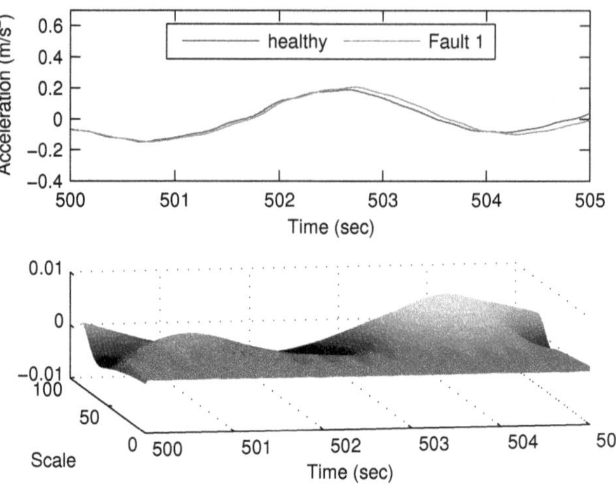

Figure 8: Illustration of the enhanced delineation of two similar signals (top) by the differential wavelet transform (bottom) which accentuates the minute differences between the two signals

If the above condition is satisfied at a pixel (t_k, s_l), then we tag the pixel as (t_k^j, s_l^j) to note its inclusion in the change signature δ_j. The inclusion of the threshold, h, in Eq. (5) is a provision to exclude pixels at zero crossings of the dynamic response wavelet coefficients. Without this provision, at zero-crossings (e.g., when the Gauss wavelet coefficients are zero due to zero slopes at the peaks of the acceleration signals) any nonzero wavelet coefficient would dominate this zero wavelet coefficient and hence include superfluous pixels in the corresponding damage signature. We have found the value of $h = 0.005$ to be sufficient as a safeguard against inclusion of zero-crossing pixels in the change signatures. The other factor in the change signature extraction routine of Eq. (5) is the dominance factor, η_d. Since higher dominance factors lead to fewer pixels in the change signature at the risk of missing minute differences between the dynamic responses, higher dominance factors correspond to higher standards of change detection.

For illustration purposes, change signatures between the two healthy signals (healthy, wind 1 and healthy, wind2) in Fig. 4 are shown in the top plot of Fig. 9. For comparison, also shown in Fig. 9 (bottom plot) are the change signatures of the faulty signal (fault 1, wind 1 and healthy, wind 1) in Fig. 4.

Shown in this figure, are two sets of signatures, marked as blue and red. The blue signatures are associated with

$$\left|\overline{W\{y_j^n\}}(t_k, s_l)\right| > h > \eta_d \left|\overline{W\{y_j\}}(t_k, s_l)\right| \Longrightarrow (t_k, s_l) \in \delta_j^n$$

and the red signatures correspond to

$$\left|\overline{W\{y_j\}}(t_k, s_l)\right| > h > \eta_d \left|\overline{W\{y_j^n\}}(t_k, s_l)\right| \Longrightarrow (t_k, s_l) \in \delta_j$$

The change signatures in Fig. 4 are clearly farther apart for the faulty signal than those for the healthy signal, represented by the larger lag caused by the damage in contrast to the wind.

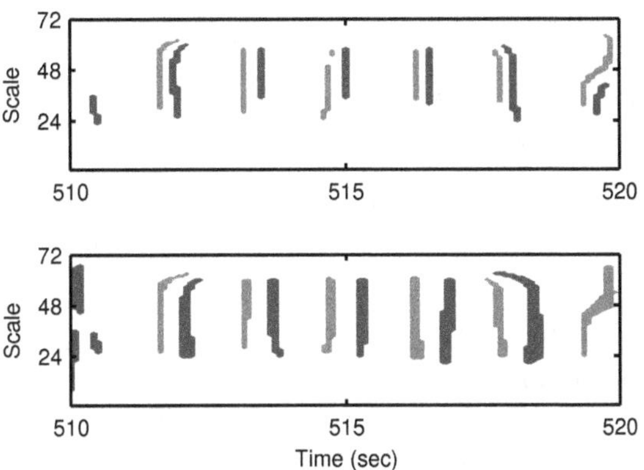

Figure 9: Change signatures between two signals from a healthy tower with different wind conditions (top) and those between a signal from a faulty tower and a healthy signal (bottom)

4.5. Image Distances of Damage Signatures

Since the change signatures only characterize the change of the signal, they are bound to be generated under normal conditions because of the wind. Accordingly, a separate measure is required to differentiate between the change signatures caused by wind and those by damage. As was shown visually in Figs. 4 and 6, the differentiating aspect of the two effects is the larger lag that is caused by the damage in contrast to the wind. To characterize this aspect, image distances are utilized to represent the distance of the change signatures.

To assess the distances between the change signatures, either the Euclidean distance or the weighted Euclidean distance (also known as image Euclidean distance (Wang, Zhang, & Feng, 2005)) can be used. However, our analysis indicates that the weighted Euclidean distance provides a more succinct distance measure because of its assignment of larger weights to pixels of higher proximity to each other. The weighted Euclidean distance, d_I, hereafter referred to as imaged distance, discounts the difference in magnitudes of wavelet coefficients according to the mutual distance between their locations on the time-scale plane, as (Wang et al., 2005)

$$d_I^2(\delta_j, \delta_j^n) = \frac{1}{2\pi\sigma^2} \sum_{k=1,\, l=1}^{N,M} \exp\{-|P_k - P_l|^2/2\sigma^2\}$$

$$(W\{\delta_j\}_k - W\{\delta_j^n\}_k)(W\{\delta_j\}_l - W\{\delta_j^n\}_l) \qquad (7)$$

where σ is a width parameter that represents the discount rate associated with the pixel distance, k and l denote the coordinates of each pixel on the time-scale or frequency-scale plane, P_k and P_l denote pixel locations, and $|P_k - P_l|$ represents the distance between two pixels on the plane lattice. According to Eq. (7), the image distance fully incorporates the difference in magnitude of wavelet coefficients with identical locations and discounts by the weight "$\exp\{-|P_k - P_l|^2/2\sigma^2\}$" the magnitude difference when the two locations do not coincide on the time-scale or frequency-scale plane (i.e., image lattice).

For the application of image distances to damage detection, consider the Gauss WT of the tower acceleration shown in Fig. 7 in the time-window of 510-520 s after the shutdown maneuver (at 500-508 s). The surface consists of a series of peaks and valleys at higher scales. This results in a recurring set of vertical signatures shown in Fig. 9. Given that these signatures may have been caused by wind or damage, the distinction between them is clarified through the distances between the signatures. However, change signatures could also be caused by noise or the edge effects of wavelet transform. Therefore, image distances should be ideally computed for the change signatures that are representative of wind or damage. Referring to Fig. 9, an image distance computed for a single pair of change signatures would be sufficient for representing the cause of change. To facilitate this focus on the signature pairs, a window in the corresponding plane is considered, as shown in Fig. 10, to isolate a few pairs of change signatures for computation of the image distance.

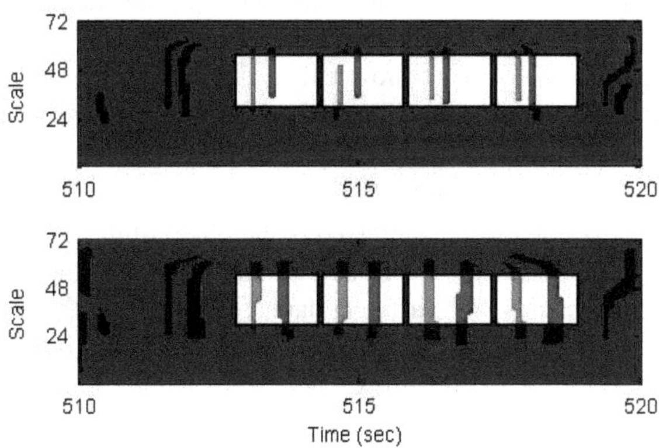

Figure 10: The change signatures in Fig. 9 enclosed by a window in the time-scale domain to better capture the area of relevance to the change

5. DAMAGE ESTIMATES

The possibility of estimating tower and/or blade damage by image distances was studied by transforming the acceleration records (the tower acceleration time histories or the blade acceleration power spectral densities) via the Gauss or Sombrero WT into their corresponding domains and then extracting their change signatures using the acceleration from the healthy turbine with the wind 1 profile as the baseline. This was also performed for the acceleration records obtained with the other wind profiles. The success of damage estimation by this method was then verified by obtaining larger image distances for the acceleration signals from the damaged turbine in comparison to those from the healthy turbine under different wind conditions. The analysis was performed using Gauss WT, for tower damage detection, and Sombrero WT, for blade damage detection. However, the analysis is not specific to the WT, as either WT could be used for detection of tower or blade damage.

5.1. Tower Damage Estimates

Image distances were obtained for the tower acceleration time histories obtained at different locations. The acceleration records of the healthy and damaged tower obtained with different wind profiles were transformed to the time-scale domain by the Gauss WT. Change signatures were then extracted at the dominance factor of $\eta_d = 6$ for the acceleration signals of the healthy and damaged tower in the time window of 510 to 520 seconds, using the acceleration of the healthy tower with wind 1 as the baseline.

For successful damage estimation, the images distances from the damaged tower need to be higher than those from the healthy tower, albeit with different wind profiles. To test this hypothesis, the image distances of the acceleration signals from the healthy tower (different winds) at its nine output locations are compared with those from the 5% damaged tower in Fig. 11. In all these cases, the wind profiles had a mean speed of 12 m/s. The image distances in Fig. 11 are clearly larger for the damaged tower than the healthy tower. This validates the premise of the method that image distances can discern the effect of damage from the wind on acceleration time histories of the tower.

The next issue to be addressed is the effect of damage level on the image distances. To address this point, the image distances obtained with different damage levels from the acceleration at output location 7 of the tower are shown in Fig.12 with different wind profiles. The results indicate that the image distances provide consistent and reliable estimates of the damage regardless of the damage level across different wind profiles. However, the magnitudes of damage distances do not seem to be affected by the damage level.

Yet another issue to be explored is the influence of the mean wind speed on damage estimation. Since wind is the single most important factor in acceleration of the wind turbine, it is likely to provide higher excitation levels at higher speeds. On the other hand, at higher than the rated wind speed, the control system will pitch the blade so as to maintain constant power. Accordingly, three wind speeds of 6, 12, and 18 m/s were selected within the operating range of the wind turbine. The image distances obtained with different wind speeds from the acceleration at output location 7 of the tower for both the 5% damaged and healthy tower are shown in Fig. 13 for different wind profiles. The image distances indicate that damage can still be reliably estimated at these other wind speeds, albeit at a lower margin than those achieved at 12 m/s. At the mean wind speed of 6 m/s, the image distance of the damaged tower is quite close to that of the healthy tower with wind profile 3. Similarly, at the mean wind speed of 18 m/s, the image dis-

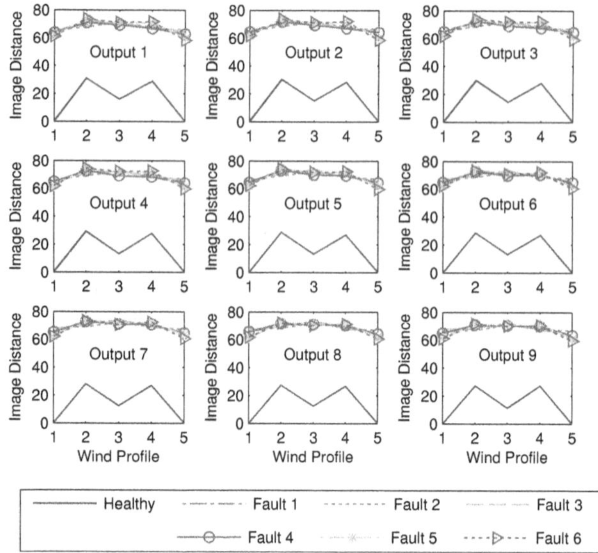

Figure 11: Image distances of the acceleration time histories at 9 output locations of the tower obtained for the healthy tower and a 5% damaged tower with five different wind profiles having a mean speed of 12 m/s

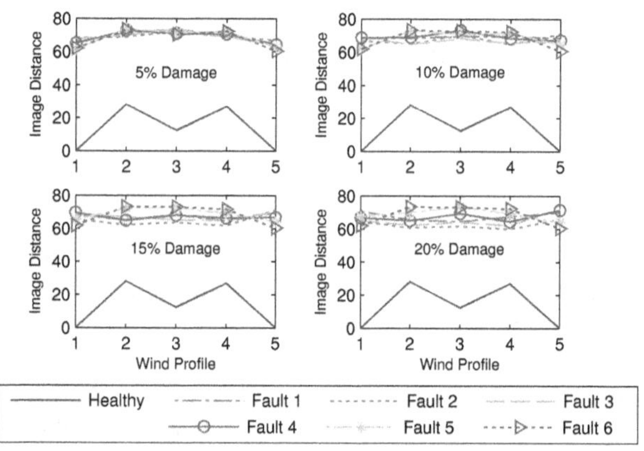

Figure 12: Image distances with varying levels of damage obtained from the acceleration time histories at output location 7 of the tower with different wind profiles having a mean speed of 12 m/s

tance of the healthy tower with wind profile 5 is within the range of those considered to be representative of the damage.

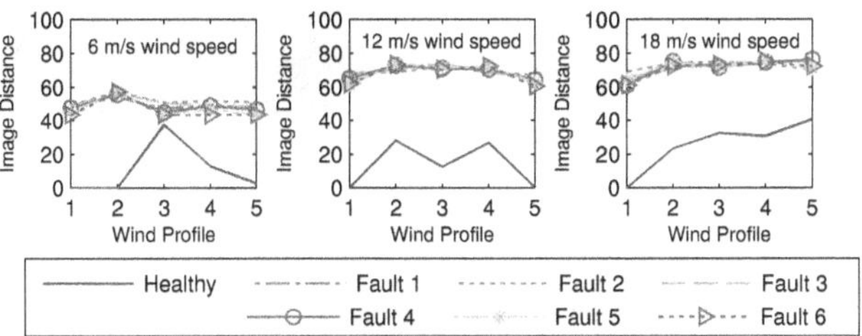

Figure 13: Image distances at different mean wind speeds obtained from the acceleration time histories at output location 7 of the tower for the healthy and 5% damaged tower with different wind profiles

5.2. Blade Damage Estimates

The acceleration used for blade damage estimation was obtained during normal operation, since the dynamic response of interest was the power spectral density of the third and fourth mode acceleration. Therefore, there was no shutdown maneuver necessary as a standardized excitation. Blade damage was estimated from the power spectral density of the blade acceleration at different blade locations. For blade acceleration, the Sombrero WT was used to provide peaks and valleys similar in location to those of the actual acceleration. Also, instead of the acceleration time history that was used for the tower, the windowed spectra of the blade acceleration were used for representation of acceleration in the frequency-scale domain. For illustration purposes, the wavelet transform of the spectrum of blade acceleration at output location 7 of the blade for the healthy blade, excited by wind profile 1, is shown in Fig. 14. The surface of the WT is shown in the top plot of Fig. 14 and its contour in the bottom plot. The wavelet transform is created using the Sombrero wavelet and results in ridges around the third and fourth flap-wise natural frequencies.

The effect of damage and wind on the Sombrero WT of the frequency spectra is shown in Fig. 15 for the blade acceleration at output location 7. The top plot corresponds to the healthy blade excited by wind profile 1 (i.e., the baseline). The middle plot corresponds to the healthy blade excited by wind profile 2, and the bottom plot is associated with a 30% damaged blade excited by wind profile 1. The baseline (top) plot is very similar to the middle plot which is associated with the healthy blade as well. But the bottom plot contains a significant off-set (shift) relative to the other two as the result of the damage. It should be noted that this off-set is also observed in the frequency spectra of Fig. 6, which are the sources of the contours in Fig. 15. Therefore, one may be tempted to perform damage estimation from the offset between the frequency spectra peaks. However, as is shown

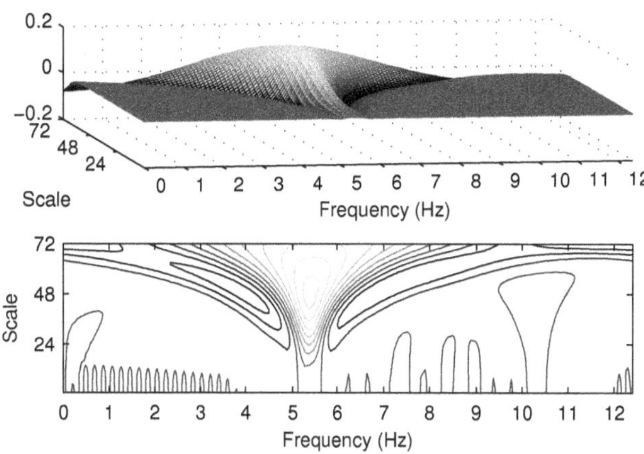

Figure 14: Sombrero WT of the healthy blade acceleration at output location 7 of the blade excited with wind profile 1 (top) and its contour in the frequency-scale plane (bottom)

later in the Discussion section, such an estimation technique does not provide nearly as robust a set of results as those by the image distances of the change signatures. The advantage of relying on the WT of the frequency spectra stems from the multi-resolution image of the signal provided at various scales, which enhances identification of the natural frequency over peak picking in the frequency domain.

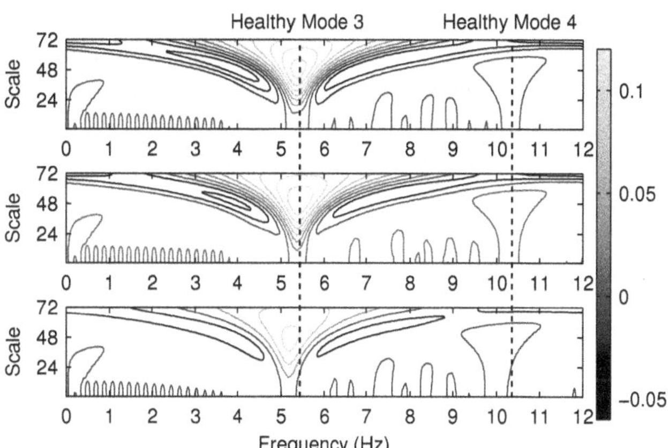

Figure 15: Contour plots of the sombrero WT of the blade acceleration at output location 7 of the blade. The top plot corresponds to the healthy blade excited by wind profile 1 (i.e., the baseline). The middle plot corresponds to the healthy blade excited by wind profile 2, and the bottom plot is associated with a 30% damaged blade excited by wind profile 1.

Change signatures were obtained from the Sombrero WTs of the blade acceleration power spectra, but at a lower dominance factor of $\eta_d = 1.4$. The change signatures obtained from the healthy (wind 2) and damaged (wind 1) blade accelerations at output location 7 of the blade, depicted in the middle and bottom plots of Figs. 6 and 15, with the acceleration of the healthy blade (wind 1) as baseline (i.e., corresponding to top plot of these figures) are shown in Fig. 16. It is clear from the results that the damaged blade produces far more change signatures.

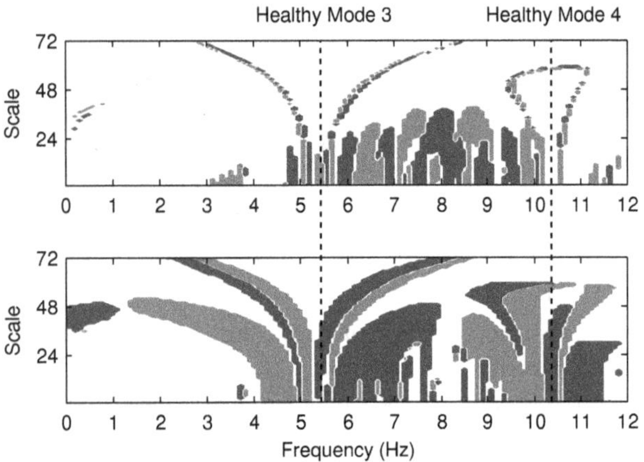

Figure 16: Change signatures of the Sombrero WT of the acceleration at blade output location7. The top plot represents the difference between the healthy blade acceleration excited by wind profile 2 relative to the acceleration by wind profile 1, used as baseline. he bottom plot represents the difference caused by 30% damage with the same wind excitation as the baseline.

To compute the image distances for the blade acceleration, a larger window than the window used for the tower acceleration was required, to encompass modes 3 and 4. Furthermore, the computation window was positioned at higher scales than the one for the tower to mitigate the effect of noise at lower scales. For illustration purposes, the windows used for computing the image distances of the blade acceleration at each of the modes 3 and 4 are shown in Fig. 17.

The image distances of the change signatures obtained separately for the third and fourth modes from each output location of the blade are shown in Figs. 18 and 19, respectively. Unlike the results for the tower, which were consistent across all output locations, the image distances for the blade are not the same for all output locations. For instance, output location 1 at the root of the blade does not seem to provide the distinction between the damaged and healthy blades that output location 7 provides, for example. Nevertheless, in most cases the damage seems to be decipherable with adequate margins between the image distances of the damaged and healthy blade.

As with the tower, image distances were obtained for different levels of blade damage, to evaluate the capacity of the image distances in characterizing the level of damage. The image distances obtained from modes 3 and 4 of output location 7 acceleration for different damage levels are shown in Fig. 20. The image distances not only provide adequate margins for

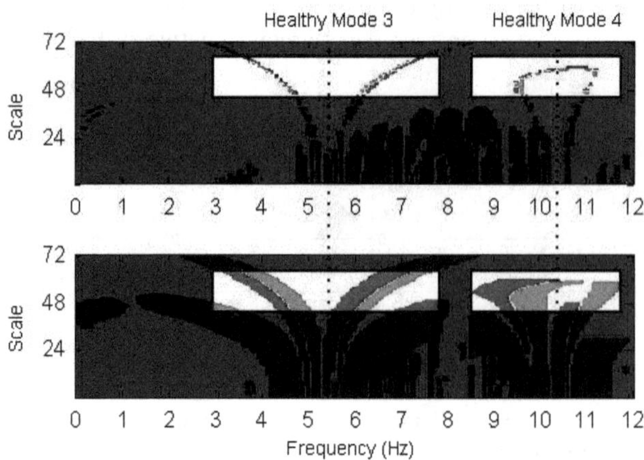

Figure 17: Windows (white areas) defined in the frequency-scale plane to focus image distance computation of the change signatures for blade acceleration around the third and fourth mode natural frequencies

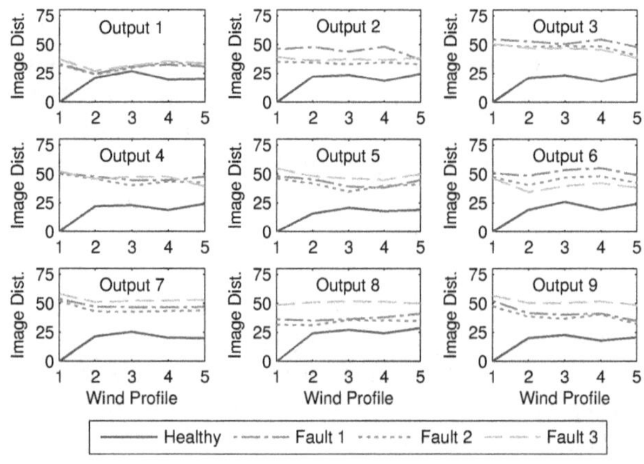

Figure 18: Image distances of the change signatures in the proximity of the third mode for the healthy and 30% damaged blade by each of the nine blade output locations

damage estimation beyond 10% damage (10% reduction in stiffness coefficient), but also indicate the level of damage. This is in contrast to the image distances from the tower in Fig. 12, which were insensitive to the damage level.

To evaluate the influence of wind speed on the damage estimation results, image distances were also obtained for the change signatures of modes 3 and 4 accelerations of blade output location 7 with two other mean wind speeds at 6m/s and 18 m/s. The image distances from the three mean wind speeds are shown in Fig. 21. The results indicate that mode 3 provides a more consistent basis for damage estimation across different wind speeds and that the largest margin exists at the lowest wind speed of 6 m/s. This could be due to the absence of pitch control at this lower speed and lower turbulence.

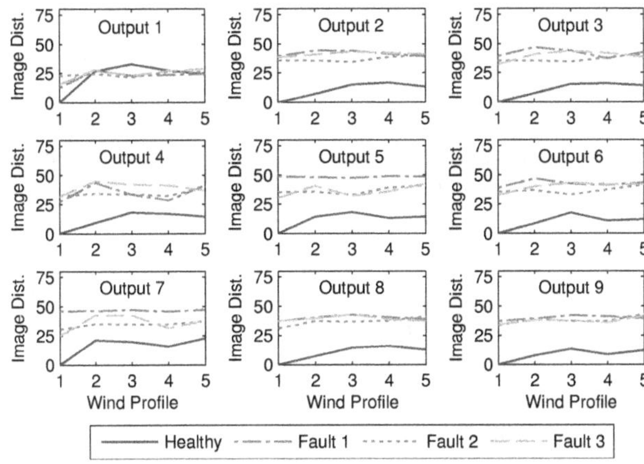

Figure 19: Image distances of the change signatures in the proximity of the fourth mode for the healthy and 30% damaged blade by each of the nine blade output locations

6. DISCUSSION

The results presented above indicate the effectiveness of the method in estimating the tower and blade damage in presence of varying wind conditions. The method relies on several factors which enhance its robustness, including its use of wavelet transforms to accentuate the delineation of vibration signals, its change signature mechanism that allows selective detection of effect levels, and the capacity to use windows for exclusion of noise effects. Nevertheless, the question remains as whether similar performance can be obtained by a far simpler method such as peak picking in either time or frequency domains. To address this point, we performed blade damage estimation by peak picking in the frequency domain using the frequency spectrum of blade acceleration containing the third and fourth modes. For this, the frequency spectra were low-pass filtered to smooth their edges and the third or fourth mode frequency was estimated by finding the local maximum of the spectrum within the frequency window associated with the mode. The distance of the estimated frequency from its baseline was then used as the damage estimate, as shown in Fig. 22. The results indicate that although the frequency distances at locations 1, 2 and 7 provide adequate differentiation between the damage and wind effects, the distances at the other locations are not as distinct. These distances were also found to be sensitive to the level of smoothing and generally less robust to fault locations and different wind speeds. The multi-scale smoothing of the signal provided by the wavelet transforms is considered to be an important factor, among others, in the added robustness of the proposed method.

The results from the proposed method also indicate that while the image distances are sensitive to the level of damage, the location of blade damage, and the operating conditions (wind speed and profile), they provide overall margins that effectively distinguish the effect of damage from the wind. The results further indicate that unlike the tower damage estimates,

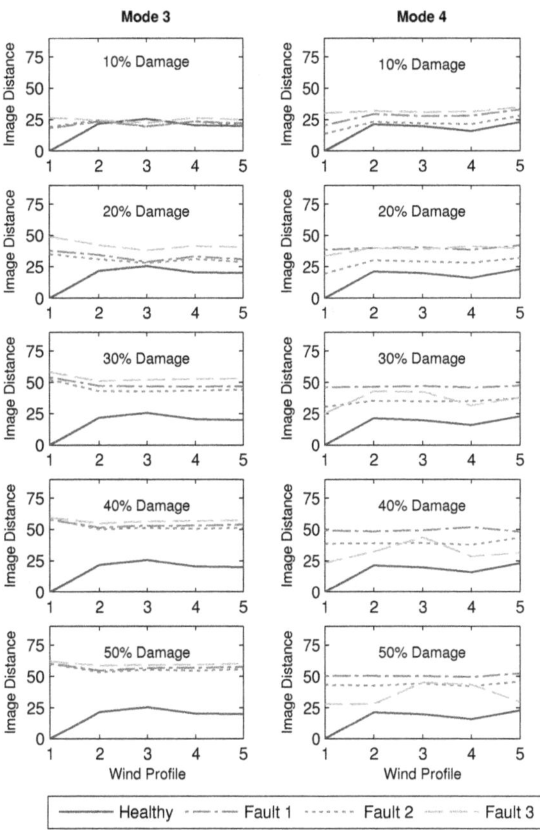

Figure 20: Image distances of the change signatures from the third and fourth modes of the healthy and different level damaged blades using the acceleration of blade output location 7 with wind profiles having the mean wind speed of 12 m/s

the blade damage estimates are sensitive to sensory location. This is due to the fact that the motion of the tower is dominated by the first mode acceleration and therefore all of the sensory locations produce similar results. The acceleration of the blade, on the other hand, is studied in proximity of the third and fourth modes. As a result, the sensors that are close to the nodes of either mode do not provide reliable information. For instance, the blade damage estimates in Figs. 18 and 19 from output location 1 of the blade, close to its root, are generally inferior to the others. This is due to the absence of acceleration at this point and the FAST model, which represents the connection between the blade and the hub as rigid. As to the algorithmic issues not studied in this paper and the concerns associated with the practical application of the method, the following points warrant further investigation.

- *Algorithmic Issues:*
  - Width parameter: Adjustment of the image distance width parameter, σ, in Eq. (7) affects how pixels are discounted due to distance. For the tower and the blade, a $\sigma = 1$ was used. However, the change

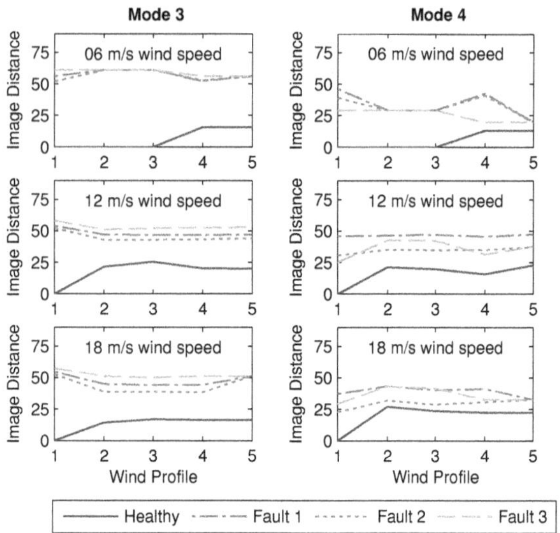

Figure 21: Image distances of the change signatures according to the third and fourth modes computed from the acceleration of blade output location 7 for the healthy and 30% damaged blade at different mean wind speeds

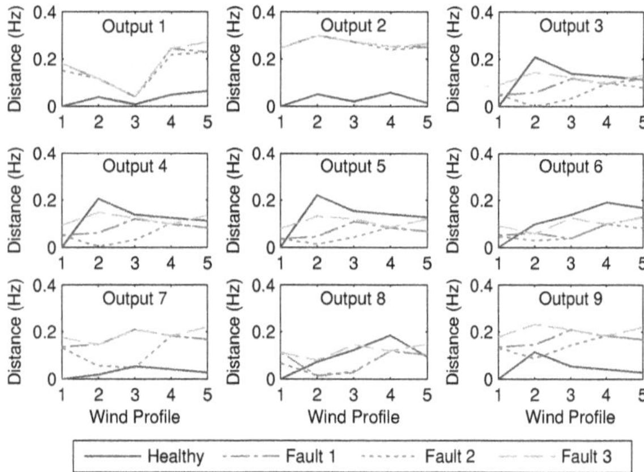

Figure 22: Frequency distances caused by different 12 m/s mean wind speed profiles and 30% blade damage according to peak picking of the power spectral density of blade acceleration at the nine output locations of the blade

signatures are further apart for the tower than the blade. Therefore, fine tuning of the width parameter might lead to improved results.

- Dominance factor: A range of dominance factors, η_d in Eq. (5), were tested for each damage estimation case. A higher dominance factor generally resulted in change signatures with fewer pixels and the corresponding image distances which were less sensitive to the wind profile.
- Windowing: Windows are used to focus image distance estimation on the areas of plane where change signatures are representative of mostly damage. In the case of the tower, the change signatures represent the cyclic nature of the response, therefore, the windows are placed to capture signature pairs. Accordingly, the location of the windows depend on the wavelet transform used for damage estimation. For the case of the blade, on the other hand, the windows need to be positioned around the third and fourth mode natural frequencies. Therefore, apart from slight adjustments for the dominance factor used, the window needs to be placed in the middle of the scale range to avoid noise, which is mostly present at low scales (high frequency), and edge effects, which usually appear at higher scales (low frequency).

- *Application Concerns:*
  - Measurement noise: In practice, measurement noise will influence the dynamic response. However, we do not expect it to significantly affect the performance of the method, because of the several provisions of the method that enable it to cope with the stochastic nature of dynamic responses. One such provision is the capacity to exclude low-scale (high-frequency) regions of the plane from the image distance window, so as to minimize the influence of noise. Another provision is the dominance factor, η_d in Eq. (5), which allows exclusion of small changes caused by noise from the change signatures. In general, the varying wind conditions with which this method is designed to cope are considered to be much more problematic than measurement noise, because of their low frequency nature that coincides with the acceleration signals of interest in the 0.3 - 12 Hz range.
  - Modal analysis of the blade vibration: These simulations have been performed using the third and fourth flap-wise blade natural frequencies. FAST is only able to model two flap-wise and one edgewise mode of vibration per simulation. In contrast, the real system will contain all modes of vibration in its dynamic response, wherein the lower modes dominate. Therefore, it may be more difficult to extract the third and fourth mode properties, hence the shifts of the third and fourth mode frequencies. To provide a preliminary evaluation of the potential challenge posed by such condition, simulation runs were conducted in FAST using the first and third modes together in one case, and the first and fourth modes together in another case. The frequency spectrum of each of these simulation runs is compared in Fig. 23 to the frequency spectrum of the dynamic response obtained using the third and fourth modes. The results in this figure do

not show a significant diminishment of the third or fourth modal estimation due to the presence of the 1st mode in the dynamic response. Another issue not considered in the present study, due to the limitation of FAST, is modal coupling due to the twist of the blades.

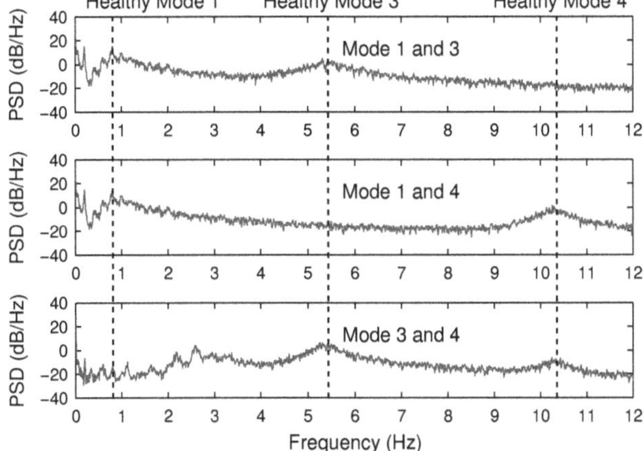

Figure 23: Comparison of the blade spectra of three different scenarios each containing only two of the flap-wise mode shapes: modes 1 and 3 in the top plot, modes 1 and 4 in the middle plot, and modes 3 and 4 in the bottom

- Sensor location: The insensitivity of tower damage estimates to sensor location suggests that a single sensor may be sufficient for tower damage detection. The blade damage estimates, however, are more nuanced. Certain sensors perform better at detecting damage using the third mode of vibration. Others perform better using the fourth mode of vibration. This warrants further analysis to ascertain the optimal location of sensors and their numbers for robust and reliable damage estimation.
- Type of sensor: Although the data used in this research is based on simulated accelerometer data, the results are not expected to be constrained by the type of data. Strain gauge data is likely to be as suitable for the analysis.
- Data acquisition constraints: Since identification of tower damage in this paper is based on the tower dynamic response time history, there will be no sampling constraints on the data acquisition system so long as the time histories capture the phase shift caused by damage. However, identification of blade damage is based on at least the third mode of vibration, so the sampling frequency needs to be high enough to capture this mode. Given that the storage requirements for such high frequency data may prove to be infeasible, one could consider analyzing the data right after data acquisition and discarding it once the system is evaluated as healthy.
- Wind profiles: This study has examined three wind speeds. At each wind speed, five full field, turbulent wind profiles were used to test the effect of wind on the dynamic response. However, this topic should be studied further to determine if the method can handle wind files which are even more dissimilar. It is also desirable to perform a full scale validation of the method.
- Model Accuracy: The results presented in this paper have been obtained with utmost attention to the intricacies of the problem. To this end, the model of the wind turbine has been configured to account for power regulation by the pitch controller and the aerodynamics of the blades as well as the 3D dynamics of the blades, the tower, the drive train, the generator, the yaw system. Beyond these results, a more realistic validation of the methods needs to rely on experimental data.

7. CONCLUSION

A unified method of damage estimation is introduced for blade and tower damage detection of operating wind turbines. For tower damage estimation, the acceleration time response of the tower to a shutdown maneuver is considered, and for blade damage estimation, the blade frequency spectrum of the third and fourth modes of acceleration. For damage estimation, pairwise change signatures are extracted from the continuous wavelet transforms of the dynamic responses and their image distances are used for damage estimation. The performance of the method is studied based on simulated acceleration of the NREL 5 MW wind turbine using the aero-elastic design code FAST. The results indicate that the method provides robust damage estimates in presence of varying wind with the acceleration records of various sensory locations of the tower and blade. Furthermore, the tower damage estimates are shown to be insensitive to the damage level, whereas the blade damage estimates represent the damage level as well.

ACKNOWLEDGEMENTS

The authors are grateful to the makers of Wavelab from the Dept. of Statistics at Stanford University who have provided a valuable platform for producing some of the results in this paper. This research is supported in part by NSF IGERT (Grant No. 1068864) on Offshore Wind Energy Engineering, Environmental Science, and Policy.

REFERENCES

Avendano-Valencia, L., & Fassois, S. (2012). In-Operation Output-Only Identification of Wind Turbine Structural Dynamics: Comparison of Stationary and Non–

Stationary Approaches. In *ISMA Conference on Noise and Vibration Engineering.* Leuven, Belgium.

Buhl, M. (n.d.). *NWTC Design Codes.* http://wind.nrel.gov/designcodes/preprocessors/modes/. (Last modified 26-May-2005; accessed 24-January-2012)

Carden, E., & Fanning, P. (2004). Vibration based condition monitoring: a review. *Structural Health Monitoring, 3*(4), 355–377.

Chang, C., & Chen, L. (2004). Damage detection for a rectangular plate by spatial wavelet based approach. *Applied Acoustics, 65,* 819-832.

Danai, K., Civjan, S. A., & Styckiewicz, M. M. (2011). Direct method of damage localization for civil structures via shape comparison of dynamic response measurements. *Computers & Structures.*

Danai, K., & McCusker, J. R. (2009). Parameter estimation by parameter signature isolation in the time-scale domain. *ASME J. of Dynamic Systems, Measurement and Control, 131*(4), 041008.

Doebling, S. W., Farrar, C. R., Prime, M. B., & Shevitz, D. W. (1996). *Damage identification and health monitoring of structural and mechanical systems from changes in their vibration characteristics: a literature review* (Tech. Rep.). Los Alamos National Lab., NM (United States).

Farrar, C. R., & Doebling, S. W. (1997). An overview of modal-based damage identification methods. In *Proceedings of DAMAS Conference* (pp. 269–278).

Friswell, M. I. (2007). Damage identification using inverse methods. *Philosophical Transactions of the Royal Society A: Mathematical, Physical and Engineering Sciences, 365*(1851), 393–410.

Jonkman, J., & Buhl Jr, M. (2005). FAST users guide. *Rep. No. NREL/EL-500-38230, NREL, Golden, Colorado, USA.*

Jonkman, J., Butterfield, S., Musial, W., & Scott, G. (2009). Definition of a 5-MW reference wind turbine for offshore system development. *National Renewable Energy Laboratory, NREL/TP-500-38060.*

Kelly, N., & Jonkman, B. (n.d.). *NWTC Design Codes.* http://wind.nrel.gov/designcodes/preprocessors/turbsim/. (Last modified 03-February-2011; accessed 17-July-2012)

Loutridis, S., Douka, E., Hadjileontiadis, L., & Trochidis, A. (2005). A two-dimensional wavelet transform for detection of cracks in plates. *Engineering Structures, 27,* 1327–1338.

Mallat, S. (1998). *A wavelet tour of signal processing. 1998.* Academic, San Diego, CA.

Mallat, S., & Hwang, W. L. (1992). Singularity detection and processing with wavelets. *Information Theory, IEEE Transactions on, 38*(2), 617–643.

Ovanesova, A., & Suarez, L. (2004). Applications of wavelet transforms to damage detection in frame structures. *Engineering Structures, 26,* 3949.

Rucka, M., & Wilde, K. (2006). Application of continuous wavelet transform in vibration based damage detection method for beams and plates. *Journal of Sound and Vibration, 297,* 536–550.

Santos, J. A. dos, Maia, N., Soares, C. M., & Soares, C. M. (2008). Structural damage identification: a survey. *Trends in computational structures technology, Saxe-Coburg Publications, Stirlingshire, UK,* 1–24.

Wang, L., Zhang, Y., & Feng, J. (2005). On the Euclidean distance of images. *Pattern Analysis and Machine Intelligence, IEEE Transactions on, 27*(8), 1334–1339.

On-Line Fault Detection in Wind Turbine Transmission System using Adaptive Filter and Robust Statistical Features

Ruoyu Li[1] and Mark Frogley[2]

[1,2] *SKF USA Remote Diagnostics Center,SKF USA Inc., Houston, TX 77086*

ruoyu.li@skf.com
mark.frogley@skf.com

ABSTRACT

To reduce the maintenance cost, avoid catastrophic failure, and improve the wind transmission system reliability, online condition monitoring system is important. Developing effective online fault detection methodology is important. In this paper, an adaptive filtering technique is applied for enhancing the fault impulse signals-to-noise ratio in wind turbine gear transmission systems. Multiple statistical features designed to quantify the impulse signals of the processed signal are extracted for rotating machine fault detection. The multiple dimensional features are then transformed into one dimensional feature. A minimum error rate classifier will be designed based on the transformed one dimensional feature to identify the gear transmission system with defect. Vibration signals collected from wind turbines in the real operation will be used to demonstrate the effectiveness of the presented methodology.

Keywords: Adaptive filtering, Fault detection, Fault diagnosis, Condition monitoring, Gear transmission system, Statistical features, Pattern classification, Wind turbine transmission system.

1. INTRODUCTION

Wind Power is the world's fastest growing renewable energy source. With the developing and growing of wind power, reducing the cost of generating the wind energy becomes a critical issue. As wind turbines are often located in remote locations, the operation and maintenance costs are usually high. According to a survey on the failures of the Swedish wind turbines (Ribrant et al., 2007), more than 30% of the failures are mechanical failures. The repair costs of the mechanical failure are relatively high comparing to other failures such as sensor issue, and electric related issues (Hyers *et al.*, 2006) (Nilsson and Bertling, 2007). Thus deploying condition based monitoring (CBM) system could prevent catastrophic machine failure, improve the reliability and decrease the maintenance costs.

For wind turbine transmission system condition monitoring, different types of signals such as vibration, acoustic emission, temperature, oil debris, power performance and so on could be used (Han & Song, 2005), (Zhu *et al.*, 2012) , (Schlechtingen *et al.*, 2013), (Abouhnik & Albarbar, 2012) Among them, vibration signals are currently widely used technique for mechanical fault detection and diagnosis. In the real applications, one type of rotating mechanical fault, for instance bearing surface defect, gear tooth crack, chipped gear tooth, generate impulse signals (McInerny and Dai, 2003) (Wang, 2001) (Endo, H., and Randall, 2007). When these faults develop inside rotating machinery, each time the rotating components pass over the damage point, an impact force will be created. The impact force will cause a ringing of the support structure at the structural natural frequency. By effectively detecting those periodic impulse signals, one group of rotating machine faults could be detected. In real complex machines it is not always possible to place the sensors directly on the rotating components. Thus the fault impulse signals collected by the sensors installed at some distance away from the rotating components are usually relatively weak and buried in the background noise and other rotating components, such as shaft, blades, and gears and so on. Moreover, wind turbine transmission systems work under dynamic operating conditions. The changing of rotating speed introduces smearing effects to the Fourier spectrum (Wang and Heyns, 2011). This will further increase the difficulties in fault detection and diagnostics. Finding a way to increase the impulse signal-to-noise ratio (SNR) is important for wind turbine transmission system fault detection. Another difficulty in online monitoring of wind turbine system is how to analyze the data in a large volume efficiently. For example, a typical condition monitoring system for wind turbine mechanical transmission consists about six to eight sensors on different locations and for each sensor, there are about two to four measurement readings with different setups. Usually every 10 to 120 minutes, different key features values will be collected and sent back to the server for trend analysis purpose and every 8-12 hours, time waveforms and spectrums will be captured and send back to the server for detail analysis purpose. For a wind farm with size of 100 turbines, when all the operating conditions met the requirement of activating the condition monitoring system, there will be about 7200 to 23040 extracted trend values for just one key feature and about 2400 to 9600 time waveforms usually with 4096 or 8192 sampling points and spectrums with 1600 to 6400 frequency lines daily for analyzing. Let us assume that it will take approximately 20 seconds to 30 seconds for an experienced human expert to analyze the spectrum for machine fault detection purpose. That means it will take a human vibration expert about 13 hours to 80 hours to analyze all the daily collected signals. Thus develop an effective automatic way on analyzing the signals and making decision will be in great need.

In this paper, an online automatic fault detection methodology has been developed. The methodology uses the adaptive filtering technique for enhancing the fault impulse signals-to-noise ratio in wind turbine gear transmission systems. Then multiple statistical features designed to quantify the impulse signals of the processed signal are extracted for detecting one type of rotating machine faults which generate impulse signals. The multiple dimensional features are transformed into one dimensional feature. A minimum error rate classifier is then designed to identify the type of faults which generate periodic impulses signals. Wind turbine vibration signals collected from the real operation will be used to demonstrate the effectiveness of the presented methodology. The remainder of the paper is organized as follows. Section 2 explains theoretical basis of the methodology. The analysis results on vibration signal collected from wind turbine systems are provided in Section 3. Section 4 summarizes the work done and concludes the paper.

2. THEORETIC BASIS

2.1. Self-adaptive Noise Cancellation

Self-adaptive noise cancellation (ANC) (Widrow *et al.*, 2005) is used in this paper. Its structure is shown in Figure 1. The technique has been widely used in biological signal processing (Rahman *et al.*, 2011), (Inan *et al.*, 2010), (Thakor & Zhu, 1991) to remove the electrical interference, in the audio signal processing (Greenberg, 1998), (Sambur, 1978) to improve the quality of interested voice signal, in the non-destructive testing area (Zhu & Weight, 1994) to improve the SNR for the acoustic emission signal, and the vibration signal based rotating machine fault detection (Li & He, 2011) , (Antoni & Randall, 2004) , and (Bechhoefer *et al.*, 2009) to enhance the impulse fault SNR.

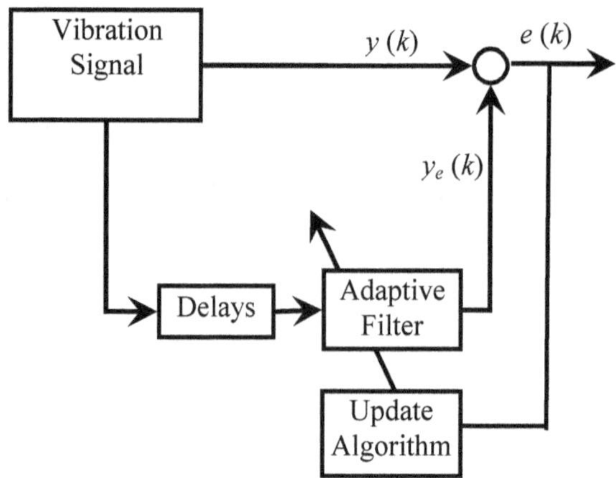

Figure 1. The scheme of the adaptive noise cancellation

The idea here is to use the adaptive filter to track the periodic components inside the signal and then remove it from the original vibration signal. The parameters of the filter are adaptively updated by using the error signal, expressed in Eq. (1). The way for calculating the optimal parameters of the prediction model is to minimize the mean squared error between the original signal and the predicted signal.

$$e_k = y(k) - y_e(k) \tag{1}$$

where $y(k)$ is the input vibration signal at time k, $y_e(k)$ is the estimated output signal, e_k is error between the estimated signal and the input signal at time k.

To adaptively adjust the coefficient of the prediction model, the least mean squared algorithm is applied to update the coefficients online. The update equations, presented in section 1 of the book (Benesty & Huang, 2003) are shown in Eq. (2) and Eq. (3).

$$y_e(n) = W_k^T X_k \tag{2}$$

$$W_{k+1} = W_k + 2\mu e_k X_k \tag{3}$$

where, $W_k = [w_1(k), w_2(k), \ldots, w_l(k)]$ is the parameters vector of the adaptive filter at time k, l is the length of the filter, $X_k = [y(k-\Delta), y(k-\Delta-1), \ldots y(k-\Delta-(l-1))]$ is the delayed version of input y(k), Δ is the delay time and is μ is the learning step size.

A simulation is used here for demonstration purpose. A sine waveform, impulses and Gaussian noise are used to compose the simulated signal. The adaptive noise cancellation was then applied to the simulated signal. Both the simulated signal and the processed signal are shown in Figure 2. In Figure 2, x-axis represents the time and y-axis represents the amplitude of the signal. The top red curve stands for the simulated signal. The blue curve stands for the processed signal. From Figure 2, one could easily see that the adaptive noise cancellation algorithm effectively remove the periodic component, the sine waveform.

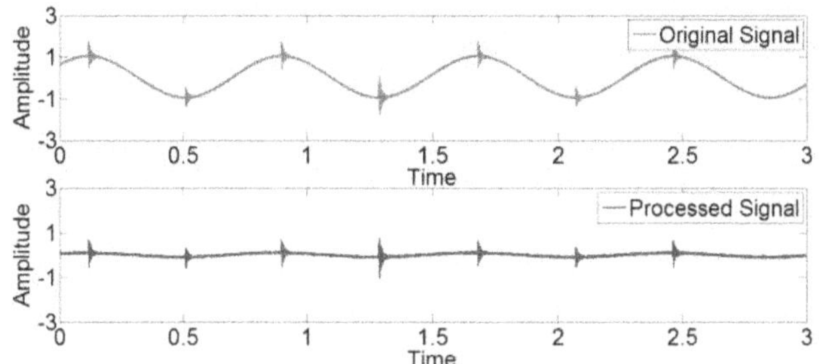

Figure 2. The simulated signal and the processed signal

2.2. Statistical Feature extraction

To design a classifier, quantification values are needed. As the purpose in this paper is for impulse signal detection, five statistical values commonly used for rotating machine fault detection and diagnosis (Lei *et al.*, 2008) (Samanta, *et al.*, 2001) (Medjaher *et al.*, 2012) , were used here to quantify the processed impulse signal enhanced signal. They are Kurtosis, Crest Factor, Root Mean Square (RMS), Impulse Factor, and Skewness (Norton *et al.*, 2003).

Kurtosis is the fourth statistical moment. It is a good indication of the "peakedness" of the signal. The Kurtosis is defined as,

$$x_{KT} = \frac{\frac{1}{n}\sum_{i=1}^{n}(x_i - \bar{x})^4}{\left(\frac{1}{n}\sum_{i=1}^{n}(x_i - \bar{x})^2\right)^2} \quad (4)$$

The Crest Factor is the ratio of the peak value to the RMS value. It is a good measurement of spikiness of a signal. The Crest Factor is defined as,

$$x_{CR} = \frac{\max|x_i|}{\sqrt{\frac{1}{n}\sum_{i=1}^{n}{x_i}^2}} \quad (5)$$

The RMS value is usually used to represent the overall energy of the signal. The RMS is defined as,

$$x_{rms} = \sqrt{\frac{1}{n}\sum_{i=1}^{n}{x_i}^2} \quad (6)$$

The Impulse factor is the ratio of the absolute peak value to the absolute mean value of the signal. It is sensitive for detection of bearing problem (Yiakopoulos *et al.*, 2011). The Impulse factor is defined as,

$$x_{IF} = \frac{\max|x_i|}{\frac{1}{n}\sum_{i=1}^{n}|x_i|} \quad (7)$$

The Skewness is a third statistical moment. Skewness is usually used to measure the symmetry of the signal. The Skewness is defined as,

$$x_{KT} = \frac{\frac{1}{n}\sum_{i=1}^{n}(x_i - \bar{x})^3}{\left(\sqrt{\frac{1}{n}\sum_{i=1}^{n}(x_i - \bar{x})^2}\right)^3} \quad (8)$$

In Eq. (4) to (8), x is the signal with n samples. $\bar{x}$ is the mean value of x .

2.3. Linear Discriminant Analysis

Linear discriminant analysis (LDA) (McLachlan, 2004) is a well known technology which is widely used for feature extraction and dimension reduction. Many successful applications are presented in various research papers (Swets and Weng, 1996) (Martinez and Kak, 2001) (Lu *et al.*, 2003). LDA projects the data sets with multiple classes into a lower-dimensional vector, which provides maximized separation between the classes. In this paper, as only two classes, health state and fault state are considered, multiple class LDA will not be discussed and only two classes LDA will be considered.

Let's assume that we have a set of d-dimensional n samples $s = [s_1, s_2, \ldots, s_n]$. N_1 of them belong to class C_1. N_2 of them belong to class C_2. If we define a transformation matrix w, then we could obtain a scalar y by projecting the s onto a line,

$$y = w^T s \quad (9)$$

y is a set of n samples $y = [y_1, y_2, \ldots, y_n]$. N_1 of them belong to class C_{T1}. N_2 of them belong to class C_{T2}.

To find the scalar which maximizes the separation between the two classes, the following objective function is adopted (McLachlan, 2004). By maximizing the objective function in Eq. (9), we could obtain the projecting which provides best separation between the two classes.

$$J(w) = \frac{|\tilde{\mu}_1 - \tilde{\mu}_2|^2}{\tilde{s}_1^2 + \tilde{s}_2^2} \quad (10)$$

Where

$$\tilde{\mu}_i = \frac{1}{N_i}\sum_{s \in C_{Ti}} w^T s = w^T \mu_i \text{ and}$$

$$\tilde{s}_i^2 = \sum_{y \in C_{Ti}} (y - \tilde{\mu}_i)^2, \ i=[1\ ,2]\ .$$

Through maximizing process, the transform matrix could be calculated as,

$$w = S_W^{-1}(\mu_1 - \mu_2) \quad (11)$$

Where

$$\mu_i = \frac{1}{N_i}\sum_{s\in C_i} w^T s \text{ , } i=[1\text{ , }2] \text{ and}$$

$$S_W = \sum_{i=1}^{2}\sum_{s\in C_i}(s-\mu_i)(s-\mu_i)^T$$

3. DATA COLLECTION AND EXPERIMENTAL RESULTS

Recently, a number of SKF IMx-W, WinCon units (SKF, 2011) were installed on a fleet of wind turbines to perform the online condition monitoring. The SKF IMx-W, WinCon unit is a modern industrial product specialized designed for on-line vibration monitoring. It has been successfully applied to online monitor the wind turbine mechanical faults all around the world. The key features of the system include (SKF, 2011):

(1) Lightning protection.

(2) Wall-mounted.

(3) Sixteen analogue inputs and two digital inputs.

(4) Simultaneous measurement of all channels.

(5) Multi-parameter gating.

The monitored wind turbine mechanical transmission systems consist of main bearing, gear transmission system and generator. A simplified transmission system with one planetary gear stage is shown in Figure 3. The monitoring system used two accelerometers on the main bearing, four on the gearbox, and two on the generator. The system also measured the rotating speed of the high speed shaft to provide the speed reference.

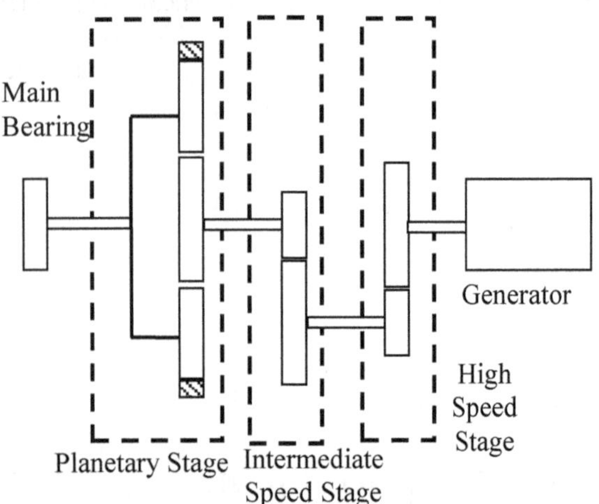

Figure 3. The simplified typical wind turbine transmission system

During our daily monitoring process, bearing defects on the intermediate shaft from four wind turbines have been captured. After the detection, the bearings have been scheduled to replace with the new bearing of the same type. In this paper, the acceleration signal collected from the intermediate shaft of four of the wind turbines with bearing issue will be investigated. The SKF series accelerometers were used for the data collection. This type of sensor has 100mv/g sensitivity and the frequency range is about 0.5 Hz to 10 kHz. The sampling frequency for the intermediate acceleration measurements were set to be 5120 Hz and about 1.6 seconds long signal was collected for each data collection. Routinely, the vibration signals will be collected every 8 hours during the normal daily collection.

For analysis purpose, 20 vibration signals before the replacement and after the replacement were randomly selected from the database of each turbine from the signals collected during the past 6 month. A typical vibration signal collected from a healthy gearbox is shown in Figure 4 and that collected from a gearbox with defect is shown in Figure 5. Figure 6 shows the picture of the damaged bearing inner race. The crack on the surface could be clearly seen from both figures.

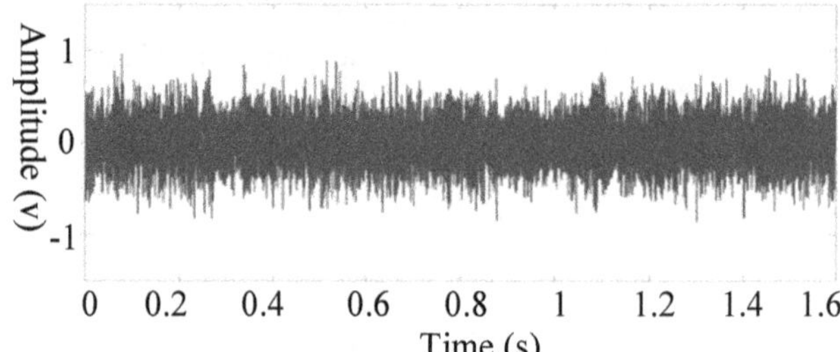

Figure 4. The vibration signal collected from a healthy gearbox

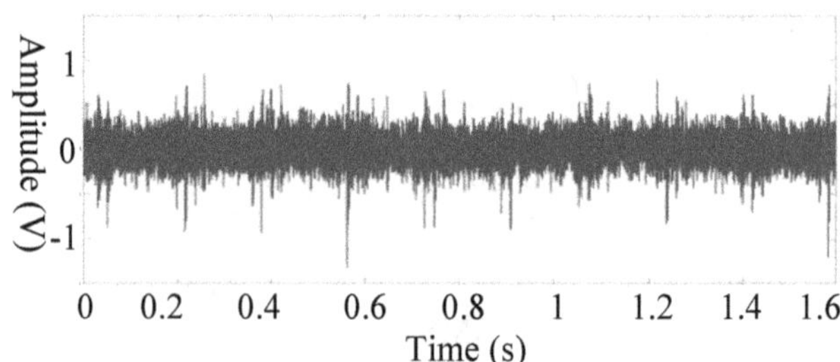

Figure 5. The vibration signal collected from a gearbox with bearing inner race defect

(a) (b)

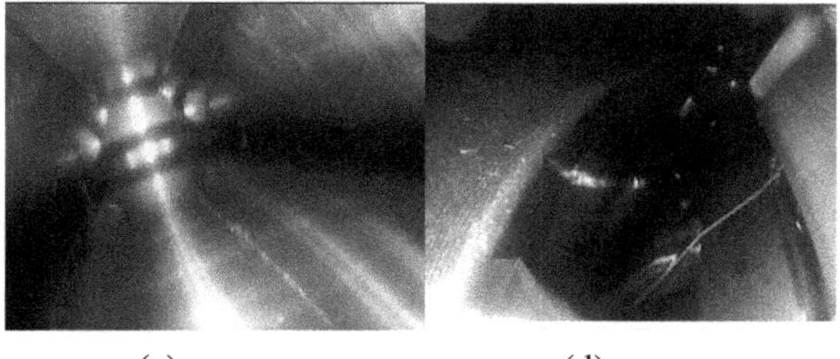

(c) (d)

Figure 6. (a) Damaged Bearing with Crack on the Whole Inner Race of Wind Turbine #1. (b) Damaged Bearing with Crack on the Surface the Inner Race of Wind Turbine #2. (c) Damaged with hair line Crack on the Surface of the Inner Race of Wind Turbine #3. (d) Damaged with Crack on the Surface of the Inner Race of Wind Turbine #4

The ANC was applied to process the vibration signals and sample results of the vibration from the healthy state and the damaged state are shown in Figure 7 and Figure 8.

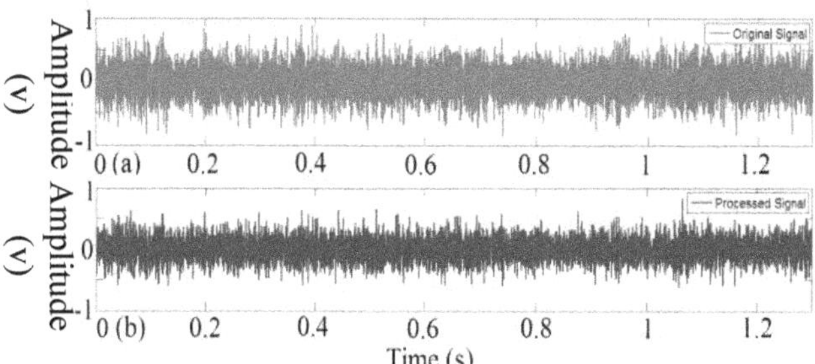

Figure 7. The processed results of the healthy gearbox

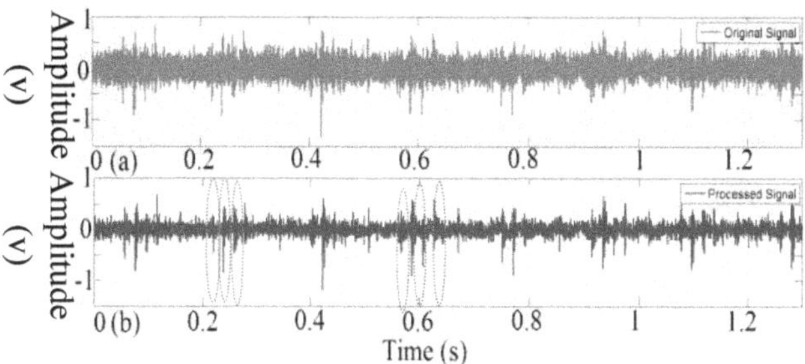

Figure 8. The processed results of the gearbox with bearing inner race defect

Comparing the results shown in Figure 7 (b) and Figure 8 (b), one could easily see the periodic fault impulses in Figure 8 (b). The periodic frequency of those impulses is related to the bearing inner race defect. This is a clear indication of there being a bearing inner race defect. In Figure 8 (a), one could see some of the impulse signals appearing in the original vibration signal, but due to the low impulse SNR, the analysis on the time waveform is hard to find the periodic behavior of the impulses. However, from the processed results shown in Figure 8 (b), the impulses signals (the red circled one) could be easily identified. By simple analysis on processed time wave form, the potential fault could be identified.

The comparison between the ANC and the widely used envelope analysis (McInerny and Dai, 2003) are investigated in this paper. The bandwidth of the band-pass filter of the envelope analysis is important and many research papers have presented ways to select the optimal bandwidth (Sawalhi *et al.*, 2007) (Eric *et al.*, 2011). However, this is beyond the research scope of this paper. In this paper, the bandwidth is set to between 500 Hz and 2 kHz. Ten vibration signals collected on the wind turbines with known bearing fault have been randomly selected. For demonstration purpose, the Fourier spectrums of both algorithms of one of the bearing fault signals were shown in Figure 9 (a) and (b). Figure 9 (a) shows the Fourier spectrum of the envelope analysis and Figure 9 (b) shows the Fourier spectrum of the ANC. From Figure 9 (a) and (b), the high peak at the bearing inner race defect frequency and its harmonics could be easily observed on both the spectrum. Both algorithms effectively capture the bearing defect.

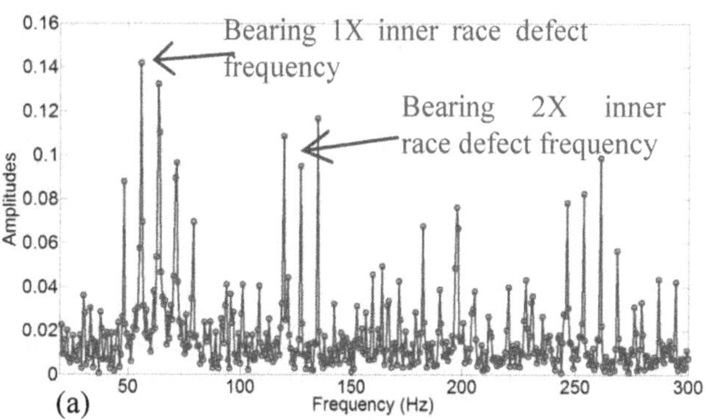

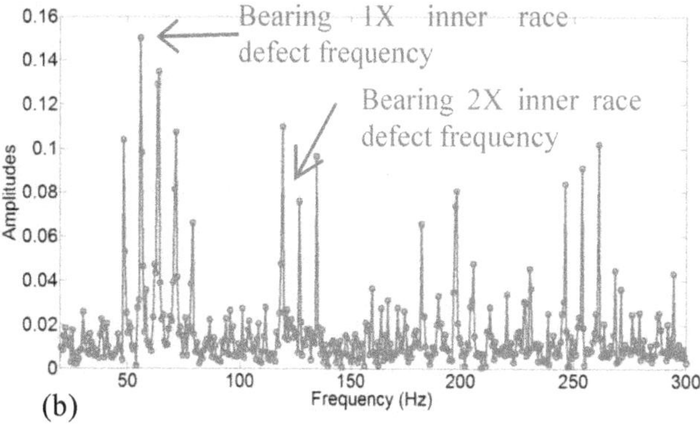

Figure 9. The spectrum of (a) the envelope analysis and (b) the adaptive noise cancellation algorithm

To quantify the performance of the two algorithms, the quantification value is calculated by the following equation.

$$I_q = \frac{\sum_{i=1}^{2} A_i}{S_{RMS}} \qquad (12)$$

where A_i is the amplitude of the ith harmonics of the bearing fault frequency of the spectrum and S_{RMS} is the RMS value of the spectrum.

In our case, as the bearing is with inner race defect, the inner race defect frequency of the bearing is used for calculation. The higher the I_q value is the better performance the algorithm has. The results of the I_q value are shown in Table 1. The percentage of the difference between I_q values is the ratio of the subtraction of I_q of the adaptive noise cancellation from that of the envelope analysis to the I_q of the envelope analysis. From the results shown in Table 1, one could see that adaptive noise cancellation algorithm has slightly higher I_q value of the traditional envelope analysis.

Table 1. The I_q value of the traditional envelope analysis and the adaptive noise cancellation

Healthy Gearbox			**Gearbox with Defect**			**Normalized Difference**
	Mean	**STD**		**Mean**	**STD**	
Kurtosis	2.9413	0.2080	**Kurtosis**	6.7379	1.8137	1.2908
Crest Factor	8.3849	1.7355	**Crest Factor**	12.8863	1.6221	0.5368
RMS	2.0668	1.0749	**RMS**	1.6640	0.6494	0.1949
Impulse Factor	10.5101	2.1233	**Impulse Factor**	17.6538	2.7323	0.6797
Skewness	-0.0181	0.0892	**Skewness**	-0.0545	0.1636	2.0110

The Kurtosis, Crest Factor, Impulse Factor, RMS, and Skewness of the processed signal were calculated for both the healthy vibration signal and the vibration signal with damaged components. The mean and the standard deviation value of the statistical values were shown in Table 2. Those feature values were then transformed into one dimensional feature vector by using the LDA. To quantify how well the individual statistical features, listed in Table 2, separate the healthy states from the fault states, the percentage of the difference between the two states are calculated. The percentage of the difference is the ratio of the absolute difference between the healthy states and the fault states to the feature value of the healthy states. The values are shown in Table 2. To obtain a general ideal on the computational requirement of the developed methodology, tests were conducted on a pc with a 2.60 GHz CPU and 3 GB RAM. By processing the 8192 point vibration signals, the processing time was approximately 0.5 seconds.

The histogram of the transformed feature of both the healthy machine and the damaged machine is shown in Figure 10. In Figure 10, the x-axis represents the amplitude of the transformed feature. The y-axis represents the numbers of the machine in the group. The purple group stands for the fault states and the red group stands for the healthy states.

Table 2. The calculated values of the processed signal of healthy gearbox and the gearbox with defect

Traditional Envelope	**Adaptive Noise Cancellation**	**Percentage of difference (%)**
16.33	16.36	0.16
15.50	16.30	5.18
12.93	14.68	13.57
11.92	12.91	8.29
16.15	17.94	11.05
11.54	13.21	14.48
17.72	19.29	8.85
17.27	18.00	4.22
14.51	15.30	5.46
13.50	14.27	5.71

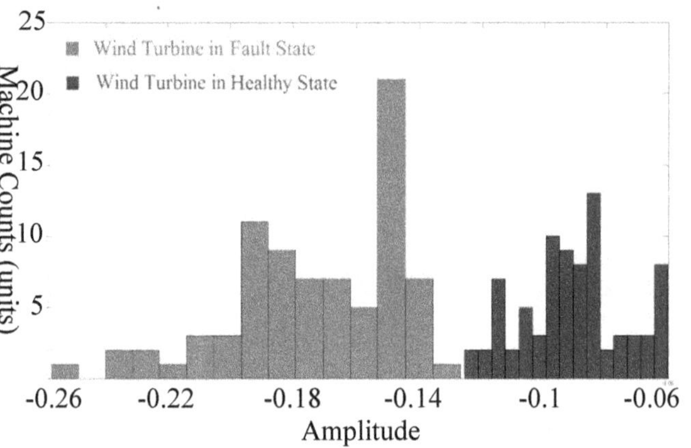

Figure 10. The histogram of the transformed feature of the healthy machine and the machine with damaged components

The decision boundary is then determined by the minimum error rate rule. As in this paper, only two classes, healthy state and damaged state are used, the prior probability determines how likely the unknown observation belongs to each state. For example, if we define the prior probability value to be 0.5. Then the probability of the unknown extracted feature belongs to healthy state is 0.5 and that of the damage state is 0.5. In this paper, three different boundaries were calculated by using three different prior probability values. They are 0.1, 0.5, and 0.9. The results are shown in Figure 11. In Figure 11, the x-axis represents the number of the samples. The y-axis represents the amplitude of the transformed feature. Four different shapes are used to represent different wind turbines. The stars, circles, diamonds, and squares represent wind turbine #1, wind turbine #2, wind turbine #3, and wind turbine #4, respectively. The color of the shapes represents the state of the wind turbine. The red represents the wind turbine is in healthy state. The purple represents the wind turbine in fault state. The blue dash line represents the decision boundary for 0.1. The value of the boundary is -0.070. The green dot line represents the decision boundary for 0.5. The value of the boundary is -0.081. The pink solid line represents the decision boundary for 0.9. The value of the boundary is -0.088. From Figure 11, one could see that the boundary for 0.1 is the most fault sensitive boundary among the three boundaries which identified several healthy states as damaged state. The boundary for 0.9 is the most fault insensitive boundary among the three boundaries which misclassified several damaged states as healthy state.

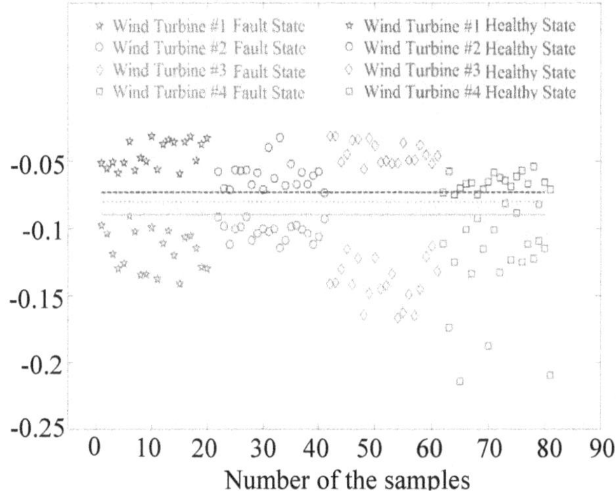

Figure 11. The transformed feature and the decision boundaries with different prior statistics (blue line: P(Healthy)=0.1, green line: P(Healthy)=0.5,pink line: P(Healthy)=0.9)

The confusion matrix of three decision boundaries is shown in Table 3. In real online condition monitoring applications with large volume of data, determination of the decision boundary is important. Fault sensitive boundary will potentially cause more false alarms while fault insensitive boundary tends to miss the true faults. Thus determination of the boundaries in real applications is based on the following rule. When the system is deployed to a newly created wind farm, the less fault sensitive boundary should be used. With aging of the wind farm, the boundary will be adjusted to more fault sensitive one.

Table 3. The confusion matrix of the different boundaries

		Calculated Healthy	Calculated Fault
0.1 Boundary	True Healthy	75	5
	True Fault	0	80
0.5 Boundary	True Healthy	80	0
	True Fault	1	79
0.9 Boundary	True Healthy	80	0
	True Fault	2	78

From Table 3, one could see that 96.875%, 99.375%, and 98.750% of accuracy are achieved for 0.1, 0.5, and 0.9 boundaries, respectively.

4. CONCLUSIONS

In real applications, effective fault detection algorithms are an essential part of the condition monitoring system, especially for the online continuous monitoring systems, like wind turbine condition monitoring systems. In this paper, an effective automatic fault detection methodology has been developed. The methodology uses adaptive filtering technique to improve the fault SNR and LDA to reduce the features' dimensions. Wind turbine vibration signals obtained in real operation were used to demonstrate the effectiveness of the presented methodology.

Currently, the developed methodology will continue testing on the vibration signals of wind turbines with different designs. In future work, it will be interesting to study the effectiveness of the developed method on a variety of rotating equipments in different industries and applications, such as ranging arms in the mining industries, pumps in the paper mill, and motors in the food industries, and so on.

REFERENCES

Abouhnik, A., and Albarbar, A., "Wind turbine blades condition assessment based on vibration measurements and the level of an empirically decomposed feature", *Energy Conversion and Management*, vol. 64, pp. 606-613, 2012.

Antoni, J., and Randall, B., "Unsupervised noise cancellation for vibration signals: part I--evaluation of adaptive algorithms," *Mechanical Systems and Signal Processing*, vol. 18, pp. 89-101, 2004.

Bechhoefer, E., Li, R., and He, D., "Quantification of condition indicator performance on a split torque gearbox", *Proceedings of the 2009 AHS Forum*, Grapevine, TX, May 27-29, 2009.

Bechhoefer, E., Menon, P., and Kingsley, M., "Bearing envelope analysis window selection Using spectral kurtosis techniques", In *2011 IEEE Conference on*

Prognostics and Health Management (PHM), pp. 1-6, 2011.

Benesty, J., and Huang, Y., *Adaptive signal processing: applications to real-world problems*: Springer, 2003.

Greenberg, J. E., "Modified LMS algorithms for speech processing with an adaptive noise canceller," *IEEE Transactions on Speech and Audio Processing*, vol. 6, pp. 338-351, 1998.

Han, Y., and Song Y.H., "Condition monitoring techniques for electrical equipment—a literature survey", *IEEE Transactions on Power Delivery*, vol. 18, pp, 4-13, 2003

Inan, O. T., Etemadi, M., Widrow, B., and Kovacs, G., "Adaptive cancellation of floor vibrations in standing ballistocardiogram measurements using a seismic sensor as a noise reference," *IEEE Transactions on Biomedical Engineering*, vol. 57, pp. 722-727, 2010.

Lei, Y., He, Z., and Zi, Y. "A new approach to intelligent fault diagnosis of rotating machinery", *Expert Systems with Applications*, vol. 35, No. 4, 1593-1600, 2008.

Li, R., and He, D., "Development of an Advanced Narrowband Interference Cancellation Method for Gearbox Fault Detection," in *AHS 67th Annual Forum and Technology Display*, Virginia Beach, Virginia, 2011.

Lu, J., Plataniotis, K. N., and Venetsanopoulos, A. N., Face recognition using LDA-based algorithms., *IEEE Transactions on Neural Networks*, vol. 14, No.1, pp. 195-200, 2003.

Martinez A., and Kak A., "PCA versus LDA", *IEEE Transactions on Pattern Analysis and Machine Intelligence*, vol. 23, no. 2, pp. 228-233, 2001

McInerny, S. A., and Dai, Y., "Basic vibration signal processing for bearing fault detection", *IEEE Transactions on Education*, vol. 46, No. 1, 149-156, 2003.

McLachlan, G. J., *Discriminant analysis and statistical pattern recognition*,vol. 544, Wiley-Interscience, 2004

Medjaher K., Camci F., and Zerhouni N., "Feature Extraction and Evaluation for Health Assessment and Failure Prognostics", In *First European Conference of the Prognostics and Health Management Society.*, pp. 111-116. 2012.

Norton, M. P., & Karczub, D. G., *Fundamentals of noise and vibration analysis for engineers*. Cambridge university press. 2003

Rahman, M. Z. U. , Rafi, A.S., and Rama, K.R., "Efficient sign based normalized adaptive filtering techniques for cancellation of artifacts in ECG signals: Application to wireless biotelemetry," *Signal Processing*, vol. 91, pp. 225-239, 2011.

Ribrant, J., and Bertling, L.M., "Survey of Failures in Wind Power Systems With Focus on Swedish Wind Power Plants During 1997–2005," *Energy Conversion, IEEE Transactions on* , vol.22, no.1, pp.167,173, 2007

Samanta, B., Al-Balushi, K. R., and Al-Araimi, S. A., "Artificial neural networks and support vector machines with genetic algorithm for bearing fault detection, *Engineering Applications of Artificial Intelligence*, vol. 16, No. 7, pp. 657-665, 2003

Sambur, M., "Adaptive noise canceling for speech signals," *IEEE Transactions on Acoustics, Speech and Signal Processing*, vol. 26, pp. 419-423, 1978.

Sawalhi, N., Randall, R. B., and Endo, H., "The enhancement of fault detection and diagnosis in rolling element bearings using minimum entropy deconvolution combined with spectral kurtosis", *Mechanical Systems and Signal Processing*, vol. 21, No. 6, pp. 2616-2633., 2007.

Schlechtingen M., Santos I.F., and Achiche S., "Wind turbine condition monitoring based on SCADA data using normal behavior models. Part 1: system description", *Applied soft Computing*, vol. 13, pp. 259-270, 2013.

SKF, SKF Multilog On-line System IMx-W, ***www.SKF.com,2011***

Swets, D. and Weng, J., "Using discriminant eigenfeatuers for image retrieval", *IEEE Transactions on Pattern Analysis and Machine Intelligence*, vol. 18, no. 8, pp. 831-836, 1996.

Thakor, N. V. and Zhu Y., "Applications of adaptive filtering to ECG analysis: noise cancellation and arrhythmia detection," *IEEE Transactions on Biomedical Engineering*, vol. 38, pp. 785-794, 1991.

Widrow, B., Glover, J., McCool, J., Kaunitz, J., Williams, C.S., Hearn, R.H., Zeidler, J.H., Dong, E., and Goodlin, R.C., "Adaptive noise cancelling: Principles and applications," *Proceedings of the IEEE*, vol. 63, pp. 1692-1716, 2005.

Yiakopoulos, C.T., Gryllias, K.C., and Antoniadis, I.A., "Rolling element bearing fault detection in industrial enviroments based on a k-mean clustering approach", *Expert Systems with Applications*, vol. 38, No. 3, pp. 2888-2911, 2011

Zhu J., He D., and Bechhoefer E., "Survey of lubrication oil condition monitoring, diagnostics, and prognostics techniques", *Proceedings of the 2012 Conference of he Society for Machinery Failure Prevention Technology*, pp. 193-212, Dayton, OH, April 24-26, 2012.

Zhu, Y., and Weight J.P., "Ultrasonic nondestructive evaluation of highly scattering materials using adaptive filtering and detection," *IEEE Transactions on Ultrasonics, Ferroelectrics and Frequency Control*, vol. 41, pp. 26-33, 1994.

Wang, W., "Early detection of gear tooth cracking using the resonance demodulation technique", *Mechanical Systems and Signal Processing*, vol. 15, No. 5, pp. 887-903, 2001

Endo, H., and Randall, R. B., "Enhancement of autoregressive model based gear tooth fault detection technique by the use of minimum entropy deconvolution filter", *Mechanical Systems and Signal Processing,* vol. 21, No. 2, pp. 906-919, 2007

Wang, K., and Heyns, P. S., "The combined use of order tracking techniques for enhanced Fourier analysis of order components", *Mechanical systems and signal processing*, vol. 25, No. 3, pp. 803-811, 2011.

Hyers, R. W., McGowan, J. G., Sullivan, K. L., Manwell, J. F., and Syrett, B. C., "Condition monitoring and prognosis of utility scale wind turbines", *Energy Materials: Materials Science and Engineering for Energy Systems*, vol. 1, No. 3, pp. 187-203, 2006.

Nilsson, J., and Bertling, L., " Maintenance management of wind power systems using condition monitoring systems—life cycle cost analysis for two case studies", *IEEE Transactions on Energy Conversion*, vol. 22, No. 1, pp. 223-229, 2007.

BIOGRAPHIES

Ruoyu Li received the B.S. degree in automatic control and M.S. degree in control theory and control engineering from Guilin University of Electronic Technology, Guilin, China, in 2002 and 2005, respectively, and the Ph.D. degree from the University of Illinois at Chicago, Chicago, IL.

He is currently a reliability engineer with the Remote Diagnostic Center, SKF USA Inc., Houston, TX. His research interests include statistical signal processing, machine learning, data mining, rotating machinery fault diagnostics and prognostics, sensor networks, and mechatronics systems design.

Mark Frogley has worked in the field of rotating machinery condition monitoring and asset preventive maintenance since 1993. He has more than 15 years' experience on setting up and operating large predictive maintenance programs. Worked with SKF for more than 15 years, he is currently a technology integration manager with SKF USA Inc. Lansdale, PA, USA.

Lubrication Oil Condition Monitoring and Remaining Useful Life Prediction with Particle Filtering

Junda Zhu[1], Jae M. Yoon[1], David He[1], Yongzhi Qu[1], and Eric Bechhoefer[2]

[1]*Department of Mechanical and Industrial Engineering, The University of Illinois at Chicago, Chicago, IL, 60607, USA*
davidhe@uic.edu

[2]*NRG Systems, Hinesburg, VT, 05461, USA*
erb@nrgsystems.com

ABSTRACT

In order to reduce the costs of wind energy, it is necessary to improve the wind turbine availability and reduce the operational and maintenance costs. The reliability and availability of a functioning wind turbine depend largely on the protective properties of the lubrication oil for its drive train subassemblies such as the gearbox and means for lubrication oil condition monitoring and degradation detection. The wind industry currently uses lubrication oil analysis for detecting gearbox and bearing wear but cannot detect the functional failures of the lubrication oils. The main purpose of lubrication oil condition monitoring and degradation detection is to determine whether the oils have deteriorated to such a degree that they no longer fulfill their functions. This paper describes a research on developing online lubrication oil condition monitoring and remaining useful life prediction using particle filtering technique and commercially available online sensors. It first introduces the lubrication oil condition monitoring and degradation detection for wind turbines. Viscosity and dielectric constant are selected as the performance parameters to model the degradation of lubricants. In particular, the lubricant performance evaluation and remaining useful life prediction of degraded lubrication oil with viscosity and dielectric constant data using particle filtering are presented. A simulation study based on lab verified models is provided to demonstrate the effectiveness of the developed technique.

1. INTRODUCTION

Lubrication oil is an important information source for early machine failure detection just like the role of the human blood sample testing in order to perform disease detection. In modern industries, lubrication oil plays a critical part in condition maintenance of complicated machineries such as wind turbines. In recent years, health condition monitoring and prognostics of lubrication oil has become a significant topic among academia and industry. Significant effort has been put into oil diagnostic and prognostic system development and research. In comparison with vibration based machine health monitoring techniques, lubrication oil condition monitoring provides approximately 10 times earlier warnings for machine malfunction and failure (Poley, 2012). The purpose of most research is, by means of monitoring the oil degradation process, to provide early warning of machine failure and most importantly extend the operational duration of lubrication oil in order to reduce the frequency of oil changes and therefore reduce maintenance costs.

For the wind industry, in order to reduce wind energy costs, there is a pressing need to improve the wind turbine availability and reduce the operational and maintenance costs. The reliability and availability of a functioning wind turbine depends largely on the protective properties of the lubrication oil for its drive train subassemblies such as gearbox and means for lubrication oil condition monitoring and degradation detection. The wind industry mostly uses offsite lubrication oil analysis. The lubrication oil in the wind turbine is normally sampled every 6 months and sent to oil analysis labs for feedback on the condition of the oil. However, the online health monitoring of functional failures of lubrication oil has been an issue that cannot be handled by such techniques and remains to be an unsolved problem. The purpose of lubrication oil condition monitoring and degradation detection is to determine whether the oil has deteriorated to such a degree that it no longer fulfills its protective function and to provide early warning of the possibility of total failure. As stated by Sharman and Gandhi (2008), and many other researchers, the primary function of lubrication oil is to provide a continuous layer of film between surfaces in relative motion to reduce friction and

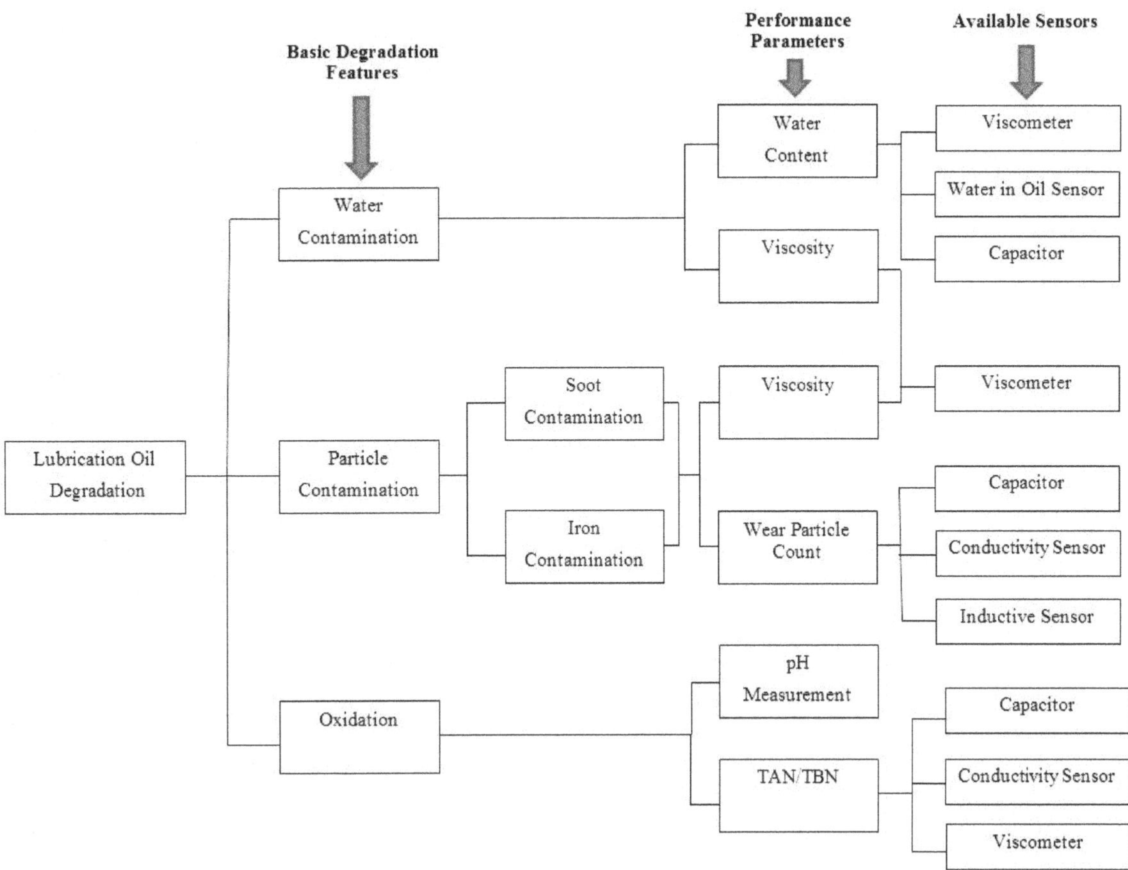

Figure 1. The relationship among the basic degradation features, performance parameters, and available oil condition sensors

prevent wear, and thereby, prevent seizure of the mating parts.

The secondary function is to cool the working parts, protect metal surfaces against corrosion, flush away or prevent ingress of contaminants and keep the mating component reasonably free of deposits. In a lubricated system, variation in physical, chemical, electrical (magnetic) and optical properties change the characteristics of the lubrication oil and lead to the degradation as its protective properties. The main causes of turbine lubricant deterioration are oxidation, particle contamination, and water contamination. These three are defined in this paper as lubrication oil basic degradation features. The parameters that describe the lubrication oil performance or level of degradation are called performance parameters. These parameters include viscosity, water content, total acid number (TAN), total base number (TBN), particle counting, pH value and so forth. Each performance parameter can be measured by certain sensing techniques. The relationship among the basic degradation features, performance parameters, and available oil condition sensors is shown in Fig. 1. Also, Table 1 shows the performance parameters for different kinds of applications and their benchmark for lubrication oil degradation. For example, for water content, it measures the water contamination percentage of the lubrication oil. This performance parameter is necessary and crucial to gearbox, hydraulic system, engine, compressor and turbine applications. Water content can be measured by a capacitance sensor, viscosity sensor, and water in oil sensor.

To find a feasible solution for online lubrication oil health condition monitoring and remaining useful life (RUL) prediction, it is necessary to conduct a comprehensive review of the current oil health monitoring techniques. The investigation on current state of the art lubrication oil monitoring techniques is reported in (Zhu *et al.*, 2012; 2013). Over the years, scientists and experts have developed sensors and systems to monitor one or more of the lubrication oil performance parameters in order to monitor the oil condition effectively. These sensors and systems can be summarized into four categories including electrical (magnetic), physical, chemical, and optical techniques. For example, the most effective electrical technique for oil health monitoring is detecting the dielectric constant change of the lubrication oil. According to recent studies, the

Performance Parameters	Measurement Function	Unit	Benchmark of Degradation	Applications					Available Measurement Approach
				Gear box	Hydraulic system	Engine	Compressor	Turbine	
Viscosity (40 °C)	Contamination of lubricant by some other oil, oxidation	Cst (mm^2/s)	≥ 55 ≤ 50	yes	yes	yes	yes	yes	Kinetic Viscometer
Viscosity (100 °C)			≥ 10 ≤ 8						Micro-acoustic Viscometer
Water Content	Presence of water	%	≤ 2	yes	yes	yes	yes	yes	Capacitance sensor (Dielectric constant) Kinetic Viscometer Water in oil sensor
TAN/TBN	Acidity/alkalinity of lubricant (oxidation level)	mgKOH/gm	≥ 0.6 ≤ 0.05	yes	yes	yes	yes	yes	Capacitance sensor (Dielectric constant) Kinetic Viscometer Conductivity Sensor
Flash point	Presence of dissolved solvents or gases in the lubricant	°C	≥ 220 ≤ 140	no	yes	yes	no	no	Thermometer
Wear Particle Count	Wear particles in parts per million	ppm	≤ 40	yes	yes	yes	yes	yes	Capacitance sensor (Dielectric constant), Kinetic Viscometer, Conductivity Sensor, Inductive Sensor
Particle Counting	Detect number of particles for sample size of 100cc	mg/L	≤ 200	no	yes	no	no	yes	

Table 1. Performance parameters, applications and their benchmark for lubrication oil degradation.

capacitance or permittivity change can be used to monitor the oxidation, water contamination, and wear particle concentration. On the other hand, for physical techniques, viscosity is commonly discussed. The lubrication oil oxidation, water contamination, particle concentration, and some other property changes all have an influence on oil viscosity. Therefore, viscosity is considered an objective mean of oil degradation detection. The final goal of all above mentioned systems is to achieve lubrication oil online health monitoring and remaining useful life prediction in industrial machineries. Note, that most sensing systems are only capable of off-line monitoring, in which oil samples are collected from the machinery by specialists and sent to laboratories for oil condition analysis. In this way, the actual condition of the lubrication oil cannot be determined online because of the sampling and analysis delay. With the deployment of online oil condition monitoring techniques, one can optimize the maintenance schedule and reduce the maintenance costs.

In this paper, based on the previous results of a comprehensive investigation of oil condition monitoring techniques reported in (Zhu *et al.*, 2012; 2013), the two most effective online lubrication oil sensors, kinematic viscometer and dielectric constant sensors, are selected to develop an online lubrication oil health monitoring and remaining useful life prediction tool. Kinematic viscosity is the absolute viscosity with respect to liquid density while dielectric constant is the relative permittivity between the lubrication oil and air.

The purpose of this paper is to present the development of an online lubrication oil condition monitoring and remaining useful life prediction technique based on a particle filtering algorithm and commercially available online sensors. This technique is developed by integrating lubrication oil degradation physical models with the particle filtering algorithm. The physical models are used to simulate the deterioration process of the lubrication oil due to water contamination in terms of the kinematic viscosity and dielectric constant. A simulation case study based on lab verified models is used to demonstrate the effectiveness of the technique.

In this paper, a particle filtering algorithm is utilized as RUL prediction tool. For oil condition monitoring, an effective and accurate state estimation tool will be beneficial to reduce machine downtime. An on-line RUL estimator includes two stages: state estimation and RUL prediction. First, in the state estimation stage, even though there are many state estimation techniques, Kalman filter and particle filter are the most utilized ones. However, Kalman filter requires many assumptions such as: 1) zero-mean Gaussian process noise, 2) zero-mean Gaussian observation noise, 3) Gaussian posterior probability density function (pdf), etc. Because nonlinear Kalman filter is linearization based technique, if the system nonlinearity grows, any of linearization (either local or statistical linearization) methods breaks down (Merwe *et al*., 2000). Second, in RUL estimation stage, particle filtering can handle statistic prediction data unlike the other methods (parameter estimation). As a result, particle filtering algorithm provides feasible solutions for a wide range of RUL predication applications. A particle filtering algorithm integrated with physics based oil degradation models will provide a basis to develop practically feasible tools for accurate RUL prediction of lubrication oil.

The remainder of the paper is organized as follows. Section 2 is focused on the development of an online lubrication oil health monitoring and remaining useful life prediction tool using a kinematic viscometer and dielectric constant sensor. In this section, physical models that simulate the kinematic viscosity and dielectric constant as a function of the water contamination level and temperature are presented. The validation of the physical models using the experimental data is performed. The developed physical models are then integrated into a particle filtering framework to develop the lubrication oil remaining useful life prediction tool. The developed tool is then illustrated with a simulation case study based on lab verified models. Finally, Section 3 concludes the paper.

2. Development of Lubrication Oil RUL Prediction Tool

2.1. Lubrication Oil Deterioration Models Due To Water Contamination

In this section, physical models that represent lubrication oil deterioration due to water contamination in terms of viscosity and dielectric constant are presented. Since experimental approach has certain disadvantages including limited degradation scenario coverage, long pre-installation training time and unavoidable test errors, the physical models can be built aiming at eliminating all those shortcomings and provide an accurate/ideal sensor output which reflects the actual health status of the lubricant oil. Water contamination is selected as a representative basic degradation feature. Using the physical models, given any temperature and a water contamination ratio, one could simulate the kinematic viscometer and the capacitance sensor outputs with maximum accuracy.

2.1.1. Kinematic Viscosity

Define:

T = temperature, in Celsius

$V_{oil,T}$ = viscosity of the healthy oil at temperature T, in Cst

$V_{water,T}$ = viscosity of the water at temperature T, in Cst

P = water volume percentage

According to Stachowiak and Batchelor (2005), water and oil mixture viscosity at a certain temperature $V_{M,T}$ can be computed as:

$$V_{M,T} = \left(V_{oil,T} - V_{water,T}\right) \times (1 - P) + V_{water,T} \quad (1)$$

where:

$$V_{water,T} = -0.451 \times \ln T + 2.3591 \quad (2)$$

Note that in Eq. (1), $V_{oil,T}$ is defined as the healthy lubrication oil information and is extracted from our initial test while $V_{water,T}$ is defined as the water physical attribute which can be considered known factors. Based on Equation (1), we can compute the degree of oil degradation as the result of water contamination in terms of viscosity as: $DD_{viscosity} = \frac{V_{M,T}}{V_{oil,T}}$.

Equation (1) represents the kinematic viscosity of the degraded oil as a function of temperature and water contamination ratio.

2.1.2. Dielectric Constant

Define:

$\varepsilon_{oil,T}$ = dielectric constant of healthy oil at temperature T

$\varepsilon_{water,T}$ = dielectric constant of water at temperature T

According to Jakoby and Vellekoop (2004), the dielectric constant of water and oil mixture at a certain temperature $\varepsilon_{M,T}$ can be computed as:

$$\varepsilon_{M,T} = \varepsilon_{oil,T} \times \left(1 + 3 \times P \times \frac{\varepsilon_{water,T} - \varepsilon_{oil,T}}{\varepsilon_{water,T} + 2 \times \varepsilon_{oil,T} - P \times \left(\varepsilon_{water,T} - \varepsilon_{oil,T}\right)}\right) \quad (3)$$

where:

$$\varepsilon_{water,T} = 80 - 0.4 \times ((T + 273) - 293) \quad (4)$$

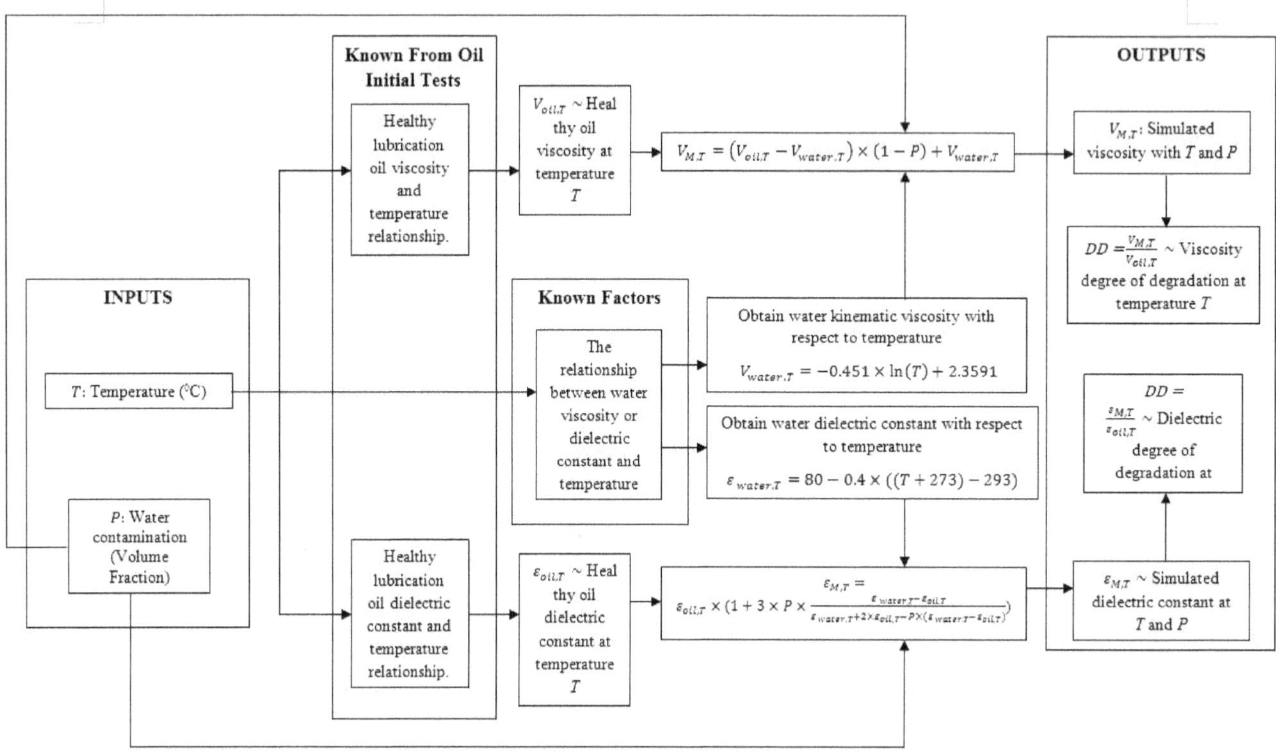

Figure 2. Lubrication oil water contamination simulation model for viscosity and dielectric constant

Note that in Eq. (3), $\varepsilon_{oil,T}$ is defined as the healthy lubrication oil information and is extracted from our initial test while $\varepsilon_{water,T}$ is defined as the water physical attribute which can be considered known factors. Based on Equation (3), we can compute the degree of oil degradation as the result of water contamination in terms of dielectric constant as: $DD_{dielectric\ constant} = \frac{\varepsilon_{M,T}}{\varepsilon_{oil,T}}$.

Equation (3) represents the dielectric constant of the degraded oil as a function of temperature and water contamination ratio.

The simulation application of the lubrication oil deterioration model due to water contamination in terms of viscosity and dielectric constant can be summarized in Fig. 2. The simulation input is the temperature and water contamination ratio. The simulation output is the degraded oil kinematic viscosity and dielectric constant. Using the simulation application, one could generate a series of viscosity and dielectric constant values accordingly to reflect the true status of the lubrication oil.Experimental Validation of the Physical Models

2.1.3. Experimental Setup

In this section, the experiment setup using both capacitance and viscosity sensors are presented. In order to obtain the viscosity and the dielectric constant data, VISCOpro2000 from Cambridge Viscosity Inc. and Oil quality sensor from GILL Sensor were used. For the kinematic viscometer, the sensor output data with a RS232 port and was connected to Window PC through a RS232 and USB converter. The software interface on the PC was HyperTerminal that comes with Microsoft Windows XP. The viscometer involves a piston that dipped into the test lubricant and the coils inside the sensor body magnetically force the piston back and forth a predetermined distance. By alternatively powering the coils with a constant force, the round trip travel time of piston is measured. An increase in viscosity is sensed as a slowed piston travel time. The time required for the piston to complete a two way cycle is an accurate measure of viscosity. The practical unit of viscosity is centipoises (Cp), which is identical to the MKS unit mPa s (The viscosity of water is approximately 1 Cp). The viscosity sensor and its data acquisition system are shown in Fig.3. As we programmed according to the use manual that comes with the sensor. The sensor will send out analogue output including absolute viscosity, temperature compensated viscosity and the according temperature along with the date and time.

The dielectric constant sensor from Gill Sensor Inc. measures the capacitance of the test liquid then calculate the dielectric constant by the equation D=Coil/Cair, which is the capacitance of the test liquid divided by the capacitance of air, then output a voltage accordingly. The output analog signal was captured by LabJack U12 which was the data acquisition unit for the sensor and the voltage signal was recorded with Logger and Scope, software that comes with the U12.

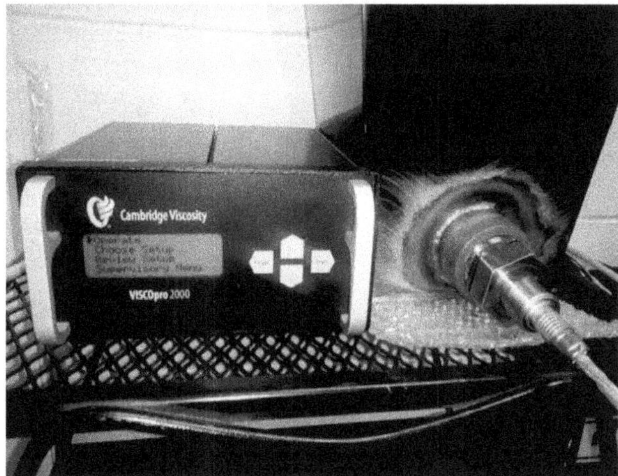

Figure 3. Viscometer and its data acquisition system

The dielectric constant sensor and its data acquisition system along with the entire experiment setup are shown in Fig. 4 and 5.

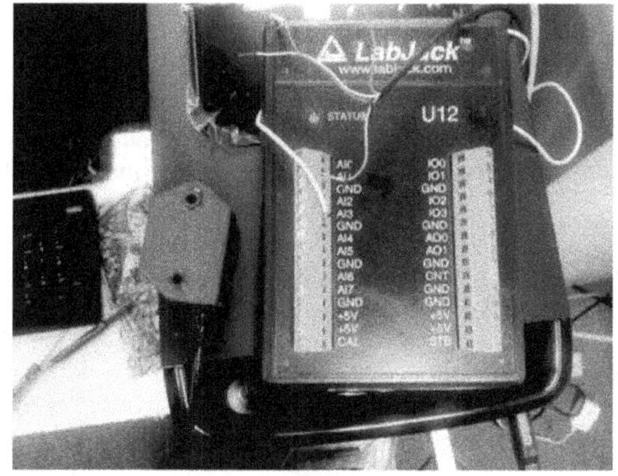

Figure 4. Dielectric constant sensor and the LabJack U12 data acquisition system

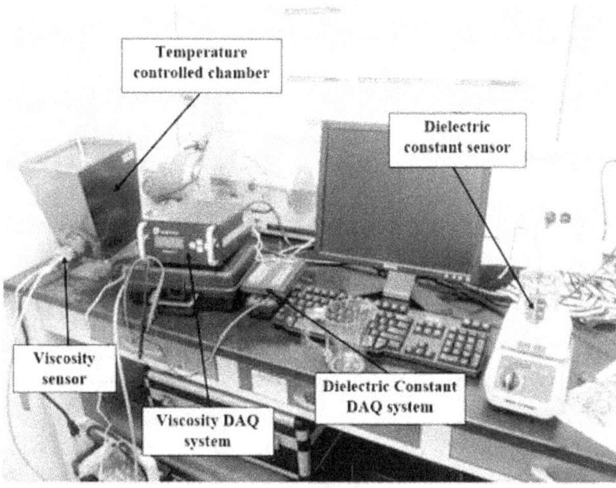

Figure 5. Experimental setup

Also needed for the tests are temperature control units. For the Dielectric constant test, we used a Hotplate from Thermo Scientific. It is a ceramic hotplate with temperature control and digital indication of temperature on the contact surface. However, since the viscometer had to be installed with sensor side facing up, we installed the sensor on a steel container and heated the oil inside with a liquid heater. In both situations, the test oils were preserved in a temperature controlled container and heated from around 25 to approximately 60 degrees Celsius. Instant temperatures were recorded along with the according viscosity and dielectric constant.

2.1.4. The Validation Results

In order to validate the physical models, viscometer and dielectric constant sensor readings under different water contamination levels with varying temperatures were compared with those computed from the physical models under the same conditions.

During the experiment, Castrol SAE 15W-20 lubrication oil was selected to perform the physical model validation. The healthy SAE 15W-20 lubrication oil kinematic viscosity in relation with temperature was obtained from the experimental tests as following:

$$V_{oil,T} = 57470.5189 \times T^{-1.935}; \quad (5)$$

Also, the healthy SAE 15W-20 lubrication oil dielectric constant in relationship with temperature was obtained from the experimental tests as following:

$$\varepsilon_{oil,T} = 4.90028 \times T^{-0.121}; \quad (6)$$

Fig. 6, 7, 8, and 9 show the plots of the kinematic viscosity obtained from the experiments and the physical models at water contamination level of 0.5%, 1%, 2%, and 3%, respectively. 40 data points were used to validate the viscosity physical model.

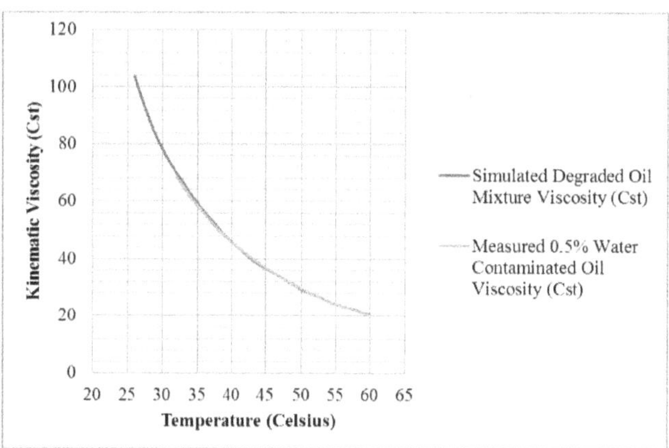

Figure 6. Kinematic viscosity comparison between simulated 0.5% water contaminated oil and measured 0.5% water contaminated oil

Judging from the kinematic viscosity curves, the experiment result validated the simulation result. For a fixed water contamination level, as temperature increases the viscosity drops, the measured viscosity variation follows the pattern of the simulated kinematic viscosity curves.

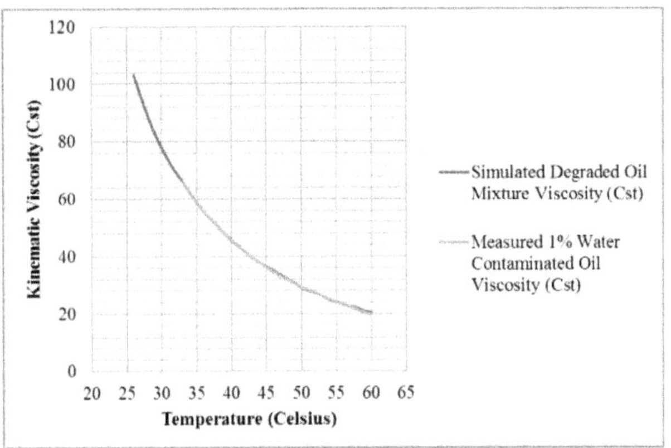

Figure 7. Kinematic viscosity comparison between simulated 1% water contaminated oil and measured 1% water contaminated oil

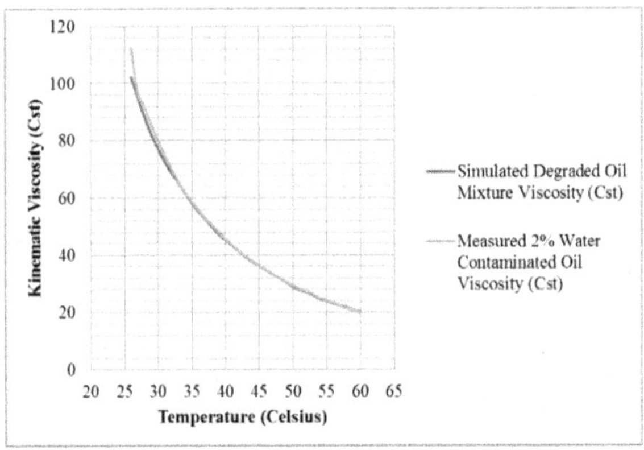

Figure 8. Kinematic viscosity comparison between simulated 2% water contaminated oil and measured 2% water contaminated oil.

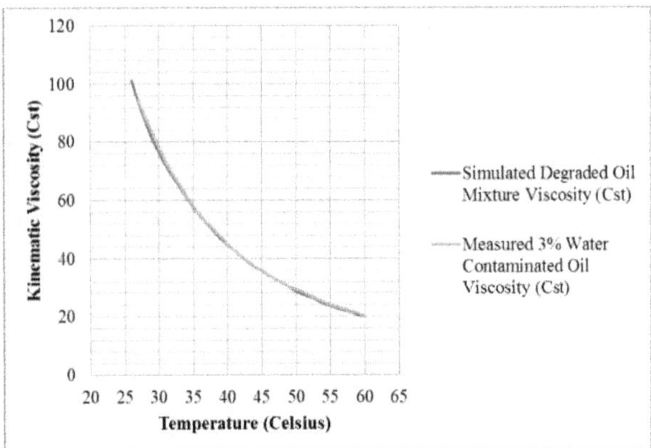

Figure 9. Kinematic viscosity comparison between simulated 3% water contaminated oil and measured 3% water contaminated oil

Fig.10 shows the plots of the dielectric constant obtained from the experiments and the physical models at water contamination level of 0.5%. 40 data points were used to validate the dielectric constant physical model. Similar to the case of kinematic viscosity, the experiment result validated the simulation result. For a fixed water contamination level, as temperature increases the dielectric constant increases, the dielectric constant variation follows the pattern of the simulated dielectric constant curves. The dielectric constant physical model has been validated by Jakoby and Vellekoop (2004) for lubrication oil applications.

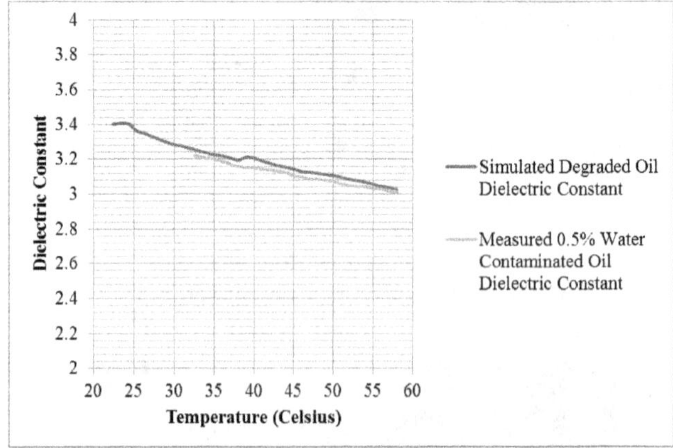

Figure 10. Dielectric constant comparison between simulated 0.5% water contaminated oil and measured 0.5% water contamination oil

2.2. Lubrication Oil RUL Prediction Using Particle Filters

Using particle filter for RUL prediction is a recent development in combining both physics based and data driven approaches for prognostics (He *et al.*, 2012). Applications of particle filters to prognostics have been reported in the literature, for example, remaining useful life predication of a mechanical component subject to fatigue crack growth (Zio and Peloni, 2011), online failure prognosis of UH-60 planetary carrier plate subject to axial crack growth (Orchard and Vachtsevanos, 2011), degradation prediction of a thermal processing unit in semiconductor manufacturing (Butler and Ringwood, 2010), and prediction of lithium-ion battery capacity depletion (Saha *et al.*, 2009). The reported application results have shown that particle filters represent a potentially powerful prognostics tool due to its capability in handling nonlinear dynamic systems and non-Gaussian noises using efficient sequential importance sampling to approximate the future state probability distributions. Particle filters were developed as an effective online state estimation tool (see Doucet *et al.*, 2000; Arulampalam *et al.*, 2002). In this paper, an integrated approach using particle filters for lubrication oil RUL prediction is presented.

2.2.1. Particle Filter for State Estimation

Applying particle filters to state estimation will be discussed first. Particle filters are used to estimate the state of a dynamic system using state and observation parameters. The state transition function represents the degradation in time of the lubrication oil. The observation or measurement represents the relationship between the degradation state of the lubrication oil and the health monitoring sensor outputs.

To apply particle filtering method, state estimation problem should be formulated first as stated by Yoon (2012). The problem of state estimation (a.k.a. filtering) is to estimate the dynamic state in terms of the posterior probability density function (pdf), based on all available information, including the sequence of measurements up to the current time step *k*. Let us introduce $x_k \in \mathbb{R}^{nx} and\ z_k \in \mathbb{R}^{nz}$ which represent system state vector and observation (or measurement) vector at the current time k respectively, where nx and nz are the dimension of the corresponding state vector and observation vector; $\mathbb{R}$ is a set of real numbers; $k \in \mathbb{N}$ is the time index; and $\mathbb{N}$ is the set of natural numbers. Consider the following discrete-time hidden Markov model (a.k.a state transition and observation model):

$$\boldsymbol{X_k}|(\boldsymbol{X_{k-1}} = x_{k-1}) \triangleq p(x_k|x_{k-1}) \quad (7)$$

$$\boldsymbol{Z_k}|(\boldsymbol{X_k} = x_k) \triangleq p(z_k|x_k) \quad (8)$$

where $\boldsymbol{X_k} = \{x_0, x_1, \dots, x_k\}$ is the sequence of the system state up to time $k \in \mathbb{N}$, and $\boldsymbol{Z_k} = \{z_0, z_1, \dots, z_k\}$ is the sequence of observation that is available up to current time k. Note that the above notation $\boldsymbol{X_k}$ is sometimes represented as $x_{0:k}$. Also, the state transition and the state observation models can be rewritten in functional form as follows:

$$x_k = f_{k-1}(x_{k-1}, v_{k-1}) \quad (9)$$

$$z_k = h_k(x_k, w_k) \quad (10)$$

where $v_k \in \mathbb{R}^{nv}$ and $w_k \in \mathbb{R}^{nw}$ denote the process noise and measurement noise at time k respectively; v_k and w_k are white noise; the initial state distribution $p(x_0) \triangleq p(x_0|z_0)$ is assumed known. Note that the state transition function is a mathematical representation of the lubrication oil degradation in time. Also, the observation model represents the health monitoring sensor outputs indicating the degradation state of the lubrication oil.

Then, the marginal pdf of the state can be recursively obtained in two steps: prediction and update. In the prediction step, suppose the state estimate at the time $k-1$ $p(x_{k-1}|\boldsymbol{Z_{k-1}})$ is known. Then the prediction (or prior) pdf of the state is obtained involving the system model via the *Chapman-Kolmogorov* equation as:

$$p(x_k|\boldsymbol{Z_{k-1}}) = \int p(x_k|x_{k-1})p(x_{k-1}|\boldsymbol{Z_{k-1}})dx_{k-1} \quad (11)$$

In the update step, the new measurement z_k becomes available and the posterior pdf can be obtained via the *Bayes* rule as follows:

$$p(x_k|Z_k) = \frac{p(z_k|x_k)p(x_k|\boldsymbol{Z_{k-1}})}{p(z_k|\boldsymbol{Z_{k-1}})} \quad (12)$$

where the normalizing constant is:

$$\mathrm{p}(\mathrm{z_k}|\mathbf{Z_{k-1}}) = \int p(x_k|\boldsymbol{Z_{k-1}})\, p(z_k|x_k)dx_k \quad (13)$$

The above obtained recursive propagation of the posterior pdf is a conceptual solution; it cannot analytically determined.

In any state estimation problem, based on the desired accuracy and processing time, a wide variety of tracking algorithms can be utilized. Especially, particle filter (a kind of suboptimal filter) increases accuracy while minimizing assumptions on the dynamic and measurement models. Due to its general disposition, particle filter became widely used in various filed. In the particle filter process, the marginal posterior density at time *k* can be approximated as follows:

$$p(x_k|\boldsymbol{Z_k}) \approx \sum_{i=1}^{N} w_k^i \delta(x_k - x_k^i) \quad (14)$$

where $\{\mathrm{x}_k^i, w_k^i\}_{i=1}^N$ represents the random measure of the posterior pdf $p(x_k|Z_k)$; $\{x_k^i, i = 1, \dots, N\}$ is a set of support points with associated weights $\{w_k^i, i = 1, \dots, N\}$; $\delta(\cdot)$ is a Dirac delta function; and sum of weights $\sum_i w_k^i = 1$. Since we are not able to directly sample from the posterior $p(x_k|\boldsymbol{Z_k})$ itself, associated weights w_k^i are computed by introducing importance density $q(x_k|\boldsymbol{Z_k})$ which is chosen easily sample from (normally transitional prior is used):

$$w_k^i \propto \frac{p(x_k^i|\boldsymbol{Z_k})}{q(x_k^i|\boldsymbol{Z_k})} \quad (15)$$

Thus, the desired posterior and weight update can be factorized in recursive forms as:

$$p(x_k|\boldsymbol{Z_k}) \propto p(z_k|x_k)p(x_k|x_{k-1})p(x_{k-1}|\boldsymbol{Z_{k-1}}) \quad (16)$$

$$\mathrm{w}_k^i = w_{k-1}^i \frac{p(z_k|x_k^i)p(x_k^i|x_{k-1}^i)}{q(x_k^i|x_{k-1}^i, z_k)} \quad (17)$$

Note that, after the weights are obtained via (11), weight normalization is required ($\sum_i w_k^i \neq 1$) to satisfy the nature of probability density function ($\sum_i w_k^i = 1$) as follows:

$$\mathrm{w}_k^i = \frac{w_k^i}{\sum_j w_k^j} \quad (18)$$

It can be shown that $\lim_{\mathrm{N}\to\infty}\{\mathrm{x}_k^i, w_k^i\}_{i=1}^{\mathrm{N}} = p(x_k|Z_k)$.

2.2.2. Particle Filter for RUL Prediction

In order to apply particle filter to estimate the remaining useful life (RUL), an l-step ahead estimator is required. An l-step ahead estimator will provide a long term prediction of the state pdf $p(x_{k+l}|\boldsymbol{Z_k}), for\ l = 1, \dots, T-k$, where T is the time horizon of interest (i.e. time of failure). In making an l-step ahead prediction, it is necessary to assume that no information is available for estimating the likelihood of the state following the future l-step path $\boldsymbol{X}_{k+1:k+l}$, that is, future measurements $z_{k+l}, for\ l = 1, \dots, T-k$ cannot be used for updating the prediction. In other word, the desired state pdf of particular future time $p(x_{k+l}|\boldsymbol{Z_k})$ can be factorized with the current posterior pdf $p(x_k|\boldsymbol{Z_k})$ to desired $p(x_{k+l}|\boldsymbol{Z_k})$ and the state transition function $p(x_k|x_{k-1})$ as $\prod_{j=k+1}^{k+l} p(x_j|x_{j-1})$. By combining Eq. (7) and (10), an unbiased l-step ahead estimator can be obtained as stated by Zio and Peloni (2011), as well as Orchard and Vachtsevanos (2009).

$$p(x_{k+l}|\boldsymbol{Z_k}) = \int \dots \int \prod_{j=k+1}^{k+l} p(x_j|x_{j-1})\, p(x_k|\boldsymbol{Z_k}) \prod_{j=k}^{k+l-1} dx_j \qquad (19)$$

Despite the fact that an unbiased estimator provides the minimum variance estimation, solving equation (13) can be either difficult or computationally expensive. Thus, a sampling based approximation procedure of the l-step-ahead estimator is provided by Zio and Peloni (2011).

Assume that the state $x_{t=k}$ represents the particle contamination level at the current time k, the particle contamination level increases by time and RUL is the object's remaining usable time before it fails (or needs maintenances). If an l-step-ahead state from the time k (i.e. $x_{t=k+l}$) goes across a pre-specified critical value λ (i.e. $x_{t=k+l} \geq \lambda$), the object's RUL at the time k can be computed as $\text{RUL}_k = (k+l) - k = l$. At each time step before its failure (i.e. $t \leq k+l$), the state $x_{t \leq k+l}$ would be projected up to the future time of failure $t = k+l$. In this manner, estimating $\text{RUL} \leq l$ is equivalent to estimating $x_{k+l} \geq \lambda$, rewriting as:

$$\hat{p}(RUL \leq l|\boldsymbol{Z_k}) = \hat{p}(\boldsymbol{X}_{k+l} \geq \lambda|\boldsymbol{Z_k}) \qquad (20)$$

When RUL (l-step-ahead prediction) is implemented using particle filter as stated by He, *et al.* (2012) corresponding weights are computed by introducing an estimated measurement $\widehat{z_{k+n}}$ according to Eq. (10) (i.e. measurement model) as:

$$\widehat{\boldsymbol{z}_{k+n}} \sim h_{k+n}(\widehat{x_{k+n}}) \qquad (21)$$

where n is a future time step $0 < n \leq l$. Then, the updating process is accomplished by Eq. (12) and (13). While RUL is computed, no measurement errors for the estimated measurements $\widehat{z_{k+n}}$ are considered. Note that the actual system has not been altered. Zero measurement errors are only applied in order to predict l-step-ahead state $\hat{p}(\boldsymbol{X}_{k+l} \geq \lambda|\boldsymbol{Z_k})$ because the future observation values are never accessible. In this paper, an integrated prognostic technique using the l-step-ahead RUL estimating particle filter is exploited.

2.3. Simulation Case Study

In order to validate and demonstrate the effectiveness of the particle filter technique based lubrication oil RUL prediction approach, a simulation case study was conducted. In this simulation case study, a scenario of lubrication oil deterioration due to water contamination was simulated with the physical models presented in Sections 2.1.1 and 2.1.2. In this scenario, a temperature template was used to simulate a daily temperature variation of the wind turbine as shown in Figure 11.

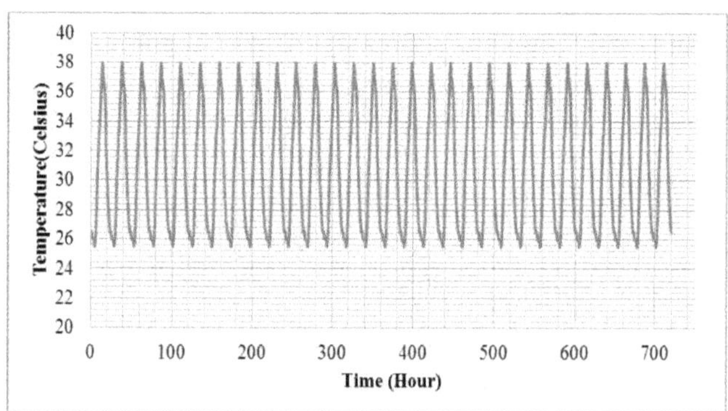

Figure 11. Temperature variation template

The other aspects of the simulation were defined as follows:

1. The deterioration state of the lubrication oil was defined as the water contamination level P.
2. The viscometer and dielectric constant sensor outputs were defined as observation data.
3. The lubrication oil deterioration process was simulated for 30 days (720 hours).
4. At the end of the simulation, the water contamination level P reached at 5%.
5. The sampling time interval was set to be every hour.
6. The failure threshold was set as 3% which was defined as the industry water contamination level limit.
7. At approximately the 525th hour, the water contamination level reached 3%.

Fig. 12 shows the water contamination propagation over 720 hours during the simulation with the given temperature.

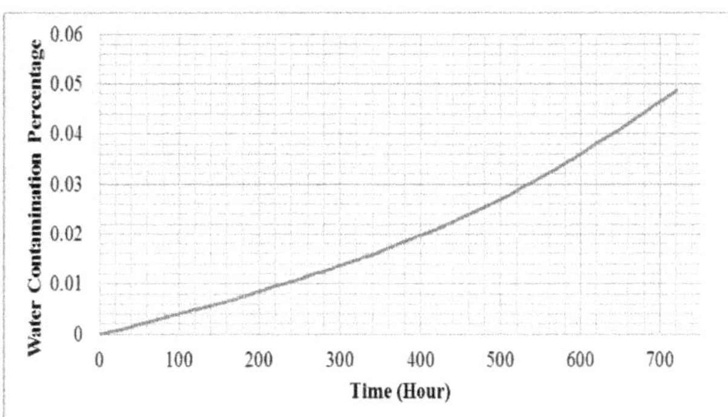

Figure 12. Water contamination propagation template

2.3.1. Particle Filter Implementation for RUL Prediction

To implement a particle filter for the RUL prediction of the lubrication oil in the simulation case study, the state transition function was defined as Eq. (22). It is generated as progression of the state of interest which in our case is the water contamination.

$$X_{k+1} = 1.0017 \times X_k + Random(0,1) \times 0.00007 \tag{22}$$

Two observation functions could be established using kinematic viscosity and dielectric constant physical models as Eq. (23) and (24).

Note that Eq. (23) is the observation function expressed in terms of kinematic viscosity and Eq. (24) the observation function expressed in terms of dielectric constant.

Generalized observation function could be established by combining kinematic viscosity and dielectric constant as Eq. (25).

$$Z_k = (5740.5189 \times T_k^{-1.935} + 0.451 \times \ln T_k - 2.3591) \times (1 - X_k) - 0.451 \times \ln T_k + 2.3591 \tag{23}$$

$$Z_k = 4.90028 \times T_k^{-0.121} \times (1 + 3 \times X_k \times \frac{-0.4 \times T_k + 88 - 4.90028 \times T_k^{-0.121}}{-0.4 \times T_k + 88 + 9.80056 \times T_k^{-0.121} - X_k \times (-0.4 \times T_k - 4.90028 \times T_k^{-0.121})}) \quad \ldots\ldots (24)$$

$$Z_k = \begin{bmatrix} Viscosity_k \\ DC_k \end{bmatrix} = \begin{bmatrix} h_1(X_k, T_k) \\ h_2(X_k, T_k) \end{bmatrix} = \begin{bmatrix} h_1(X_k, T_k) \\ h_2(X_k, T_k) \end{bmatrix} = \begin{bmatrix} (57470.5189 \times T_k^{-1.935} + 0.451 \times \ln \mathrm{T_k} - 2.3591) \times (1 - X_k) - 0.451 \times \ln \mathrm{T_k} + 2.3591 \\ 4.90028 \times T_k^{-0.121} \times (1 + 3 \times X_k \times \frac{-0.4 \times T_k + 88 - 4.90028 \times T_k^{-0.121}}{-0.4 \times T_k + 88 + 9.80056 \times T_k^{-0.121} - X_k \times (-0.4 \times T_k - 4.90028 \times T^{-0.121})}) \end{bmatrix} \tag{25}$$

In the implementation of the particle filter, number of particles was fixed as 50 and the prediction started at time point 425th hour during the simulation with l being 100 time steps. The reason for selecting 50 particle populations is to balance accuracy and processing time. The particle population impact will be discussed in the Section 2.4.2.

In order to reduce observation data fluctuation and RUL prediction variation, a temperature compensation module was integrated into the physical models. With a reference to 30 degree Celsius, which was the median temperature of the operating condition over a 24 hours cycle, the observation data was adjusted according to viscosity or dielectric constant functions with respect to the temperature. For example, at a certain temperature, the temperature compensated viscosity was the true value of the viscosity plus the theoretical viscosity difference between 30°C and current temperature. The compensated value can be obtained from the following equations:

$$\varepsilon_{compensate,T} = \varepsilon_T + (\varepsilon_{30}' + \varepsilon_T')$$

$$= \varepsilon_T + (\varepsilon_{30}' - (0.0001529 \times T^2 - 0.02241 \times T + 3.901)); \tag{26}$$

$$\varepsilon_T' = 0.0001529 \times T^2 - 0.02241 \times T + 3.901; \tag{27}$$

$$V_{compensate,T} = V_T + (V_{30}' + V_T')$$

$$= V_T + (V_{30}' - (0.21565 \times T^2 - 18.225 \times T + 431.5)); \tag{28}$$

$$V_T' = 0.21565 \times T^2 - 18.225 \times T + 431.5; \tag{29}$$

Fig. 13 and Fig. 14 present the observation variation before the temperature compensation.

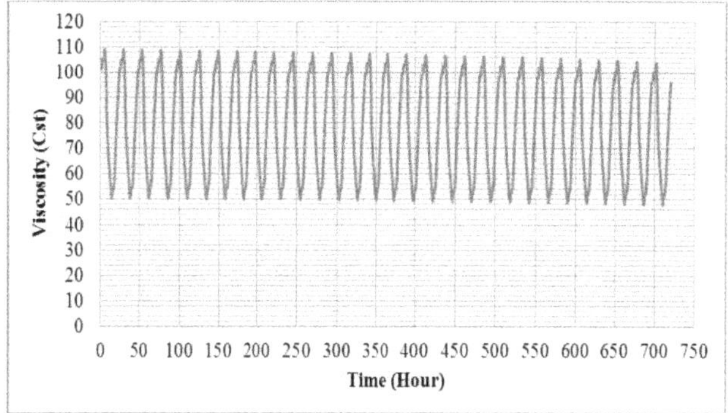

Figure 13. Observation data (kinematic viscosity) fluctuation before temperature compensation

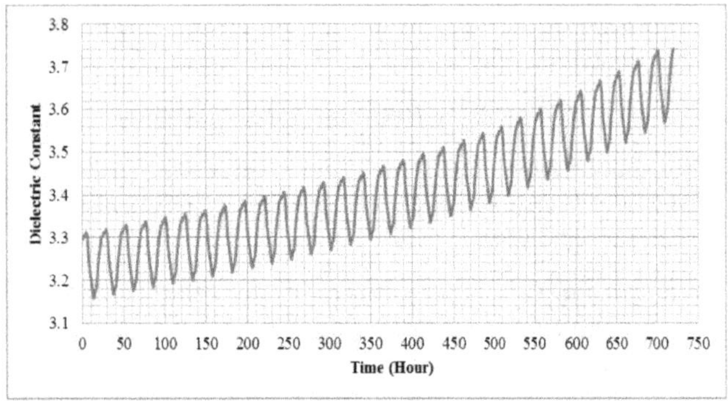

Figure 14. Observation data (dielectric constant) fluctuation before temperature compensation

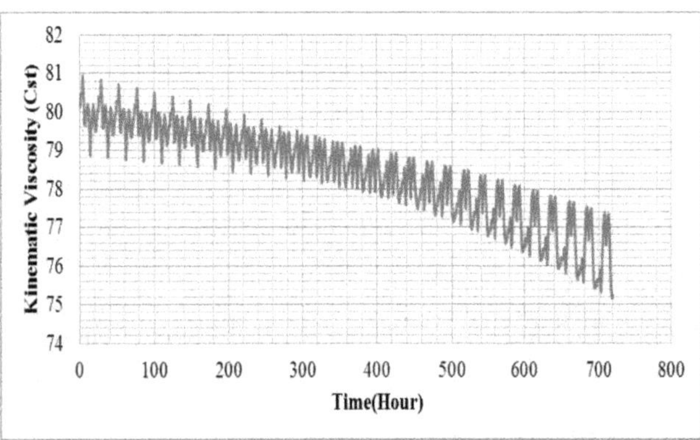

Figure 15. Observation data (kinematic viscosity) fluctuation after temperature compensation

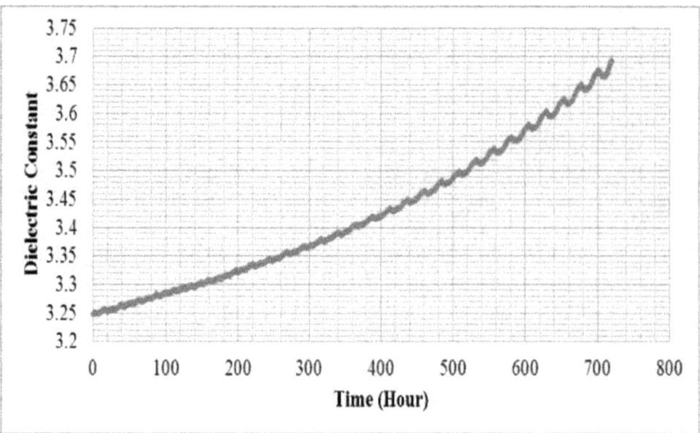

Figure 16. Observation data (dielectric constant) fluctuation after temperature compensation

Fig. 15 and Fig. 16 present the observation data variation after the temperature compensation.

In comparison of Fig. 13 with Fig.15, and Fig. 14 with Fig.16, it is obvious that the observation data fluctuation is greatly reduced after the temperature compensation and the data are ready for RUL prediction. Figure 17 summarizes the implementation of particle filter technique for lubrication oil RUL prediction.

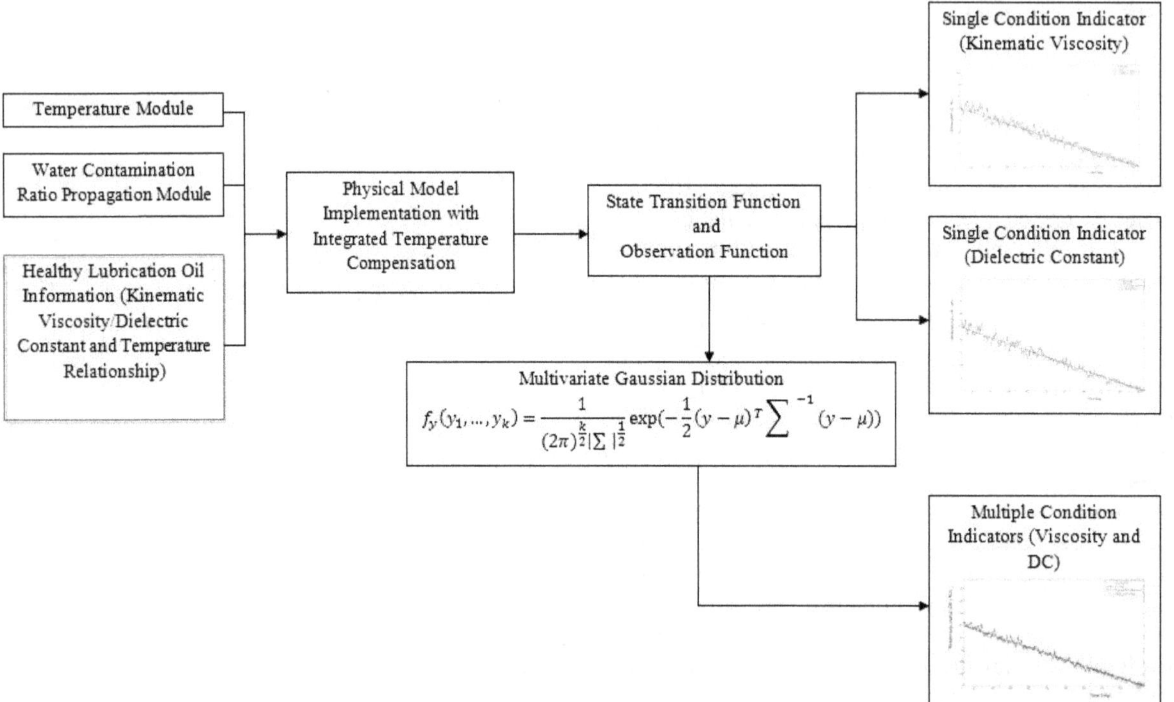

Figure 17. Particle filtering technique implementation.

2.3.2. RUL Prediction Results Using one Sensor Observation

Using the particle filter technique, RUL of the lubrication oil was predicted with either the viscosity or dielectric constant sensor observation. The prediction results are provided in Fig. 18 and Fig. 19, respectively. The x axis represents the true simulation time step. The y axis represents the time steps until failure. The blue line is the true remaining useful life and the red dots are our prediction mean while the vertical red bars are the 90% confidence intervals. From Fig. 18 and Fig. 19, one can see that with a certain degree of fluctuation at the beginning, the prediction becomes more and more accurate towards the end for both predictions. For a comparison purpose, the RUL prediction results with 200 particles are provided in Fig. 20. As one can observe, using the same dielectric constant sensor observation under the same condition, a larger particle population provide better accuracy. However, larger particle population requires more processing times. The relationship between particle population and processing times is shown in Table 2.

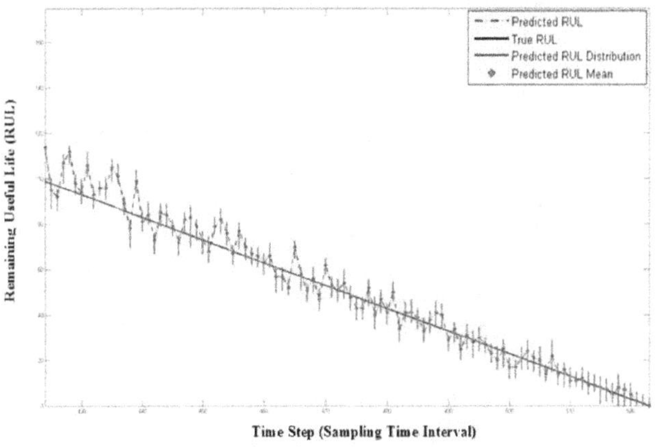

Figure 18. RUL prediction with only kinematic viscosity observation data (particle population=50)

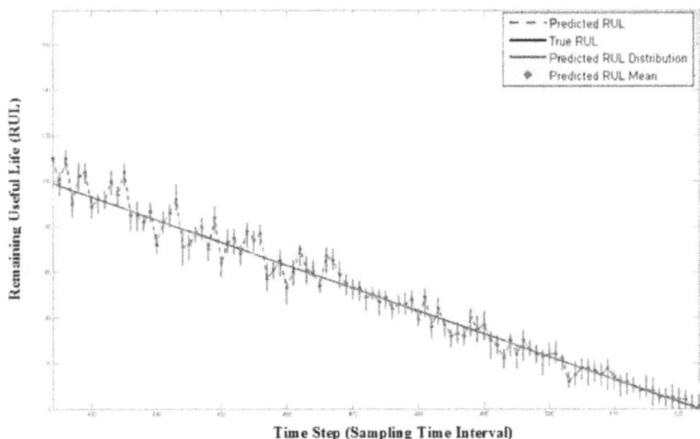

Figure 19. RUL prediction with only dielectric constant observation data (particle population=50)

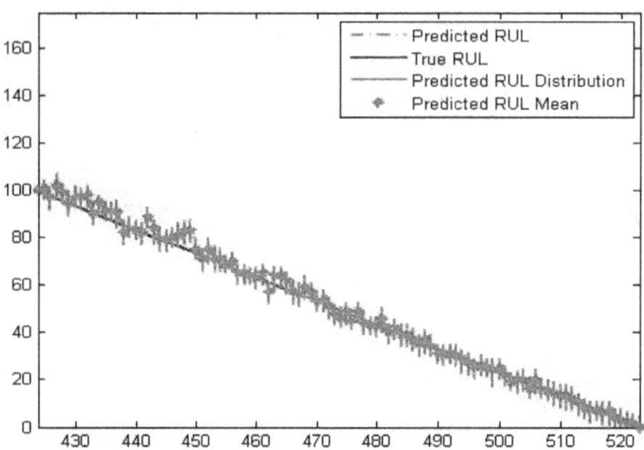

Figure 20. RUL prediction with only dielectric constant observation data (particle population=200)

Particle Population (N)	Prediction Time
50	3 minutes 49 seconds
75	4 minutes 40 seconds
100	5 minutes 47 seconds
150	7 minutes 59 seconds
200	10 minutes 16 seconds

Table 2. Particle population and prediction time relationship with only dielectric constant observation data

2.3.3. RUL Prediction Results Using Multiple Sensor Observation

The RUL prediction results presented in previous section were obtained using only one sensor. In order to combine the two sensors into a particle filter based RUL prediction, a multivariable Gaussian distribution is used:

$$f_y(y_1, \dots, y_k) = \frac{1}{(2\pi)^{\frac{k}{2}}|\Sigma|^{\frac{1}{2}}} \exp\left(-\frac{1}{2}(y-\mu)^T\Sigma^{-1}(y-\mu)\right) \quad (30)$$

where Σ is the covariance matrix of observations, $|\Sigma|$ is the determinant of Σ. Note that y_k in Equation (30) represents the sensor output data Z_k.

By applying the probability density function, each particle will be assigned a weight according to its observation and updated similarly. The RUL prediction results of combining two sensors are provided in Fig. 21. As one can see from Figure 18, Figure 19, and Figure 21, in comparison with the RUL prediction results using only one sensor, the RUL prediction variation in combining two sensors has been reduced from the beginning until the end. Moreover, the accuracy of the prediction has also been improved significantly. The shortcoming of utilizing particle filtering algorithm is that it is considered a computational expensive algorithm. However, using particle filtering algorithm in combination with viscosity and dielectric constant based physical models would provide a feasible and effective solution for RUL predication of lubrication oil.

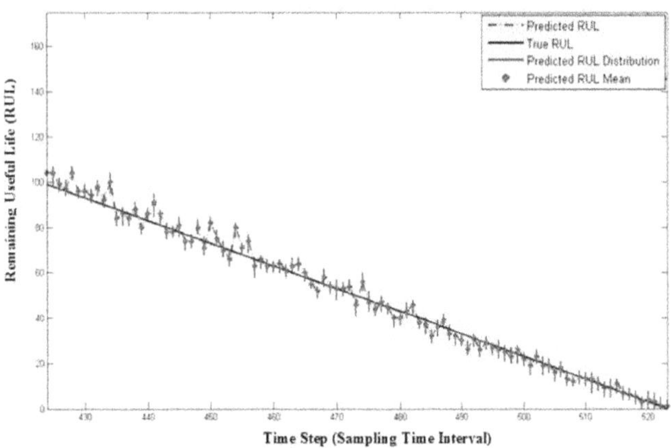

Figure 21. RUL prediction with both kinematic viscosity and dielectric constant observation data (particle population=50)

3. CONCLUSIONS

Lubrication oil condition monitoring and RUL prediction is important for reliability and availability improvement of the wind turbines and reduction of maintenance costs for the wind industry. However, up to today, no effective physics based tools for online condition monitoring of lubrication oil and RUL prediction using viscosity and dielectric constant sensors have been reported. In this paper, a solution for online lubrication oil condition monitoring and RUL prediction using viscosity and dielectric constant sensors along with the particle filtering technique is presented. In particular, physics models for lubrication oil degradation with both viscosity and dielectric constant as performance parameters have been developed and validated with lab oil test data. These lab validated oil deterioration models were integrated with a particle filtering algorithm to develop an effective RUL prediction tool for water contaminated lubrication oil. The effectiveness of developed RUL prediction tool was validated with a simulation case study based on lab verified models.

The RUL prediction results of the simulation case study showed that when only one sensor was utilized, the RUL prediction with particle filtering had a slight fluctuation around the true RUL at the beginning of the prediction process. When both viscosity and dielectric sensors were used, the prediction fluctuation at the beginning was reduced and the RUL prediction accuracy was greatly improved throughout the entire prediction process. Also, larger particle population increase prediction accuracy. However, as particle population increases, the computational time for RUL prediction increases along with it.

REFERENCES

Agoston, A., Otsch, C., and Jakoby, B., 2005, "Viscosity sensors for engine oil condition monitoring-Application and interpretation of results.", *Sensors and Actuators A*, Vol. 121, pp. 327 - 332.

Agoston, A., Dorr, N., and Jakoby, B., 2006, "Online Application of sensors monitoring lubricating oil properties in biogas engines.", *IEEE Sensors 2006, EXCO, Daegu, Korea, October 22 - 25, 2006*

Agoston, A., Otsch, C., Zhuravleva, J., and Jakoby, B., 2004, "An IR-absorption sensor system for the determination of engine oil deterioration.", *Proceeding of IEEE, sensors*, Vienna, Austria, Otc. 24 – 27, 2004, Vol. 1, pp. 463 – 466.

Agoston, A., Schneidhofer, C., Dorr, N., and Jakoby, B., 2008, "A concept of an infrared sensor system for oil condition monitoring", *Elekrotechnik & Informationstechnik.*, Vol. 125/3, pp. 71 - 75.

Arulampalam, S.M., Maskell, S., Gordon, N., and Clapp, T., 2002, "A Tutorial on Particle Filters for Online Nonlinear/Non-Gaussian Bayesian Tracking", *IEEE Transactions on Signal Processing*, Vol. 50, No. 2, pp. 174 – 188.

Benner, J.J., Sadeghi, F., Hoeprich, M.R., and Frank. M.C., 2006, "Lubricating properties of water oil emulsions." Journal of Tribology, Transaction of ASME, April 2006. Vol. 128, pp. 296 - 311.

Beran, E., Los M. and Kmiecik, A., 2008, "Influence of thermo-oxidative degradation on the biodegradability of lubricant base oils", *Journal of Synthetic Lubrication*, Vol. 28, pp. 75 - 83.

BHRA, 1988, *Condition Monitoring Supplement*, Cranfield, Bedfordshire: BHRA.

Bozchalooi, I.S. and Liang, M., 2009, "Oil Debris Signal Analysis Based on Empirical Mode Decomposition for Machinery Condition Monitoring", *2009 American Control Conference*, June 10 - 12, 2009, St. Louis, MO, USA

Butler, S. and Ringwood, J., 2010, "Particle Filters for Remaining Useful Life Estimation of Abatement Equipment used in Semiconductor Manufacturing", *Proceedings of the First Conference on Control and Fault-Tolerant Systems*, Nice, France, pp. 436 - 441.

Byington C., Palmer C., Argenna G., and Mackos N., 2010, "An integrated, real-time oil quality monitor and debris measurement capability for drive train and engine systems" *Proceedings* of *American Helicopter Society 66th Annual Forum and Technology Display*, Pheonix, Arizona, May 11 – 13, 2010

Byington, C., Mackos, N., Argenna, G., Palladino, A., Reimann, J., and Schmitigal, J., 2012, "Application of symbolic regression to electrochemical impedance spectroscopy data for lubricating oil health evaluation", *Proceedings of Annual Conference of Prognostics and Health Management Society 2012*, Minneapolis, Minnesota, September 23 – 27, 2012

Doucet, A., Godsill, S., and Andrieu, C., 2000, "On Sequential Monte Carlo Sampling Methods for Bayesian Filtering", *Statistics and Computing*, Vol. 10, pp. 197 – 208.

Doucet, A., Godsill, S., and Andrieu, C., 2000, "On Sequential Monte Carlo Sampling Methods for Bayesian Filtering", *Statistics and Computing*, Vol. 10, pp. 197 – 208.

Durdag, K., 2008, "Solid state acoustic wave sensors for realtime in_line measurement of oil viscosity.", *Sensor Review*, Vol. 28/1, pp.68 – 73.

Guan, L., Feng, X.L., Xiong G., and Xie, J.A., 2011, "Application of dielectric spectroscopy for engine lubricating oil degradation. ", *Sensors and Actuators A*, Vol. 168, pp. 22 - 29

Halderman, J. D., 1996, *Automotive Technology*, New York: McGraw-Hill.

He, D., Bechhoefer, E., Dempsey, P., and Ma, J., 2012, "An Integrated Approach for Gear Health Prognostics", *Proceedings of the 2012 AHS Forum*, Fort Worth, TX, April 30 – May 3, 2012.

JJakoby, B., Buskies, M., Scherer, M., Henzler, S., Eisenschmid, H., and Schatz, O., 2001, "Advanced Microsystems for Automotive Applications", Springer, Berlin/Heidelberg/New York, 2001, pp. 157 – 165.

Jakoby B. and Vellekoop, M.J., 2004, "Physical sensors for water-in-oil emulsions." *Sensors and Actuators A,* Vol. 110, pp. 28 - 32.

Jakoby, B., Scherer, M., Buskies, M., and Eisenschmid, H., 2002, "Micro viscosity sensor for automobile applications. ", IEEE Sensors, Vol. 2, pp. 1587 - 1590.

Jakoby, B., Scherer, M., Buskies, M., and Eisenschmid, H., 2003, "An automotive engine oil viscosity sensor", *IEEE Sensors. J*, Vol. 3, pp. 562 – 568.

Kittiwake Developments. Ltd. 2011, "Monitoring water in lubricant oil - maintain equipment & reduce downtime*", Critical Things to Monitor, Water in Lube Oil.*

Kumar, S., Mukherjee, P. S., and Mishra, N. M., 2005, "Online condition monitoring of engine oil", *Industrial Lubrication and Tribology*, Vol. 57, No. 6, pp. 260 - 267.

Merwe, R., Doucet, A., Freitas, N., and Wan E., 2000, "Unscented particle filter", *Cambridge University Engineering Department Technical Report cued/f-infeng/TR 380.*

Orchard M.E. and Vachtsevanos, G.J., 2011, "A Particle-filtering Approach for Online Fault Diagnosis and Failure Prognosis", *Transactions of the Institute of Measurement and Control*, Vol. 31, pp. 221 – 246.

Poley,J., 2012, "The metamorphosis of oil analysis", Machinery Failure Prevention Technology (MFPT) Conference, Condition Based Maintenance Section 1, Conference Proceedings, Dayton, Ohio, April 24 – 26, 2012.

Raadnui, S. and Kleesuwan, S., 2005, "Low-cost condition monitoring sensor for used oil analysis", *Wear*, Vol. 259, pp. 1502 - 1506.

Saha, B., Goebel, K., Poll, S., and Christophersen, J., 2009), "Prognostics Methods for Battery Health Monitoring Using a Bayesian Framework", *IEEE Transactions on Instrumentation and Measurement*, Vol. 58, No. 2, pp. 291-296.

Schmitigal, J. and Moyer, S., 2005, "Evaluation of sensors for on-board diesel oil condition monitoring of U.S. Army ground equipment", *TACOM/TARDEC*, Report No. 14113.

Sharma, B.C. and Gandhi, O.P., 2008, "Performance evaluation and analysis of lubricating oil using parameter profile approach", *Industrial Lubrication and Tribology*, Vol. 60, No. 3, pp. 131 - 137.

Stachowiak, G.W. and Batchelor, A.W., 2005, "Physical Property of Lubricants", *Engineering Tribology*, 3rd edition, pp. 3, ISBN-13: 978-0-7506-7836-0, ISBN-10: 0-7506-7836-4

Turner, J.D. and Austin, L., 2003, "Electrical techniques for monitoring the condition of lubrication oil.", *Measurement Science and Technology*, Vol. 14, pp. 1794 - 1800.

Yoon, J., 2012, "A Comparative Study of Adaptive MCMC Based Particle Filtering Methods", M.S. Thesis, University of Florida

Zio, E. and Peloni, G., 2011, "Particle Filtering Prognostic Estimation of the Remaining Useful Life of Nonlinear Components", *Reliability Engineering and System Safety*, Vol. 96, pp. 403 - 409.

Zhu, J, He, D., and Bechhoefer, E, 2012, "Survey of lubrication oil condition monitoring, diagnostics, and prognostics techniques and systems", *Machine Failure Prevention Technology (MFPT)*, Conference Proceeding, April 24 – 26, 2012, Dayton, Ohio, USA

Zhu, J, He, D., and Bechhoefer, E, 2013, "Survey of lubrication oil condition monitoring, diagnostics, and prognostics techniques and systems", *Journal of Chemical Science and Technology (JCST)*, (To appear)

BIOGRAPHIES

Junda Zhu received his B.S. degree in Mechanical Engineering from Northeastern University, Shenyang, China, and M.S. degree in Mechanical Engineering from The University of Illinois at Chicago in 2009. He is a Ph.D. candidate at the Department of Mechanical and Industrial Engineering. His current research interests include lubrication oil condition monitoring and degradation simulation and analysis, rotational machinery health monitoring, diagnosis and prognosis with vibration or acoustic emission based signal processing techniques, physics/data driven based machine failure modeling, CAD and FEA.

Jae Myung Yoon received his B.E degree in control engineering from Kwangwoon University, Seoul, Republic of Korea, He then worked as an equipment development engineer at the Memory business of Samsung Electronics Co. Ltd from 2006 through 2008. He then received M.S degree in Mechanical and Aerospace engineering from the University of Florida, Gainesville, FL. He is currently pursuing the Ph.D. degree in industrial engineering with the department of Mechanical and Industrial engineering, University of Illinois at Chicago, Chicago. His current research interests include: machinery health monitoring, diagnostics and prognostics, non-linear filtering, artificial neural networks encompassing reliability engineering.

David He received his B.S. degree in Metallurgical Engineering from Shanghai University of Technology, China, MBA degree from The University of Northern Iowa, and Ph.D. degree in Industrial Engineering from The University of Iowa in 1994. Dr. He is a Professor and Director of the Intelligent Systems Modeling & Development Laboratory in the Department of Mechanical and Industrial Engineering at The University of Illinois-Chicago. Dr. He's research areas include: machinery health monitoring, diagnosis and prognosis, complex systems failure analysis, quality and reliability engineering, and manufacturing systems design, modeling, scheduling and planning.

Yongzhi Qu received his B.S. in Measurement and Control and M.S. in Measurement and Testing from Wuhan University of Technology, China. He is a PhD candidate in the Department of Mechanical and Industrial Engineering at The University of Illinois Chicago. His research interests include: rotational machinery health monitoring and fault diagnosis, especially with acoustic emission sensors, embedded system design and resources allocation and scheduling optimization.

Eric Bechhoefer received his B.S. in Biology from the University of Michigan, his M.S. in Operations Research from the Naval Postgraduate School, and a Ph.D. in General Engineering from Kennedy Western University. His is a former Naval Aviator who has worked extensively on condition based maintenance, rotor track and balance, vibration analysis of rotating machinery and fault detection in electronic systems. Dr. Bechhoefer is a board member of the Prognostics Health Management Society, and a member of the IEEE Reliability Society.

Targeting Faulty Bearings for an Ocean Turbine Dynamometer

Nicholas Waters[1], Pierre-Philippe Beaujean[2], and David J. Vendittis[3]

[1,2,3]*Florida Atlantic University, 777 Glades Rd, Boca Raton, FL, 33431, USA*

ncwaters1@gmail.com
pbeaujea@fau.edu (corresponding author)
dvendittis@aol.com

ABSTRACT

A real-time, vibrations-based condition monitoring method used to detect, localize, and identify a faulty bearing in an ocean turbine electric motor is presented in this paper. The electric motor is installed in a dynamometer emulating the functions of the actual ocean turbine. High frequency modal analysis and power trending are combined to assess the operational health of the dynamometer's bearings across an array of accelerometers. Once a defect has been detected, envelope analysis is used to identify the exact bearing containing the defect. After a brief background on bearing fault detection, this paper introduces a simplified mathematical model of the bearing fault, followed with the signal processing approach used to detect, locate, and identify the fault. In the results section, effectiveness of the methods of bearing fault detection presented in this paper is demonstrated through processing data collected, first, from a controlled lathe setup and, second, from the dynamometer. By mounting a bearing containing a defect punched into its inner raceway to a lathe and placing an array of accelerometers along the length of lathe, the bearing fault is clearly detected, localized, and identified as an inner raceway defect. Through retroactively trending the data leading to the near-failure of one of the electric motors in the dynamometer, the authors identified a positive trend in energy levels for a specific frequency band present across the array of accelerometers and identify two bearings as possible sources of the fault.

1. INTRODUCTION

The Southeast National Marine Renewable Energy Center (SNMREC) is developing an Ocean Turbine (OT) that is capable of harnessing some of the energy contained in the Florida Current (Driscoll, 2008). Autonomy and minimal maintenance are of the utmost importance due to the high cost of accessing the OT for maintenance. By creating a self-diagnostic program capable of not only identifying the occurrence of a fault, but evaluating its severity and localizing it, said maintenance costs could be significantly reduced.

In addition to proper prognoses, early detection is critical due to the rapid deterioration of mechanisms under unexpected loads. Through vibrational analysis, which has been used to effectively detect and diagnose faults within other rotating machinery, one can achieve a level of awareness associated with the type of fault occurring, its location, and a relative level of severity. This information can then be used to determine whether maintenance is pertinent to prevent further damage to the system (Jayaswal, Wadhwani, & Malchandani, 2008).

SNMREC has also developed a dynamometer to emulate some of the electrical and mechanical features of the OT. To do so, the dynamometer mimics the loads of the ocean current acting on the shaft connected to an electric motor/generator. A program, the Smart Vibrations Monitoring System (SVMS), was created by Mjit, Beaujean, and Vendittis (Mjit, Beaujean, & Vendittis, 2011) for autonomous online monitoring of the dynamometer. SVMS is implemented using LabVIEW. The program uses such methods of fault detection and identification as Power Spectral Density (PSD) analysis, fractional octave analysis, kurtosis, cepstrum, and time waveform analysis (Mjit, 2009). The SVMS program was designed following the standards outlined by the International Standards Organization (ISO) on vibration condition monitoring (ISO, 2002; ISO, 2005) and, as such, the procedure for the high frequency analysis of the vibration signal outlined in this paper is carried out following the same guidelines.

The methods presented in this paper are implemented on an OT dynamometer, which operates on the same principal as wind turbines. Not unlike OTs, accessing wind turbines for maintenance can be difficult and costly. Thus, application of the methods outlined in this paper could greatly serve to minimize maintenance costs in the wind turbine sector of engineering. The methodology of detecting, localizing, and identifying is described in a manner to allow for straightforward application to wind turbines.

The terms detection, localization and identification are defined as follows: detection refers to detecting the presence of a bearing with a raceway defect as the defect forms, localization refers to localizing the faulty bearing location with respect to the position of the array of accelerometers placed along the length of the dynamometer, and identification refers to identifying the exact defective bearing and raceway.

The approach outlined in this paper relies on analysis of the high-frequency component of vibrations over an array of accelerometers placed along the length of the dynamometer. The vibrations (measured as an acceleration) caused by a defect in the inner or outer raceway of a given bearing is an amplitude-modulated signal (McFadden & Smith, 1984). As each ball within a given bearing rolls over the defect, it causes a spike in the acceleration signal, which can be unidentifiable in the time waveform due to background noise and vibrations caused from other rotating components (gears, other bearings, etc.). This spike is modulated by the modes of vibration of the structure and the uneven load on the shaft (McFadden & Smith, 1984).

Ideally, the resulting vibrations from the bearing defect alone would appear as an exponentially decreasing sinusoidal curve (Sheen, 2007). The difficulty in application of this faulty bearing detection method to the dynamometer mainly resides in identification of the faulty bearing, in the overall structural vibrations due to the multiple rotating components contributing to the vibration signature and slow rotational speeds leading to low signal to noise ratios. To do so, narrow-band envelope analysis is used to isolate several high-frequency modes of vibration (Sheen, 2004). This envelope analysis is performed on the vibration data collected across the array of accelerometers.

Bearing fault detection is carried out by trending the average power levels for specific frequency bands and for each accelerometer. In addition to trending average power levels, envelope power levels are computed across the array of accelerometers. An envelope power threshold is determined for each accelerometer's envelope power signal to further detect the presence of a faulty bearing.

Localization of the faulty bearing with respect to placement of each accelerometer is performed by comparing the average power levels computed over specific frequency bands for the entire array of accelerometers. The accelerometer with the highest average power level for a given frequency band is selected as the closest to the faulty bearing location, if the power level for this frequency band across the remaining accelerometers is significantly lower.

Identification of a bearing raceway defect is accomplished through demodulation of each accelerometer's signal about predetermined frequency bands, which allows for periodic impacts in the vibration signal to be clearly observed in the presence of a bearing with a raceway defect.

In the following section, this paper introduces a simplified mathematical model to provide a rudimentary understanding of the physics involved in the vibrations caused by ball bearings periodically coming in contact with a raceway defect. This model is also used to justify the assumption that the acquired vibrations due to bearing defects are the result of modulation. This model is also used to explain how the process of demodulation will be utilized to indicate which bearing is defective. Next, the signal processing approach used to detect, localize, and identify the fault is presented. Lastly, the approach is illustrated through a series of experimental data collected from a controlled lathe setup with a faulty bearing and from data collected from the dynamometer over the course of several weeks, leading up to a bearing failure.

2. SCIENTIFIC APPROACH

In this section, the authors introduce a simplified mathematical model describing the vibrations caused by a bearing with an inner raceway defect. This model relies on many assumptions that are summed up in the concluding section. This model is introduced solely to provide a very rudimentary understanding of the vibrations caused by the periodic impact of a ball coming in contact with a defect along the inner raceway of a bearing. It is also introduced to justify the use of the Hilbert transform demodulation technique used to identify the faulty bearing (Section 3.3).

As a first approach to modeling the vibrations induced by bearing raceway defects, we first consider the vibrations caused by a single ball bearing rolling over a single defect in the inner raceway. The forces acting on a single roller bearing are depicted in Figure 1.

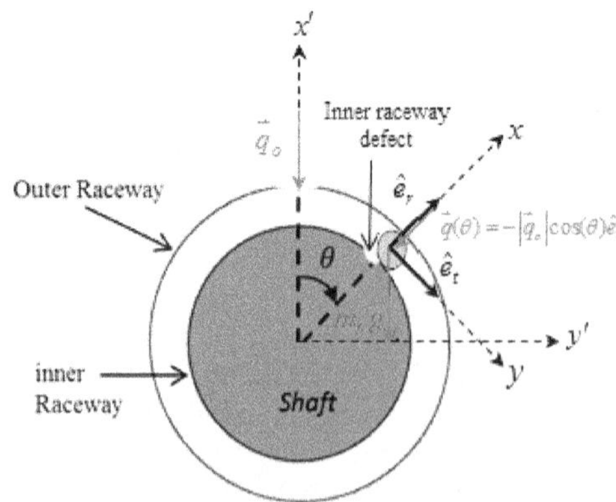

Figure 1. A shaft with a single ball bearing and a single defect on the inner raceway.

The vertical force vector $\vec{q}_0$ represents the loading due to the mass of the shaft and bearing. θ is the angle measured from the vertical to the center of the defect. The vector $m_b\vec{g}$ represents the force due to gravity acting on the ball. The vector $\vec{q}(\theta)$ represents the x-component of the shaft and bearing loading acting on the ball as it passes over the inner raceway defect. The vectors $\hat{e}_r$ and $\hat{e}_t$ represent the radial and tangential unit vectors along the x and y axes, respectively. Through analysis of the equivalent lumped parameter model, the displacement for a single impact and mode of vibration n is described by,

$$x_n(t) = X_n e^{-\omega_{0,n}\zeta_n t}\cos(\omega_{d,n}t - \phi_{0,n})u(t), \tag{1}$$

where X_n is the amplitude of the displacement for mode n at time $t = 0$. $\omega_{0,n}$ is the natural angular frequency. ζ_n is the damping ratio (we assume the system is underdamped, so that ζ_n <<1). $\omega_{d,n}$ is the damped natural angular frequency. $\phi_{0,n}$ is the initial phase for this mode and $u(t)$ is a unit step function introduced into the equation, as this system is causal.

Assuming that every ball is identical and accounting for all the modes of vibration, multiple balls passing over the defect, and all the modes of vibration,

$$x(t) = \sum_{n=1}^{N_s}\sum_{k=-\infty}^{\infty}\begin{pmatrix} X_n(kT_d)e^{-\omega_{0,n}\zeta_n(t-kT_d)}u(t-kT_d) \\ \times\cos\left[\omega_{d,n}\left(t-kT_d\right)-\phi_n\left(kT_d\right)\right]\end{pmatrix}. \tag{2}$$

Here, T_d is the period of rotation for a single ball, KT_d is the time delay associated with the kth impact, and N_s is the total number of modes of vibration. Differentiating Eq. (2) twice yields the equation that describes these vibrations in terms of acceleration.

$$a(t) = \sum_{n=1}^{N_s}\sum_{k=-\infty}^{\infty}\begin{pmatrix} A_{n,k}e^{-\omega_{0,n}\zeta_n(t-kT_d)}u(t-kT_d) \\ \times\cos\left[\omega_{d,n}\left(t-kT_d\right)-\phi_n'\left(kT_d\right)\right]\end{pmatrix}. \tag{3}$$

Equation (3) assumes that the accelerometer is placed directly over the source of vibrations. Since there is some signal loss in the vibrations as they propagate through the structure, new constants are introduced into Eq. (3) to account for this signal attenuation: $\tilde{H}_{C_1}$, $\tilde{H}_T$, $\tilde{H}_{C_2}$, and $\tilde{H}_G$ (Figure 2) represent the complex loss coefficients (or transfer functions) associated with the faulty bearing, first coupler, torque meter, second coupler, and gearbox, respectively. Note that these coefficients change with the natural angular frequency $\omega_{0,n}$.

Also, the symbol $a^{(1)}$ is used to denote the acceleration measured over the faulty bearing. $a^{(m_i)}$ represents the acceleration measured at a distance d from the faulty bearing location.

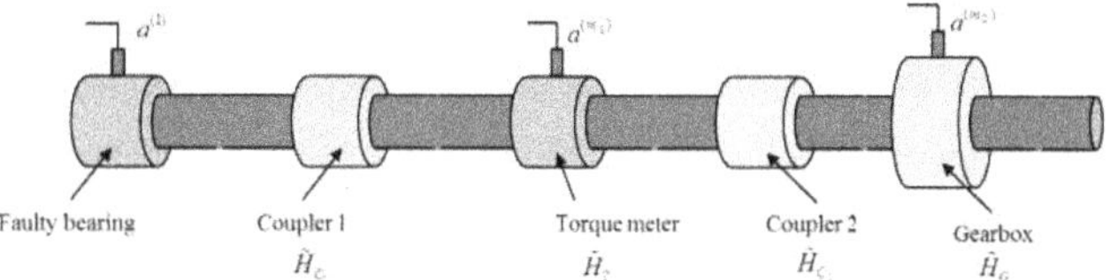

Figure 2. Simplified diagram of dynamometer components and associated loss coefficients.

For the acceleration measured at location m_i on the dynamometer, $\beta^{(m_i)}$ represents the modal attenuation of the vibrations,

$$\beta^{(m_i)} = e^{-\omega_{d,n}\zeta_n\tau_{d,n}^{(m_i)}}\tilde{H}^{(m_i)}. \tag{4}$$

The parameter $\tilde{H}^{(m_i)}$ is defined according to Eq. (5),

$$\tilde{H}^{(m_i)} = \begin{cases} \tilde{H}_{C_1}\cdot\tilde{H}_T, & \text{for } i=1 \\ \tilde{H}_{C_1}\cdot\tilde{H}_T\cdot\tilde{H}_{C_2}\cdot\tilde{H}_G, & \text{for } i=2 \end{cases}. \tag{5}$$

Thus, accounting for signal attenuation due to the vibrations propagating through the structure, the acceleration at location m_i is defined according to Eq. (6),

$$a^{(m_i)}(t) = \sum_{n=1}^{N_s}\sum_{k=-\infty}^{\infty}\begin{pmatrix} \beta^{(m_i)}A_{n,k}^{(1)}e^{-\omega_{0,n}\zeta_n(t-kT_d)}u\left(t-kT_d\right) \\ \times\cos\left[\omega_{d,n}\left(t-kT_d\right)-\phi_n^{(1)}\left(kT_d\right)\right]\end{pmatrix}. \tag{6}$$

According to Eq. (6), the maximum observed acceleration for an accelerometer placed along the length of the dynamometer will be $\beta^{(m_i)}A_{n,k}^{(1)}$. In order to determine which bearing is the source of the fault, demodulation of the acquired vibration signal is needed, which yields the envelope. The envelope of Eq. (6), is defined by the leading terms $\beta^{(m_i)}A_{n,k}^{(1)}e^{-\omega_{0,n}\zeta_n(t-kT_d)}u\left(t-kT_d\right)$. From the envelope, the period between impacts can be determined, indicating which bearing is the source of the defect and whether or not it is within the inner or outer raceway of this bearing.

3. SIGNAL PROCESSING

The methodology used to detect, localize and identify bearing raceway faults are presented in section 4. This methodology refers to several signal processing tools presented in this section.

3.1. PSD and Coherency

We define $r^{(m_i)}(t)$ as output of accelerometer m_i collected on a specific day. $r^{(m_i)}(t)$ represents the measurement equivalent of $a^{(m_i)}(t)$. We estimate the one-sided PSD of $r^{(m_i)}(t)$ as $\hat{G}_{r^{(m_i)}}(\omega)$ of the accelerometer output $r^{(m_i)}(t)$ through block averaging (Ifeachor & Jervis, 2002), using Blackman windows and 50% overlap. The window length depends on the mode of vibration studied. $\hat{G}_{r^{(m_i)}}(\omega)$ is used to identify the bands that contain the natural angular frequencies described in the previous section.

The coherency estimate $\hat{\gamma}_{r^{(m_i)}r^{(m_j)}}(\omega)$ between the accelerometer outputs $r^{(m_i)}(t)$ and $r^{(m_j)}(t)$ indicates, as a function of the angular frequency, the fraction of power originating from the same physical process,

$$\hat{\gamma}_{r^{(m_i)}r^{(m_j)}}(\omega) = \frac{\left|\hat{G}_{r^{(m_i)}r^{(m_j)}}(\omega)\right|^2}{\hat{G}_{r^{(m_i)}}(\omega)\hat{G}_{r^{(m_j)}}(\omega)}. \tag{7}$$

$\hat{G}_{r^{(m_i)}r^{(m_j)}}(\omega)$ is the one-sided Cross Spectral Density (CSD) between the accelerometer outputs $r^{(m_i)}(t)$ and $r^{(m_j)}(t)$. The coherency is a dimensionless, frequency-dependent function that varies between 0 and 1. For a specific angular frequency, a coherency value of 1 indicates that the vibration measured by the two accelerometers originates from the same physical process. $\hat{G}_{r^{(m_i)}r^{(m_j)}}(\omega)$ is estimated with the same block averaging routine used for the PSD estimate.

3.2. Power and Confidence Interval

Power-trending is performed across the period of data acquisition (an illustration of the trending process is shown in section 4.2). For each day of data acquisition, the estimated power is calculated within every predetermined frequency band $\omega_n - \frac{\Delta\omega_n}{2} \le \omega \le \omega_n + \frac{\Delta\omega_n}{2}$ and for every accelerometer m_i,

$$\hat{\Pi}_{m_i,n} = \int_{\omega_n - \frac{\Delta\omega_n}{2}}^{\omega_n + \frac{\Delta\omega_n}{2}} \hat{G}_{r^{(m_i)}}(\omega)d\omega. \tag{8}$$

The confidence interval for a given confidence level for this power level estimate is calculated as shown in Eq. (9),

$$1-\alpha = \Pr\left[\frac{v\hat{\overline{\Pi}}_{m_i,n}}{\chi^2_{\frac{1-\alpha}{2},v}} \le \overline{\Pi}_{m_i,n} \le \frac{v\hat{\overline{\Pi}}_{m_i,n}}{\chi^2_{\frac{\alpha}{2},v}}\right]. \tag{9}$$

$\overline{\Pi}_{m_i,n}$ is the true average power across the same frequency band, $v = K'_w Q$ is the degree of freedom (K'_w is the window scaling factor; Q is the number of block averages used to calculate $\hat{G}_{r^{(m_i)}}(\omega)$) and χ^2 is the chi-square distribution).

3.3. Envelope Analysis

Theoretically, the complex envelope of the acceleration measured at accelerometer m_i for mode n is of the form shown in Eq. (10).

$$\tilde{a}_n^{(m_i)}(t) = \Gamma^{(m_i)} A^{(1)}_{n,k} e^{\square\square_{0,n}\square_n(t\square kT_d)} u(t \square kT_d). \tag{10}$$

This complex envelope contains valuable information regarding the time signature of every mode as a function of the sensor location.

As mentioned earlier, the actual measured acceleration is $r^{(m_i)}(t)$. In this paper, the authors use the Hilbert transform of the measured acceleration to isolate the envelope of individual modes of vibration (demodulation of the signal). To do so, discrete Finite Impulse Response (FIR) filters are designed for every frequency band $\omega_n - \frac{\Delta\omega_n}{2} \le \omega \le \omega_n + \frac{\Delta\omega_n}{2}$. The discrete filter response for each band is labeled $h_n^{(R)}(t)$ (for the real part) and $h_n^{(I)}(t)$ (for the imaginary part). The estimated complex envelope at accelerometer m_i for mode n, denoted by $\hat{a}_n^{(m_i)}(t)$, is the circular convolution between $\left(h_n^{(R)}(t) + jh_n^{(I)}(t)\right)$ and $r^{(m_i)}(t)$

$$\hat{a}_n^{(m_i)}(t) = r^{(m_i)}(t) \square \left(h_n^{(R)}(t) + jh_n^{(I)}(t)\right). \tag{11}$$

4. METHODOLOGY

In this section, the staged approach to detecting, localizing, and identifying a growing raceway defect in a bearing is presented. It is important to note that these methods are performed under steady state conditions and at constant shaft rotational speeds for the dynamometer.

4.1. Dynamometer's impulse response and frequency band selection

The dynamometer's impulse response is measured to isolate the best frequency bands for the fault detection process. Here, the dynamometer does not rotate. Vibrations are produced across the array of accelerometers by striking the dynamometer with a calibrated hammer directly next to one of the accelerometers. The hammer strikes are administered such that vibrations are observed across the entire array of accelerometers without signal clipping. Also, the period between the hammer strikes is large enough to prevent signal overlap between vibrations due to subsequent impacts.

As the dynamometer is struck, structural resonances are excited, leading to peaks in the PSD of the vibration signals acquired by the accelerometers (estimated as described in section 3.2). In order to maximize the signal-to-noise ratio (SNR) in this PSD estimate, the dead time between strikes is minimized by cropping and concatenating the collected data about sections of the signal containing each impact and the resulting vibrations produced. As a result, the strikes appear at even intervals (approximately 0.2 seconds in length). The PSD is estimated for each accelerometer, using the same window length of 0.2 seconds.

Frequency bands in the PSD that preferably show a single peak for two accelerometers are initially considered for analysis. Each frequency band identified for modal analysis is selected for trending of average power levels based on the following criteria:

- The frequency band is preferably located in the upper frequency range to (a) avoid low frequency interferences transmitted through the floor and conduits; (b) accentuate the high frequency attenuation of a signal due to the accelerometers spacing.
- The frequency band shows high coherency (Eq. (7)) between accelerometers (i.e. the coherency is close to 1).
- If possible, the frequency band isolates a single peak in the PSD (no modal overlap).
- We can clearly observe the decrease in in-band power (Eq. (8)) across the array of accelerometers as the distance from the strike to the accelerometer increases.
- The frequency band contains sufficient damping to prevent overlap between decaying impulse responses.

4.2. Detection: Envelope Power

Detection refers to determining whether the vibrations propagating through the structure are characteristic of the presence of a faulty bearing. The process of detecting a faulty bearing is carried out at constant shaft rotational speed, once the system has reached steady state conditions. Thresholds are established for each accelerometer's envelope power signal, as a cap for healthy vibration signals. As a result, vibration levels exceeding the threshold would be indicative of a fault occurring within the system. In order to establish thresholds, each accelerometer's signal is demodulated by applying Hilbert filters about frequency bands satisfying the frequency band criteria and the envelopes are extracted for each set of data. For each accelerometer's demodulated signal, the envelope power is computed as an instantaneous power according to Eq. (12),

$$\square_n^{(m_i)}(\mathrm{t}) = \frac{\left|\tilde{r}_n^{(m_i)}(t)\right|^2}{T_s}, \tag{12}$$

where T_s is the sampling period. When thresholding the subscript *healthy* is introduced to the envelope power and envelope signals, i.e. $\square_{n,healthy}^{(m_i)}(t)$ and $\tilde{r}_{n,healthy}^{(m_i)}(t)$, to denote that the data used for thresholding is acquired while the system is operating under healthy operating conditions. The mean of each accelerometer's envelope power is computed and thresholds are set for each accelerometer's healthy envelope power such that inequality (13) holds.

$$\square_{n,healthy}^{(m_i)}(t) < Thr_n^{(m_i)} \square \bar{\square}_{n,healthy}^{(m_i)}, \quad \square t \tag{13}$$

$\bar{\square}_{n,healthy}^{(m_i)}$ is the mean value of the healthy envelope power for the isolated mode(s) of vibration n and accelerometer m_i,

$$\bar{\square}_{n,healthy}^{(m_i)} = E\left[\frac{\left|\tilde{r}_{n,healthy}^{(m_i)}(t)\right|^2}{T_s}\right]. \tag{14}$$

$Th_n^{(m_i)}$ is a constant chosen based on the accelerometer and mode(s) of vibration for the given envelope signal such that the healthy envelope power, when normalized with respect to the mean of the healthy envelope power, does not exceed this constant value for all time.

The constant $Th_n^{(m_i)}$ is also chosen such that inequality (13) is valid for the smallest value of $Th_n^{(m_i)}$ that allows for slight transients in the healthy vibrations envelope power, but vibrations due to any sort of defect, exceed the threshold $Thr_n^{(m_i)} \square \bar{\square}_{n,healthy}^{(m_i)}$ in envelope power.

Once thresholds have been determined for each accelerometer and isolated mode(s) of vibration, each accelerometer's envelope power is plotted against its respective envelope power threshold for the isolated mode(s) of vibration, $Th_n^{(m_i)} \square \bar{\square}_{n,healthy}^{(m_i)}$.

Envelope power levels that exceed the thresholds are averaged to obtain the average excess power level per time that exceeds the threshold for the accelerometer and mode(s) of vibration. This average envelope power level is denoted by $\bar{\square}_{n,exceed}^{(m_i)}$.

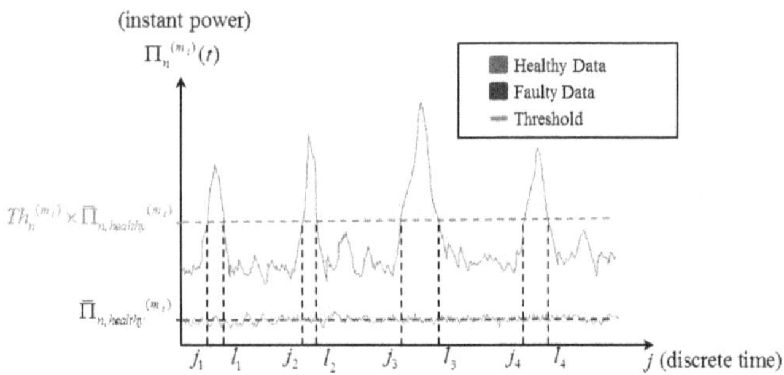

Figure 3. Zoomed in portion of an accelerometer's envelope power for mode of vibration *n*.

For the envelope power signals depicted in Figure 3, $\bar{\Pi}^{(m_i)}_{n,exceed}$ is defined according to Eq. (15)

$$\bar{\Pi}^{(m_i)}_{n,exceed} = \frac{\sum_{j=j_1}^{l_1} \left|\tilde{r}^{(m_i)}_n(t)\right|^2 + \ldots + \sum_{j=j_4}^{l_4} \left|\tilde{r}^{(m_i)}_n(t)\right|^2}{\left[(l_1 - j_1 + 1) + \ldots + (l_4 - j_4 + 1)\right] T_s}. \quad (15)$$

For each accelerometer's envelope power, if there is no fault occurring on the dynamometer, then $\bar{\Pi}^{(m_i)}_{n,exceed} = 0$ indicating that the system is healthy.

It is important to note that effective thresholds are established over some period of observation with the system operating under healthy operating conditions and once it has reached steady state conditions for a given shaft rotational speed. In order to avoid false alarms due to transients in the vibration signal resulting from impacts unrelated to mechanical failure (e.g., fish impacts) it is important to monitor the average envelope power level over multiple data sets. An alarm will be triggered assuming that the average envelope power is greater than zero for consecutive data sets.

4.3. Detection: Trending

In addition to thresholding the envelope power levels for detecting the presence of a fault, average power levels across predetermined frequency bands are trended to indicate a worsening condition:

1. The dynamometer is operated at various shaft rotational speeds. For each rotational speed, the power level within every predetermined frequency band is computed across the array of accelerometers, along with the confidence interval for each power level.
2. Each average power level is compared to its respective baseline average power level for the given shaft rotational speed.
3. Measurements are repeated over the course of weeks to determine whether the power levels across any of the frequency bands increases for any of the accelerometers over time.
4. If an upward trend in the power levels for any of the frequency bands appears across at least the majority of accelerometers and if the power levels approach or exceed an upper threshold, then there is a fault occurring (Figure 4).

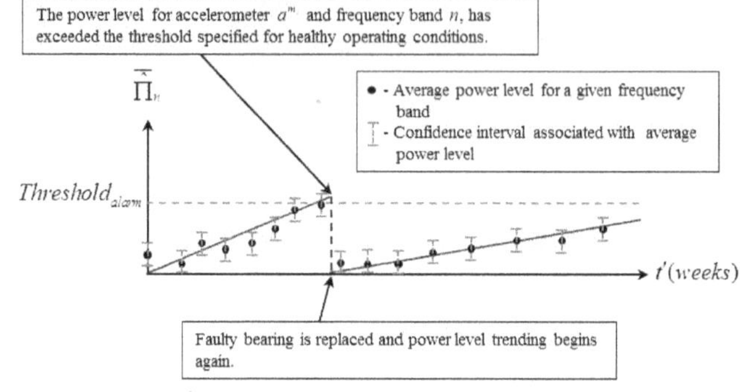

Figure 4. Trending of the average power levels over the course of weeks to detect presence of a faulty bearing.

4.4. Localization

Localization refers to determining the proximity of the faulty bearing location to the accelerometer placement (Figure 5). It is carried out at constant shaft rotational speed, under steady state conditions, and once the presence of a faulty bearing has been detected.

1. The power levels for specified frequency bands are computed for each accelerometer.
2. If an accelerometer acquires data with the highest power level within the frequency band and if for the same frequency band, the remaining accelerometers show a decrease in the power level as their distance increases, then this accelerometer is closest to the defect.

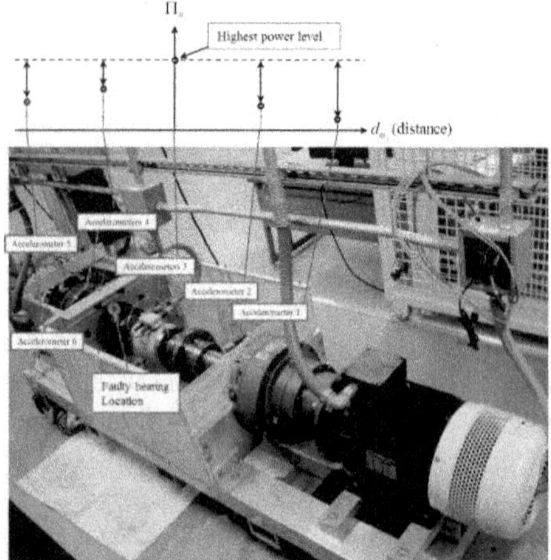

Figure 5. Localization of a faulty bearing based on comparing power levels for one frequency band and RPM across the array of accelerometers.

4.5. Identification

Identification refers to determining the exact bearing containing the raceway defect. Again, the following procedure (also depicted in Figure 6) is performed under constant shaft rotational speeds, under steady state conditions, and once a faulty bearing has been detected:

1. Envelope analysis on the acquired vibration signals is performed within the specified frequency bands and across the array of accelerometers. From the envelope the period between impacts due to balls coming in contact with the defect is determined.
2. The period between impacts, denoted T_{impact}, should correspond to the Ball Pass Frequency for the inner raceway (BPFI) or outer raceway (BPFO) of one of the dynamometer's bearings (Randall & Antoni, 2011).

These frequencies are defined as,

$$f_{o,b_i} = \frac{N_{B,b_i}}{2} f_r \left(1 \square \frac{D_{B,b_i} \cos(\square_{b_i})}{D_{P,b_i}} \right) \tag{16}$$

$$f_{i,b_i} = \frac{N_{B,b_i}}{2} f_r \left(1 + \frac{D_{B,b_i} \cos(\square_{b_i})}{D_{P,b_i}} \right) \tag{17}$$

f_{o,b_i} is the impact frequency for an outer raceway defect (BPFO) in bearing b_i. f_{i,b_i} is the impact frequency for an inner raceway defect (BPFI) in bearing b_i. N_{B,b_i} is the number of balls within the bearing. f_r is the rotational speed (in Hz) of the inner raceway, relative to the outer raceway. D_{B,b_i} is the ball diameter. D_{P,b_i} is the ball pitch diameter. $\square_{b_i}$ is the ball contact angle (which is zero). The ball pitch diameter is calculated as,

$$D_{P,b_i} = \left(\frac{I_{i,b_i} + I_{o,b_i}}{2} \right) \tag{18}$$

I_{i,b_i} and I_{o,b_i} are the maximum diameter of the inner race and maximum diameter of the outer race, respectively.

Although this paper focuses on detecting bearing raceway faults, application of this methodology to identify a bearing cage faults or ball defects, simply involves calculating the Fundamental Train Frequency (FTF) or Ball Spin Frequency (BSF) and comparing the impact frequency extracted from the analytic signal to these two values. Without providing any experimental evidence to validate the claim, the authors anticipate the application of this bearing fault detection technique to work for bearing cage faults and ball defects. However, experimental justification is left for future work.

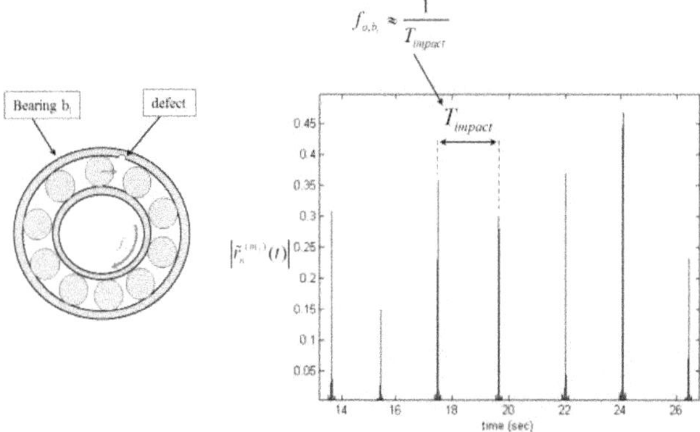

Figure 6. The period between subsequent impacts found from the modulus of the analytic signal formed by Hilbert filtering around a specified frequency band.

Equations 19 and 20 describe calculation of the FTF and BSF.

$$f_{C,b_i} = \frac{f_r}{2} \left(1 \square \frac{D_{B,b_i} \cos(\square_{b_i})}{D_{P,b_i}} \right). \tag{19}$$

$$f_{BS,b_i} = \frac{D_{P,b_i}}{2 D_{B,b_i}} \left\{ 1 \square \left(\frac{D_{B,b_i} \cos(\square_{b_i})}{D_{P,b_i}} \right)^2 \right\}. \tag{20}$$

f_{C,b_i} is the impact frequency for a cage defect in bearing b_i and f_{BS,b_i} is the impact frequency for a ball defect within bearing b_i (Randall & Antoni, 2011).

The assumptions that are made for the methodology described in this section are as follows:

- The system is operating at a constant shaft rotational speed and there is minimal speed fluctuation.
- The system has reached steady state conditions.
- Thresholds for alarms are established while the system is operating void of any defects.

A flowchart describing the overall process of detection, localization, and identification of a bearing fault is provided in Figure 7.

5. RESULTS

In the first set of experimental results, effectiveness of the methods outlined in this thesis are demonstrated by detecting, localizing, and identifying a bearing with a raceway defect punched into its inner raceway. Each of the steps involved in the process outlined in this thesis are clearly demonstrated for straightforward application to the dynamometer. In Experiment II, data leading up to a yet to be determined fault occurring on the dynamometer was retroactively trended. Prior to currently ongoing maintenance on the dynamometer, it was possible to trend data previously acquired and observe an increase in average power levels for one of the frequency bands identified from studying the impulse response of the system.

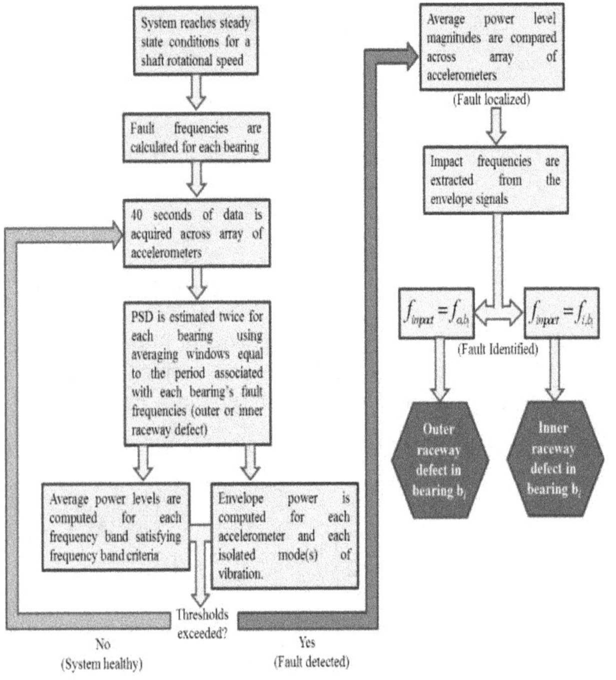

Figure 7. Bearing fault detection, localization, and identification flowchart.

5.1. Experiment I: Controlled Lathe Setup

A tachometer, as well as a plastic collar with an accelerometer fixed to its top surface that allowed for interchanging of bearings, were fixed to the shaft of the lathe (Figure 8). Four accelerometers were used to acquire vibrations along the length of the lathe. Accelerometers 1 to 3 were high frequency accelerometers with frequency responses ranging from 0.5 to 10,000 Hz (at ±3 dB), while accelerometer 4 was a low frequency accelerometer with a frequency response ranging from 0.2 to 3,000 Hz (at ±3 dB).

By collecting the vibrations across an array of accelerometers with a healthy bearing (void of known defects) in place, a baseline was established. A second identical bearing was carefully disassembled and had a defect punched into its inner raceway (referred to as the bad bearing). Due to the large manufacturer tolerances for the two bearings, even the healthy bearing produced a large amount of vibrations. In order to get a better idea of the background noise in the acquired signals for the healthy and bad bearings, a steel bearing (referred to as the ideal bearing) with much tighter manufacturer tolerances was mounted to the lathe and the vibrations were acquired along the array of accelerometers. The bearings used for the healthy and bad bearings in this experiment were stainless steel ball bearings with polymer ball cages, which allowed for simple disassembly with nothing more than hand pressure.

Figure 8. Lathe experimental setup.

The system was excited using calibrated hammer strikes administered next to accelerometer 1 and the impulse response for the system was studied (Figure 9).

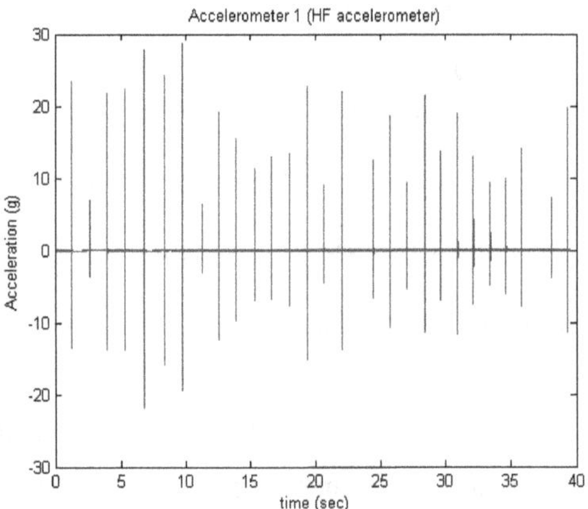

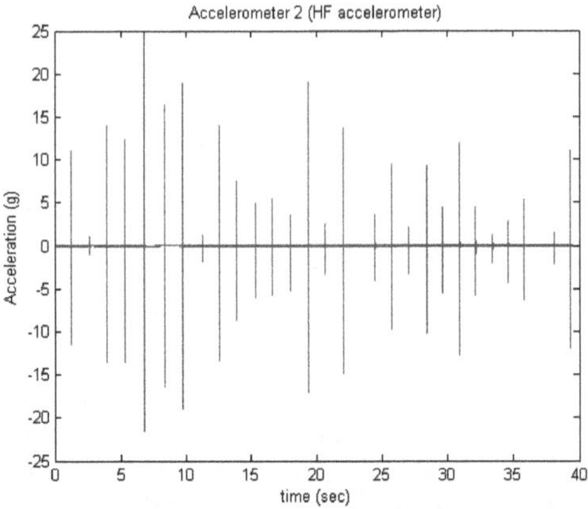

Figure 9. Time waveform of hammer impacts for accelerometers 1 (top) and 2 (bottom).

Time windows of length 0.182 seconds (4096 points), were taken about each hammer impact to form the impacts signal for the PSD and coherency estimates (Figure 10). The frequency band 439-1126 Hz, proved to have the steepest modal attenuation for the envelope with the most separation between impact vibrations and background noise.

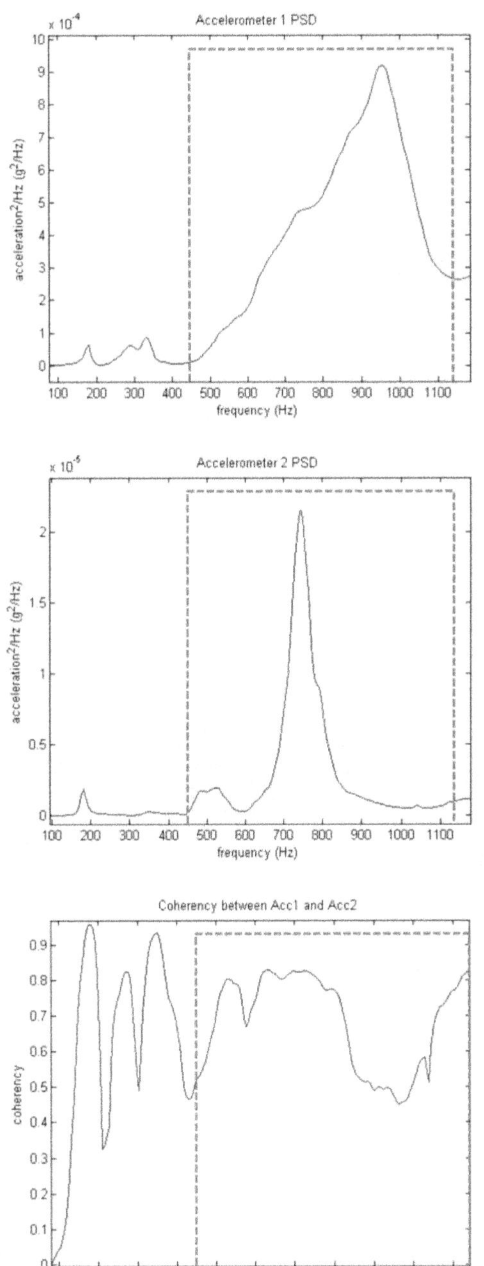

Figure 10. PSD for accelerometers 1 (top) and 2 (middle) and coherency between accelerometers 1 and 2 (bottom).

Following analysis of the PSD and coherency between accelerometers 1 and 2, the time waveform of the vibration signals was demodulated for the two accelerometers to check the modal decay for this frequency band (Figure 11).

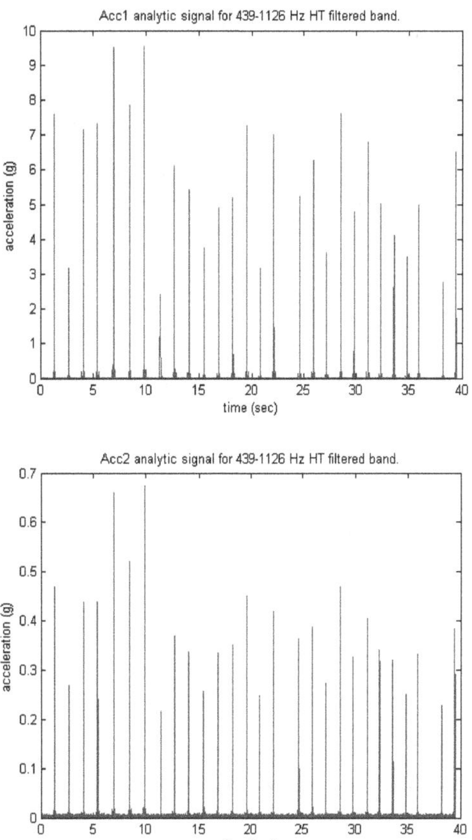

Figure 11. Demodulated signals for accelerometers 1 (top) and 2 (bottom) for the frequency band of 439-1126 Hz.

Following acquiring the vibrations across the array of accelerometers for the impulse response of the system, vibrations were collected for the healthy bearing, bad bearing, and the ideal bearing for a shaft rotational speed of 120 RPM (Figure 12). Data are sampled at a sampling rate of 22.5 kHz and each set of processed data was of 39.5 seconds in length. An 8-pole low-pass filter was applied to the data with an edge frequency of 10 kHz.

With the healthy bearing in place, each accelerometer's envelope power was plotted for the Hilbert filtered frequency band of 439-1126 Hz and thresholds were set at $Th_i \square\square\overline{}^{(m_i)}_{n,healthy}$ (for i = 1, 2, 3, or 4) above the maximum observed envelope power level. The constant Th_i varied depending on the accelerometer and was selected such that the threshold was above the highest observed envelope power level for each accelerometer and this frequency band, as shown in Figure 13.

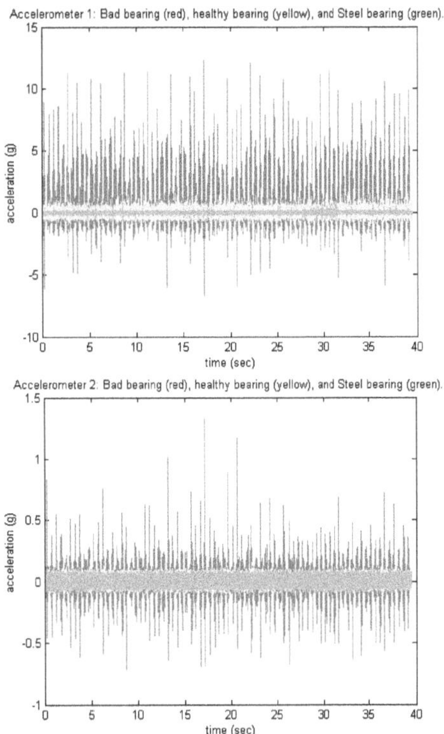

Figure 12. Time waveforms for the ideal bearing (green), healthy bearing (yellow), and bad bearing (red) for accelerometers 1 (top) and 2 (bottom).

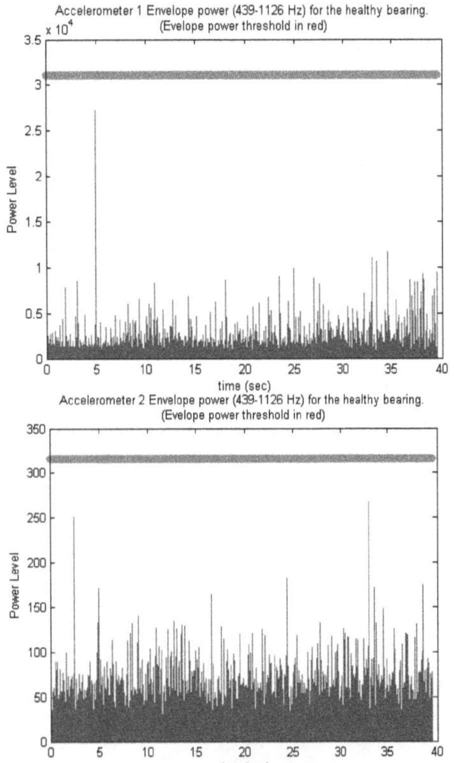

Figure 13. Envelope power and envelope power thresholds for accelerometers 1(top) and 2 (bottom) for the healthy bearing.

Data was then acquired from the bad bearing and the acquired vibration envelope power signals were plotted against their respective envelope power levels. Envelope power levels in excess can clearly be observed in Figure 14.

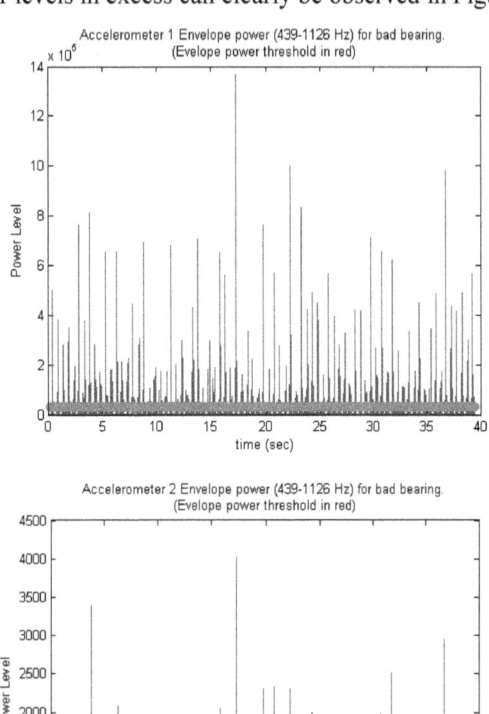

Figure 14. Envelope power levels acquired from the bad bearing for accelerometers 1 (top) and 2 (bottom).

The results of summing the excess envelope power levels across the array accelerometers for this demodulated frequency band are summed up in the Table 1.

Accelerometer number	**Threshold** $Th_n^{(m_i)} \square \overline{\square}_n^{(m_i)}$ $\left(g^2/\sec\right)$	**Excess Average Envelope Power** $\overline{\square}_{n,exceed}^{(m_i)}$ $\left(g^2/\sec\right)$
1	3.110 x 10^4	1.089 x 10^5
2	3.159 x 10^2	6.663 x 10^2

Table 1. Average envelope power levels and thresholds for bad bearing vibrations and accelerometers 1 and 2.

For the dynamometer, this would be the indication of the presence of a fault occurring in the system. As a further indicator of the presence of a fault occurring on the lathe, an increase in the average power level for this frequency band was observed across the array of accelerometers between the healthy bearing data set and the bad bearing data set.

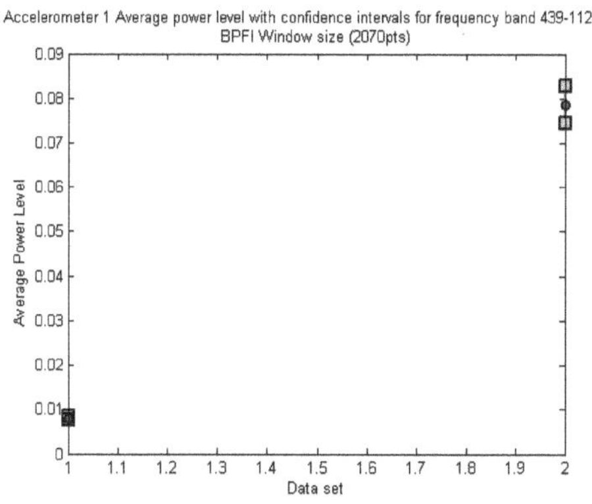

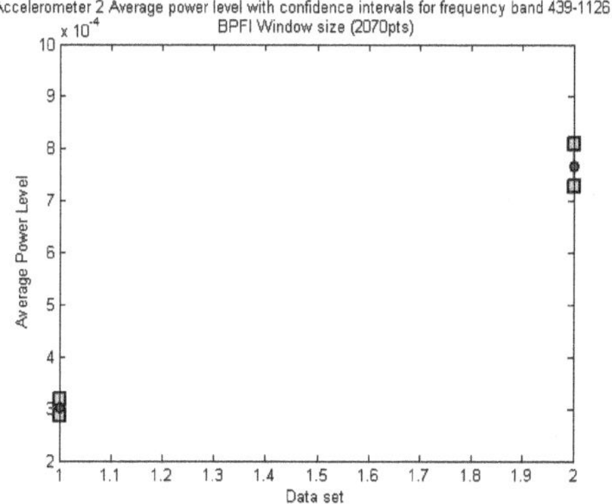

Figure 15. Average power levels across the frequency band of 439-1126 Hz for accelerometers 1 (top) and 2 (bottom).

The increase in average power levels are summed up in Table 2.

Accelerometer number	Percent increase in average power level for frequency band 439-1126 Hz
1	876 %
2	155%

Table 2. Increase in average power levels across the frequency band of 439-1126 Hz for accelerometers 1 and 2.

Having detected the presence of a defect occurring on the lathe, the next step was to localize this defect with respect to the accelerometer placement. By plotting the increase in average power levels across the array of accelerometers in a single figure, localization of the fault to the position of accelerometer 1 becomes apparent. The average power levels across the frequency band 439-1126 Hz for the faulty bearing data, along with their respective confidence intervals are shown in Figure 16.

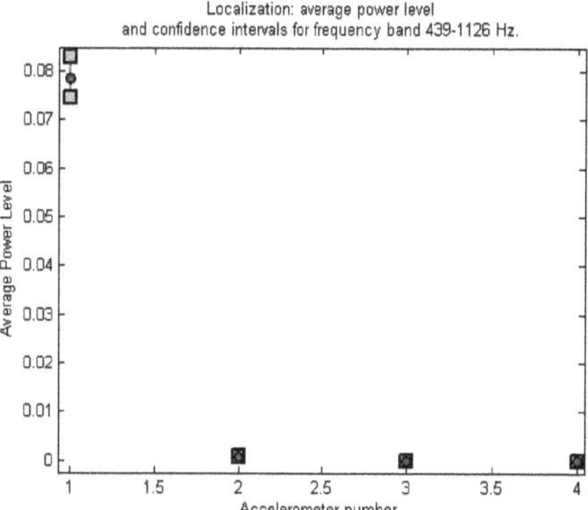

Figure 16. Average power levels with confidence intervals for faulty bearing data and frequency band 439-1126 Hz for accelerometers 1-4.

Accelerometer 1 showed the highest average power level for this frequency band, with accelerometers 2-4 showing a downward trend in power levels for this frequency band as their distance from accelerometer 1 increased. Without prior knowledge of the placement of the faulty bearing, Table 2 and Figure 16 make it apparent that the position of the faulty bearing was closest to the placement of accelerometer 1. Thus, by comparing the average power levels across the array of accelerometers for a given frequency band, it was possible to localize the position of the faulty bearing with respect to the accelerometer placement to the position of accelerometer 1.

Having detected the presence of a defect and localized its position with respect to the accelerometer placement, the next step was to identify the exact bearing and raceway the defect corresponded to. From the envelope for accelerometer 1 (Figure 17), the average period between impacts was found to be 0.0914 seconds which corresponds to an impact frequency of 10.94 Hz.

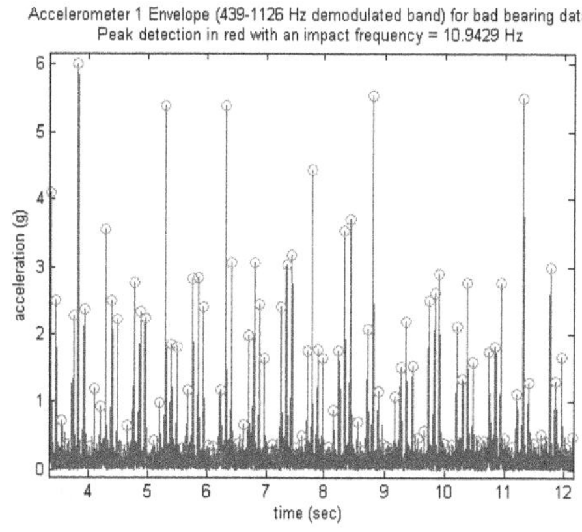

Figure 17. Envelope for bad bearing data acquired by accelerometer 1 for with demodulation about the frequency band of 439-1126 Hz.

The theoretical fault frequency for an inner raceway defect within the bearing used for the experiment and for a shaft rotational speed of 120 RPM was 10.86 Hz. Thus, the defect was identified as an inner raceway defect with 0.74% error.

5.2. Experiment II: Trending Data Leading to Fault on the Dynamometer

The approach of detecting, localizing, identifying bearings with raceway defects is illustrated through the analysis of a series of experimental data collected over the course of a month leading up to a fault in the dynamometer. By retroactively trending the data leading to the near-failure of one of the electric motors in the dynamometer, a positive trend in energy levels for a specific frequency band was observed, which was present across the array of accelerometers and two bearings were identified as possible sources of the fault. Shortly before the electric motor was dismounted, excessive vibrations and sound radiation were observed. The fault was then localized via a calibrated stethoscope to the location of two of the dynamometer's bearings.

Data collected over the course of a month were processed for accelerometers 3 and 5 (Figure 18) before the dynamometer was disassembled and the electric motor was sent for maintenance and repair.

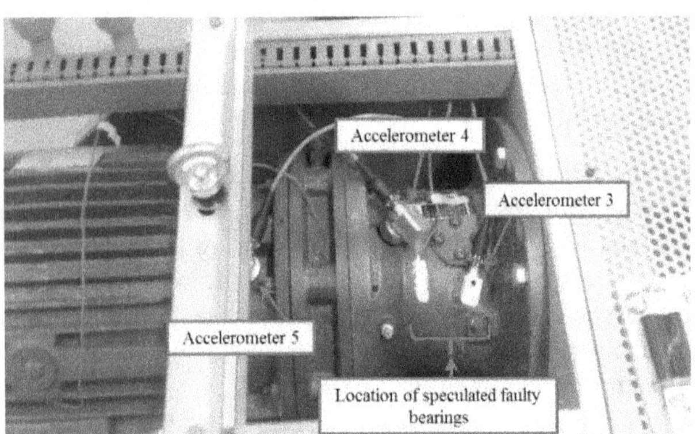

Figure 18. Accelerometer placement on the dynamometer and the location of the speculated faulty bearings.

The impulse response of the dynamometer was studied and the frequency band of 278-463 Hz was identified as satisfying the frequency band criteria (section 4.1). In processing the data collected prior to detecting the fault, an upward trend in the average power across the frequency band of 278-463 Hz was observed (Figure 19).

Within the 278-463 Hz band, the data acquired by accelerometer 3 showed an increase in averaged power level of approximately 297% in four weeks, while accelerometer 5 showed an increase of approximately 321% over the same time period. Given the confidence intervals shown in Figure 19, these increases in power level are very significant and clearly indicate a change in the structural response of the dynamometer.

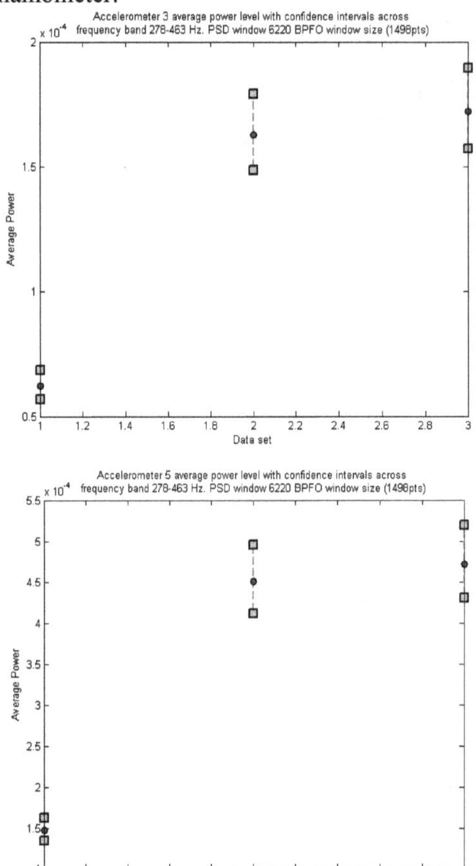

Figure 19. Average power levels across the frequency band of 278-463 Hz for accelerometers 3 (top) and 5 (bottom).

The signal acquired by accelerometer 3 on the last day was then demodulated about the same frequency band of 278-463 Hz (Figure 20) and a periodic impact was noticed from the envelope. The period between impacts was determined to be 0.2761 seconds, which corresponds to an impact frequency of approximately 3.62 Hz and closely matches the impact frequency for the two motor bearings located at position of accelerometer 3 (Figure 18).

Indeed, for a shaft rotational speed of 50 RPM, one bearing would theoretically produce vibrations at an impact frequency of 3.41 Hz (outer raceway defect), while the second bearing would produce vibrations with an impact frequency of 3.8 Hz (outer raceway defect). This is very close to the impact frequency caused by a defect on the outer raceway of two specific bearings. This periodicity in the vibration signal was only observed in the data acquired by accelerometer 3 which also indicated that one of the two bearings in question contained a defect, since accelerometer 3 was placed directly over the speculated fault location (Figure 18).

The maintenance on the dynamometer confirmed the measurements. The two faulty bearings have since been replaced.

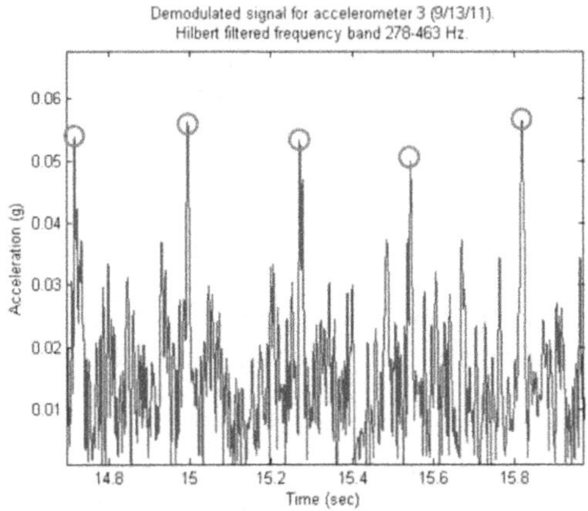

Figure 20. Envelope for data acquired by accelerometer 3 from last day of data acquisition.

6. CONCLUSION

In this paper, a method of using an array of accelerometer to detect the presence of bearing with a raceway defect, localize its position with respect to the placement of accelerometers, and identifying the exact bearing and raceway to which the defect corresponds to has been presented. The methodology outlined in this paper is described for application to an ocean turbine dynamometer, but is applicable to any rotating machinery containing bearings.

The method outlined in this paper was carried out on a lathe with a faulty bearing in order to demonstrate implementation on a more ideal system, for straight forward application to the more complex system, such as the dynamometer. Following detecting the presence of the defect, its position on the lathe was localized with respect to the accelerometer placement. Lastly, the faulty bearing was identified, in addition to the raceway the defect corresponded to (inner raceway), with 0.74% error.

Next, the process of analyzing vibrations across an array of accelerometers in order to specifically target bearing raceway defects occurring on a dynamometer was demonstrated. Through trending of the average power within across a specific frequency band satisfying the frequency band criteria, a growing defect occurring in the dynamometer was detected. Following this, the defect was localized relative to the accelerometer placement by comparing the average power levels for the same frequency band across the array of accelerometers. In addition, an envelope analysis within this frequency band showed periodic impacts, whose frequency of occurrence closely matched the impact frequency (outer raceway) for two of the motor bearings. The fault was also localized via a calibrated stethoscope: this test supports the conclusions obtained with the approach presented in this paper.

ACKNOWLEDGEMENT

This research was funded by the Southeast National Marine Renewable Energy Center (SNMREC) at Florida Atlantic University.

REFERENCES

Fernandez, A., Bibao, J., Bediaga, I., Gaston, A., & Hernandez, J., (2005). Feasibility study on diagnostic methods for detection of bearing faults at an early stage. *Proceeding of WSEAS International Conference*, November 2-4, Venice, Italy, pp.113-118. http://www.wseas.us/elibrary/conferences/2005venice/papers/508-367.pdf

Muszynska, A., (1995). Vibrational Diagnostics of Rotating Machinery Malfunctions. *International Journal of Rotating Machinery*, vol. 1, No. 3-4, pp. 237-266. doi:10.1155/S1023621X95000108

International Standards Organization (ISO) (2002). Condition monitoring and diagnostics of machines-Vibration condition monitoring part 1: General Procedures. In ISO 13373-1: 2002.

International Standards Organization (ISO) (2005). Condition monitoring and diagnostics of machines-Vibration condition monitoring part 2: Processing, analysis and presentation of vibration data. In ISO 13373-2: 2005.

Ho, D., & Randall, R. B., (2000). Optimisation of Bearing Diagnostic Techniques Using Simulated and Actual Bearing Fault Signatures. *Journal of Mechanical Systems and Signal Processing*, vol. 14, Issue 5, pp. 763-788. doi: 10.1006/mssp.2000.1304

Ifeachor, E. C., & Jervis, B. W., (2002). Spectrum estimation and analysis. *In Digital Signal Processing 2nd ed* (pp. 704-705). Harlow, England: Pearson Education Limited.

Driscoll, F. R., Alsenas, G. M., Beaujean, P. P., Ravenna, S., Raveling, J., Busold, E., & Slezycki, C. (2008). A 20 kW Open Ocean Current Test Turbine. *Proceedings of the MTS/IEEE Oceans '08*. September. Quebec City, Quebec, Canada. http://209-20-84-91.slicehost.net/assets/2009/3/4/A_20_kW_Open_Ocean_Current_Test_Turbine.pdf

Marichal, G. N., Arte´ s, M., & Garcı´a-Prada, J. C., (2010) An intelligent system for faulty-bearing detection based on vibration spectra. *Journal of Vibration and Control*, vol. 17, no. 6, pp. 931–942. doi: 10.1177/1077546310366264

Cooper, G. R., & McGillem, C. D., (1999). Spectral Density. *In Probabilistic Methods of Signal and System Analysis 3rd ed.* (pp. 257–271). Oxford, New York: Oxford University Press, Inc.

Konstantin-Hansen, H., & Herlufsen, H., (2003). Envelope Analysis for Diagnostics of Local Faults in Rolling Element Bearings. *Bruel & Kjaer Sound and Vibration Measurement*, pp. 1-8. http://www.bksv.com/doc/bo0501.pdf

Konstantin-Hansen, H., & Herlufsen, H., (2010). Envelope and Cepstrum Analysis for Machinery Fault Identification. *Bruel & Kjaer Sound and Vibration*, pp. 10-12. http://www.sandv.com/downloads/1005hans.pdf

Onel, I. Y., Dalci, K. B., & Senol, I., (2005). Detection of Outer Raceway Bearing Defects in Small Induction Motors Using Stator Current Analysis. *Sadhana*, vol. 30, part 6, pp.716. doi: 10.1007/BF02716705

Courrech, J., & Gaudet, M.,. Envelope Analysis – The Key to Rolling-Element Bearing Diagnosis. *Bruel & Kjaer Vibration and Sound.* http://www.bksv.com/doc/BO0187.pdf

Renwick, J. T., & Babson, P. E., (1985). Vibration Analysis-A Proven Technique as a Predictive Maintenance Tool. *IEEE Transactions on Industry Applications*, vol. IA- 21, no. 2. doi: 10.1109/TIA.1985.349652

Mjit, M. M., (2009). *Methodology For Fault Detection and Diagnostics in an Ocean Turbine Using Vibration Analysis and Modeling*. Master's thesis, Florida Alantic University, Boca Raton, FL. http://snmrec.fau.edu/sites/default/files/research/theses/DT-09-335.pdf

Mjit, M., Beaujean, P. J., & Vendittis, D. J., (2011). Comparison of Fault Detection Techniques for an Ocean Turbine. *Annual Conference of the Prognostics and Health Management Society*. September 25-29, Portland, OR. https://www.phmsociety.org/sites/phmsociety.org/files/phm_submission/2011/phmc_11_023.pdf

Mjit, M., Beaujean, P. J., & Vendittis, D. J., (2010). Remote Health Monitoring for Offshore Machines, using Fully Automated Vibration Monitoring and Diagnostics. *Proceedings of Annual Conference of the Prognostics and Health Management Society*, October 10-16, Portland, OR. https://www.phmsociety.org/sites/phmsociety.org/files/phm_submission/2010/phmc_10_006.pdf

Norton, M. P., & Karczub, D. G., (2003). The Analysis of Noise and Vibration Signals. In, *Fundamentals of noise and vibration analysis for engineers, 2nd ed.* (pp. 353–357). Cambridge, UK: Cambridge University Press.

Valente, M., et al. (2008). Hearing Protection. In, *Audiology Treatement, 2nd ed.* (pp. 371-373). New York, NY: Theime Medical Publishers.

Thrane, N., Wismer, J., Konstantin-Hansen, H., & Gade, S., (1995) Practical use of the "Hilbert transform", *Bruel & Kjaer Application Notes.* http://www.bksv.com/doc/bo0437.pdf

Thrane, N., (1984). The Hilbert Transform. *Bruel & Kjaer, Techical Review,* No. 3 (pp. 3-15). http://www.bksv.com/doc/bv0015.pdf

McFadden, P. D., & Smith, J. D., (1984). Model for the Vibration Produced by a Single Point Defect in a Rolling Element bearing. *Journal of Sound and Vibration*, vol. 96, no. 1, pp. 69-82. doi: 10.1016 /0022-460X(84)90595-9

Jayaswal, P., Wadhwani, A. K., & Mulchandani, K. B., (2008). Machine Fault Signature Analysis. *International Journal of Rotating Machinery*. vol. 2008. doi:10.1155/2008/583982

McInerny, S. A., and Dai, Y., (2003). Basic Vibration Signal Processing for Bearing Fault Detection. *IEEE Transactions on Education*, vol. 46, no. 1. doi: 10.1109/TE.2002.808234

Wang, W.-Y., & Harrap, M. J., (1996). Condition Monitoring of Ball Bearings Using Envelope Autocorrelation Technique. *Journal of Machine Vibration*, vol. 5, pp. 34- 44. http://cat.inist.fr/?aModele=afficheN&cpsidt=3150743

Wang, Y.-F., & Kootsookos, P. J., (1999). Modeling of Low Shaft Speed Bearing Faults For Condition Monitoring. *Mechanical Systems and Signal Processing*, vol. 12, no. 3, pp. 415-426. http://espace.library.uq.edu.au/eserv.php?pid=UQ:10968&dsID=MechSysAndSigPro.pdf

Sheen, Y. -T., (2004). A complex filter for vibration signal demodulation in bearing defect diagnosis. *Journal of Sound and Vibration*, vol. 276, pp.105-119. doi: 10.1016/j.jsv.2003.08.007

Sheen, Y.-T., (2007). An impulse-response extracting method from the modulated signal in a roller bearing. *Journal of the International Measurement Confederation*, vol. 40, Issue 9-10, pp. 868-875. http://ir.lib.stut.edu.tw/bitstream/987654321/8162/2/paper4.pdf

Su, Y.-T., & Lin, S.-J., (1992). On Initial Fault Detection of a Tapered Roller Bearing: Frequency Domain Analysis. *Journal of Sound and Vibration*, vol. 155, no. 1, pp. 75-84. doi: 10.1016/0022-460X(92)90646-F

BIOGRAPHIES

Nicholas C. Waters earned his bachelor's degree from the University of California, Davis in Applied Mathematics. In 2012, he received his master's degree from Florida Atlantic University in Ocean Engineering under the advisement of Dr. Pierre-Philippe Beaujean. His graduate career field of study was focused on machine condition monitoring and its application to an ocean turbine. Nicholas is an active member of the Institute of Electrical and Electronic Engineers (IEEE) and the Marine Technology Society (MTS).

Pierre-Philippe J. Beaujean received the Ph.D. degree in ocean engineering from Florida Atlantic University in 2001. He is an Associate Professor at the Department of Ocean and Mechanical Engineering, Florida Atlantic University. He specializes in the field of underwater acoustics, signal processing, sonar design, data analysis, machine health monitoring, and vibrations control. Dr. Beaujean is an active member of the Acoustical Society of America (ASA), the Institute of Electrical and Electronic Engineers (IEEE) and the Marine Technology Society (MTS).

Dr. David J. Vendittis [Ph.D. (Physics) - American University, 1973] is a Research Professor (part time) at the Department of Ocean and Mechanical Engineering, Florida Atlantic University. Additionally, he is the Technical Advisory Group (ASA/S2) chairman for an International Standards Organization subcommittee, ISO/TC108/SC5 - Machinery Monitoring for Diagnostics. This committee writes international standards that support Machinery Condition Monitoring for Diagnostics. He was appointed to this position by the Acoustical Society of America (ASA)

Effect of Different Workscope Strategies on Wind Turbine Gearbox Life Cycle Repair Costs

A. Lesmerises[1], and D. Crowley[2]

[1,2] *StandardAero Engineering Services, 3523 General Hudnell Dr., San Antonio, TX, USA 78226*
Alan.Lesmerises@StandardAero.com
David.Crowley@StandardAero.com

ABSTRACT

The wind turbine industry is beginning to establish orthodoxies governing the repair of gearboxes, including policies governing the replacement of bearings during gearbox heavy maintenance events. Some maintainers recommend replacing all of the bearings, every time, regardless of condition or age. At the same time, others prefer to only replace the failed bearing. The former rationale achieves availability by spending more money than absolutely necessary; the latter sacrifices reliability in exchange for a lower shop visit cost. Even though neither approach results in the lowest Life Cycle Cost, no standard practice has yet been implemented to methodically determine what would be the best approach. Furthermore, as gearboxes approach the end of their planned service lives, a different strategy may be called-for. This paper presents an illustrative example of using a reliability-based statistical analysis to determine which strategy will yield the lowest Life Cycle Cost for wind turbine gearboxes.

1. INTRODUCTION

Many Wind Turbine asset owners have been faced with the question of whether to reuse or replace bearings while a wind turbine gearbox is undergoing heavy maintenance. Some owners believe that only bearings with extensive damage should be replaced (commonly referred to as "On-Condition" Maintenance or OCM), while others prefer to proactively replace all bearings to avoid unplanned failures. The difference in shop visit costs associated with these two strategies can be significant; shop visit costs can be less than $10K (when using a condition-based approach) to over $100K (material cost to replace all bearings).

From a reliability standpoint, bearings present a rather complex problem. Standard life expectancies for bearings are typically stated as "L_{10}" (or B_{10}) lives. They represent the total service time by which 10% of a population of bearings can be expected to fail. However, their actual service lives can vary significantly depending on the design of the system they're used in, their duty cycle, and the condition of the lubrication system used to support them.

The Aviation and Gas Turbine industries have developed clear definitions for inspection and reuse of bearings used on aircraft and in powerplants. These industries typically follow a "condition-based" approach. If the bearings meet defined acceptance criteria, they are reused regardless of the economics involved or time in service. In providing these services, StandardAero complements these condition assessments with economic evaluations that relate the estimated remaining useful life to the cost of parts being replaced and the life expectancy of the unit being repaired (based on which parts will be replaced), enabling the asset owner to realize a lower long-term cost per service hour.

Just as it is for wind turbine asset owners, maintainers of many other types of complex rotating equipment face the same fundamental question of whether to repair only what is broken (an OCM strategy) and potentially live with short service lives until the next maintenance event, or should parts be refurbished or replaced every time regardless of condition (a pure overhaul strategy), potentially incurring costs without extending time in service or receiving the full benefit from the increased costs.

Answering this dilemma requires a clear methodology to determine when one strategy or another is more economical. The goal of this paper is to lay out the foundation for determining when it is more economical to proactively replace the bearings versus an "inspect and reuse" approach when a gearbox is undergoing heavy maintenance. A new SAE Aerospace Recommended Practice, JA6097 – Using a System Reliability Model to Optimize Maintenance Costs,

A Best Practices Guide (2013) describes one such methodology that can be applied to virtually any type of system, including wind turbine gearboxes. This methodology can be explained by walking the reader through an analysis of both the OCM and pure overhaul strategies, and then comparing those results with an even more optimal maintenance approach.

One significant drawback of maintenance approaches based on a single policy for all assets (such as the pure OCM or pure overhaul approaches) is that they don't take into consideration that each asset comes in for maintenance in a different "state". The particular components installed in the system frequently have different service histories, and hence make different reliability contributions to the system's future performance. This is especially significant when it comes to complex, heavily integrated repairable systems – what constitutes the "best" decisions to be made at each shop visit will be different based on the state of each asset. The principal advantage of the approach described in JA6097 is that it simultaneously addresses both the cost and reliability impact of various corrective maintenance actions being considered at a given shop visit, a major shortcoming of many other optimization techniques (Wang, 2002).

2. ASSUMPTIONS AND COST DATA

For this analysis, we wanted to consider as many costs that an asset owner will incur over the lifetime of a gearbox as possible. There may be other costs not shown here, and some of these costs will vary based on location, the make and model of units involved, source of supply, and economies of scale.

We realize that some operators may disagree with some of the particular values assumed below. At the same time, we also had to protect the confidentiality of data provided to us by other wind turbine operators. However, for the purposes of this analysis, the following costs were judged sufficient to provide a reasonable assessment and comparison. More importantly, it will still illustrate an objective data-driven methodology for determining which maintenance strategy is more cost effective. In practice, values derived from the individual asset owner's actual equipment and experiences would be used for these computations.

The following assumptions and costs used for this study:

Operating Hours/Yr	3,000	Hrs *
Total Hrs over 20 Yrs	60,000	Hrs
Total Hrs over 25 Yrs	75,000	Hrs

Table 1. Gearbox Usage

* – An approximate value reported by some wind turbine operators. While this may be representative in general, individual operators can and do experience significant seasonal variations due to geographic location (weather, topography, etc.).

Crane to RR Gearbox	$60,000	per event
RR Gearbox 3 days/2 techs	48	Hrs Labor
Shipping (in and back)	$ 6,000	per event
Misc. Material	$ 4,000	per event
Avg. cost/bearing	$ 8,000	ea
Qty Bearings/Gearbox	14	ea
Field Labor Costs	$ 100	/ Hr
Shop Labor Costs	$ 100	/ Hr
Gearbox Major Repair	450	Hrs Labor
Gearbox Med Repair	350	Hrs Labor
Gearbox Minor Repair	350	Hrs Labor
Revenue	$ 0.06	/ KWH

Table 2. Costs and labor per event

Crane	$ 60,000
Labor to RR Gearbox	$ 4,800
Lost Sales (1 week)	$ 5,000
Shipping	$ 6,000
Shop Repair Cost	$ 47,000
Total	$ 122,800

Table 3. Costs to replace 1 bearing off-tower

These also assume that there will be one (1) crane visit for each gearbox remove & replace (RR) event, and that a serviceable gearbox will always be available.

Crane	$ 60,000
Labor to RR Gearbox	$ 4,800
Lost Sales (1 Week)	$ 5,000
Shipping	$ 6,000
Shop Repair Cost	$ 157,000
Total	$ 232,800

Table 4. Costs to replace all bearings off-tower

2.1. Expected Lives for Wind Turbine Bearings

Currently, there is limited data on the times to failure for gearbox bearings. However, for the purposes of this analysis, we used the minimum required L_{10} lives specified by ANSI/AGMA/AWEA 6006-A03, Standard for Design and Specification of Gearboxes for Wind Turbines (2003).

These standard values represent life expectancies, without regard for the particular equipment in which they're installed, usage rates, environmental conditions (such as weather), and other factors that can affect bearing life. In fact, feedback from the Wind Turbine industry indicates that actual service lives do vary significantly from the L_{10} values given in the standard. However, lacking specific data from actual units in service, we assumed they represent a reasonable assessment of likely bearing lives that were sufficient for this study.

Gearbox Size	1500	KW
Total Availability	$\frac{3000 \text{ hrs/yr}}{365\text{x}24 \text{ hrs/yr}}$ = 34%	
Average Revenue	$ 0.06	/ KWH
Average Revenue @ 34% Availability	$ 734	/ day
Maximum Revenue @ 100% Availability	$ 2,160	/ day
Lost Sales Rate	$ 1,000	/ day
Lost time/Gearbox failure	5	days
Average Revenue @ 34%	$ 268,056	/ yr
KWH per Yr	4,467,600	
KWH over 20 Yrs	89,352,000	
KWH over 25 Yrs	111,690,000	
Typical Site	100	Turbines
Revenue per Site @ 34% Availability	$ 26,805,600	/ yr

Table 5. Opportunity Costs

To perform our analysis, this data was converted to equivalent Weibull Parameters using an assumed slope of 3.43 (heavy wear out). This value was chosen based on StandardAero's experience with similar equipment. While some believe a Lognormal distribution would be more appropriate for bearings, our experience overhauling aerospace gearboxes indicates that the bearings and gears are very effective at transferring debris to one another and actual failures tends to exhibit more of a wear-out behavior. The value of the Weibull slope can vary significantly depending on the system design, applied loads, and other factors. As with the cost data above, actual bearing failure history data would need to be used to determine the true bearing service characteristics (in terms of the Weibull slope and characteristic life, or equivalent if another statistical distribution was more appropriate).

To convert the values above to equivalent Weibull parameters, start with the Cumulative Distribution Function (CDF):

$$F(t) = 1 - e^{-\left(t/\eta\right)^{\beta}} \tag{1}$$

By rearranging the terms to solve for η, we get

$$\eta = \frac{t}{\left[-\ln(1 - F(t))\right]^{1/\beta}} \tag{2}$$

where β = 3.43, F(t) = 0.10 (10% for the L_{10} life), and t is the L_{10} life (in hours) value taken from the AGMA Standard (2003), listed below. The equivalent Weibull parameters for the applicable L_{10} Lives are given in Table 6.

Bearing Position	L_{10} Life (Hrs)	Characteristic Life (Eta, η)	Slope (Beta, β)
High Speed Shaft	30,000	57,816	3.43
Intermed. Speed Shaft	40,000	77,089	3.43
Low Speed Shaft	80,000	154,178	3.43
Planet Carrier	100,000	192,723	3.43
Planet Gears	100,000	192,723	3.43

Table 6. L_{10} ratings and equivalent Weibull Parameters

3. ANALYSIS AND RESULTS

For this study, two types of analysis were carried out. For the first analysis, we built a reliability model to determine event occurrence rates for each of the strategies of interest. The 2nd analysis looked at the life cycle impact at specific points along the planned life.

In this case, since the failure of any single component in the gearbox would render the entire system unserviceable, a simple series reliability model was used. The overall system reliability was computed as follows:

$$R_s(t) = \prod_{i}^{n} R_i(t \mid t_{i0}) \tag{3}$$

where the system 's' consists of a set of 'n' components, and time-continued components would contribute a conditional reliability based on any operating time accumulated to-date (t_{i0}). New or restored components use a t_{i0} value of zero for their reliability contribution.

3.1. Reliability Model – OCM vs. 100% Replacement

Using the minimum bearing lives from AGMA 6006-A03 (2003), a system level reliability model was created for a typical gearbox (bearings only). The model was constructed using Raptor reliability modeling software. The resulting reliability model was analyzed under two scenarios.

1. All bearings replaced when any one bearing fails
2. Only the failed bearing is removed and replaced. Other bearings are allowed to continue in service

The results showed that if all bearings are replaced at each visit, the Mean Time Between Failure (MTBF) of the gearbox would be 37,691 hours. Over a planned 20 year life, this would result in 1.6 expected events per gearbox, and 1.9 expected events occurring over a 25 years life.

However if only the failed bearings were removed and replaced (the OCM strategy), the subsequent MTBF would drop to 8,096 hours. This would result in 3.8 events over 20 years, and 4.6 events over 25 years.

Based on the assumed cost of each event, the 20 year life cycle cost for replacing all bearings at each heavy maintenance event appears to be less than replacing only the failed bearings (the same would also be true over a 25 year life cycle). A summary of the results is provided below.

3.2. Cost Benefit Summary

Replace All Bearings 100%		LCC / Hr of Use	LCC / KWH
Events over 20 yrs	1.59		
Events over 25 yrs	1.99		
Cost per event	$ 232,800		
Total Cost 20 Yrs	$ 370,592	$6.18	$ 0.00415
Total Cost 25 Yrs	$ 463,241	$6.18	$ 0.00415

Table 7. Costs for a "Replace 100%" strategy

Replace Only Failed Bearing & Continue		LCC / Hr of Use	LCC / KWH
Events over 20 yrs	3.76		
Events over 25 yrs	5.61		
Cost per event	$ 122,800		
Total Cost 20 Yrs	$ 461,183	$ 7.69	$ 0.00516
Total Cost 25 Yrs	$ 688,702	$ 9.18	$ 0.00617

Table 8. Costs for an OCM strategy

Replace 100% vs. Only Failed Bearing		LCC / Hr of Use	LCC / KWH
Delta @ 20 Yrs	$ 90,590	$ 1.51	$ 0.00101
Delta @ 25 Yrs	$ 225,462	$ 3.01	$ 0.00202

Table 9. Comparison of replacement strategies

3.3. Workscope Cost Impacts at Different Points in a Gearbox's Life

While reviewing the data for the analysis above, we noted that many failures are likely to occur late in a unit's planned life. Under a "replace 100% strategy", these units would receive a workscope that restored its reliability past the unit's planned service life. Based on this, we thought it important to look at how the economics of the replacement strategy change along the unit's life.

To perform this analysis, a simplified gearbox reliability model was constructed including only the gearbox bearings. By integrating the system reliability function in Eq. (3), we can determine the system life expectancy *E(T)* after repair based on any previously accumulated operating times (if any) of individual bearings installed in that gearbox.

$$E_s(T) = \int_0^{\infty} \prod_i^n R_i(t \mid t_{i0})\, dt \tag{4}$$

The parameters for the minimum bearing lives (shown above) were used to calculate the system life expectancy after each repair over the planned 20 (or 25) year lives of the gearboxes, for the following four scenarios:

(1) Replacing only the failed bearing

(2) Replacing all bearings

(3) Replacing only the minimum number of bearings to reach a planned life

(4) Replacing only the HSS and ISS bearings (the least reliable bearings)

Using the assumed cost data given above, the resulting life expectancy was divided into the workscope cost, and the results were output in terms of the expected Cost per Hour of Reliable Life (Cost/Hr) from the build. The results of this are summarized in Table 10 and shown graphically in Figure 1 and Figure 2 below.

Based on these results, the lowest cost strategy falls under either Scenario 3 or 4 where only a few select bearings are replaced. While the "Replace 100%" strategy tended to have a lower LCC cost after 9-12 years (when compared to the OCM strategy), neither were better than a strategy that continually evaluated the optimum build over the unit's life.

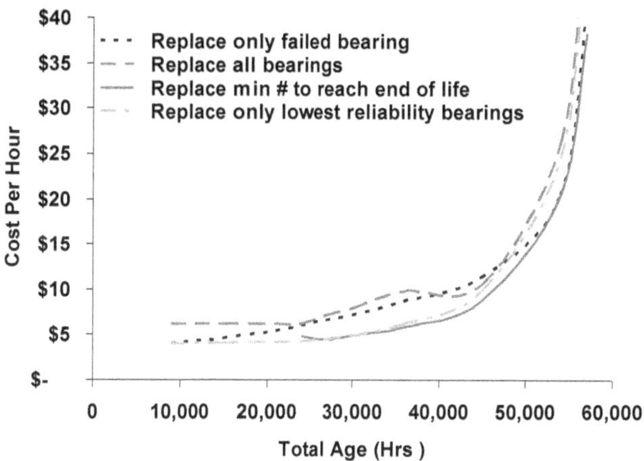

Figure 1. Cost Per Hour of Use (20 Yr Planned Life)

The cost spread between the least to best strategy ranged from approximately $2 per hour (for gearboxes workscoped in the 1st 10 years of its life), to more than $10 an hour (for gearboxes workscoped in the last 10 years of its life). The resulting savings from using the optimum strategy is approximate $100,000 (over the planned life) or $.001/KWH ($ 2/hr over last 50,000 hours, or $10/Hr over last 10,000 hours).

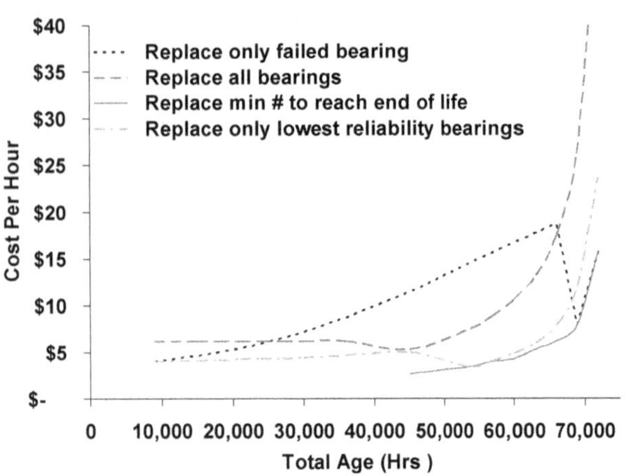

Figure 2. Cost Per Hour of Use (25 Yr Planned Life)

4. DISCUSSION

Based on the analysis above, the optimum workscope depends upon the age of the unit, the accumulated operating time and reliability of its sub-assemblies, and the planned life of the unit.

It should also be noted that only 4 workscopes were considered. If the replacement or re-use of each bearing is considered a different workscope, then there are potentially 2^{14} (16,384) workscopes, one for each possible combination of replacing or re-using any of the 14 bearings in the gearbox. This complexity (i.e., number of workscopes) will increase significantly as gears and other major components are included in the optimum cost analysis. Because of this, the example discussed here only partially answers the question of the optimum workscope for a given gearbox at a given point in time. In reality, a customized software tool would be needed to evaluate all the other permutations of possible workscopes.

The analysis performed used a very generic reliability model. In reality, each of the primary failure modes (for all components, not just bearings) that could cause a gearbox to be removed from service should be included such a model. For many gas turbine engines repaired by StandardAero, we do use a more complete reliability model that accounts for the effect of 50 to 100 distinct failure modes (and the corresponding statistical failure distributions), as part of a software tool that optimizes the workscope for each engine. This allows us to make much more accurate assessments of the costs and relative benefits of different workscopes, and tailor the workscope at each shop visit.

It is also important to reiterate that this study focused on the reliability of bearings alone for the sake of simplicity, but also because this is a particular area of concern for wind power asset owners. We recognize that the root causes of bearing failures can be driven by a number of factors including the cleanliness of oil, duty cycles, the specific design, and the final manufacturing dimensions of the gearbox. Furthermore, there are other components that can and do drive gearbox removals. Therefore, a more accurate and comprehensive reliability model that includes individual failure distributions for each root cause failure mechanism would be needed to find the true optimum maintenance workscope during each gearbox shop visit.

5. CONCLUSION

For this study, two types of analysis were carried out.

		Scenario 1 OCM (Replace only 1 failed bearing)		Scenario 2 Replace 100% (all 14 bearings)		Scenario 3 Replace minimum number of bearings to reach planned life				Scenario 4 Replace only HSS and ISS bearings	
Age (Yrs)	Age (Total Hours)	Cost/Hr 20 yr life	Cost/Hr 25 yr life	Cost/Hr 20 yr life	Cost/Hr 25 yr life	# Brgs to Reach 20 yrs	Cost/Hr 20 yr life	# Brgs to Reach 25 yrs	Cost/Hr 25 yr life	Cost/Hr 20 yr life	Cost/Hr 25 yr life
3	9000	$ 3.90	$ 3.90	$ 6.13	$ 6.13	N/A	N/A	N/A	N/A	$ 4.01	$ 4.01
5	15000	$ 4.58	$ 4.58	$ 6.13	$ 6.13	N/A	N/A	N/A	N/A	$ 4.09	$ 4.09
6	18000	$ 5.01	$ 5.01	$ 6.13	$ 6.13	N/A	N/A	N/A	N/A	$ 4.13	$ 4.13
9	27000	$ 6.85	$ 6.85	$ 7.05	$ 6.13	4	$ 4.45	N/A	N/A	$ 4.51	$ 4.32
10	30000	$ 7.68	$ 7.68	$ 7.76	$ 6.13	4	$ 4.89	N/A	N/A	$ 4.96	$ 4.40
12	36000	$ 9.75	$ 9.75	$ 9.70	$ 6.13	3	$ 5.78	N/A	N/A	$ 6.20	$ 4.58
15	45000	$ 14.02	$ 14.02	$ 15.52	$ 7.76	2	$ 8.72	4	$ 4.89	$ 9.92	$ 4.96
18	54000	$ 20.47	$ 19.85	$ 38.80	$ 11.09	1	$ 20.47	4	$ 6.99	$ 24.80	$ 7.09
19	57000	$ 40.93	$ 22.18	$ 77.60	$ 12.93	0	$ 35.60	4	$ 8.16	$ 49.60	$ 8.27
20	60000		$ 25.04		$ 15.52			4	$ 9.79		$ 9.92
21	63000		$ 27.37		$ 19.40			3	$ 11.57		$ 12.40
22	66000		$ 30.34		$ 25.87			3	$ 15.42		$ 16.53
23	69000		$ 36.45		$ 38.80			2	$ 21.80		$ 24.80
24	72000		$ 40.93		$ 77.60			1	$ 40.93		$ 49.60
25	75000										

N/A Cannot reach planned life under any Workscope

Lowest Cost/Hour Strategy

Table 10. Analysis Results: Cost per hour across planned gearbox life

The first analysis involved the construction of a system reliability model and simulating the effect of two different maintenance scenarios over planned lives of 20 and 25 years to determine the expected number of maintenance events under each strategy.

This initial analysis indicated that a strategy based on replacing all the bearings at each shop visit (Replace 100%) would result in less than two shop visits on average over the planned service life (20 and 25 years) of a gearbox. Conversely, an OCM strategy, where only the failed bearing is replaced at each event, would result in an average of close to 4 events over the planned gearbox life. When the costs of each shop visit were considered against the number of events, the "100% Replacement" strategy had a lower overall life cycle costs ($5.70/hr versus more than $7.60/hr).

The second analysis showed the optimum workscope (as measured by a combination of the shop cost and the resulting mean time to next failure) is determined by a number of factors including (1) the age of the unit, (2) the age of the other bearings in the unit and (3) the planned retirement age of the unit. It also showed that neither the OCM nor the 100% Replacement strategy was optimum. In fact, it showed that the optimum build varied significantly over the life of a unit, and the cost impact could be as much as $15 per hour of use, per workscope.

The conclusion from our analysis is that the optimum build can vary significantly over a unit's planned life and neither an "OCM" nor a "100% Replacement" strategy is optimum. The cost consequences are approximately $100,000 over the lifetime of a single unit, or $.001 per KWH.

Furthermore, asset owners need tools that can determine the optimum build throughout the service life of their equipment and, quantify the benefit in terms of cost and reliability; as well as a maintenance plan that allows them to act on the data in a manner that minimizes the asset owner's long-term costs.

NOMENCLATURE

β	"Slope" (or shape parameter) of a Weibull Cumulative Distribution Function (CDF)
η	"Characteristic Life" (or scale parameter) for a Weibull Cumulative Distribution Function
KWH	Kilowatt-Hour
L_{10} life	Time by which 10% of a population would be expected to have failed (also called B_{10} life)
LCC	Life Cycle Cost
MTBF	Mean Time Between Failure
OCM	"On-Condition Maintenance" – a maintenance philosophy of only replacing parts that have failed
RR	Remove & Replace

REFERENCES

1. American Gear Manufacturers Association (AGMA) (2003). Standard for Design and Specification of Gearboxes for Wind Turbines. In *ANSI/AGMA/AWEA 6006-A03*, Alexandria, VA, American Gear Manufacturers Association.
2. Society of Automotive Engineering (SAE) (2013). Using a System Reliability Model to Optimize Maintenance Costs, A Best Practices Guide. In *SAE ARP JA6097 MAY2013*, Warrendale, PA, Society of Automotive Engineers.
3. Wang, H. (2002). Invited Review – A survey of maintenance policies of deteriorating systems. *European Journal of Operational Research*, vol. 139 (2002), pp. 469–489, ISSN 0377-2217.

BIOGRAPHIES

Alan Lesmerises is the lead Reliability and Life Cycle Management Engineer for StandardAero Engineering Services in San Antonio, TX. He holds a Bachelor's degree in Aerospace Engineering from the University of Oklahoma, a Master's Degree in Astronautical Engineering from the Air Force Institute of Technology (AFIT), where he completed their post-graduate Reliability and Maintainability program. During nearly 21 years in the US Air Force, he was a jet engine mechanic, performed research on gas turbine engine combustion, and was a systems engineer on various Air Force acquisition programs. Since joining StandardAero in 1999, he has supported the T56 Engine overhaul operation and is one of the original members of their Reliability Engineering team. Alan is a member of the SAE, a senior member of both the AIAA and ASQ, and an ASQ Certified Reliability Engineer. He is currently the chairman of the SAE G-11M Maintainability, Supportability, & Logistics Committee, a member of the SAE Integrated Vehicle Health Management (IVHM) committee, and advises the SAE Aerospace Council as a member of the Integrated Vehicle Health Management Steering Group.

David Crowley is Director of Technology Development, StandardAero Engineering Services, San Antonio, TX. Mr. Crowley holds a Bachelor of Science Degree in Mechanical Engineering from Texas Tech University, and a Masters of Science Degree in the Management of Technology from The University of Texas at San Antonio. Prior to joining StandardAero, he worked at the US Air Force's Propulsion Directorate, San Antonio Air Logistics Center, performing various duties in support of the USAF's fleet of T56 and TF39 Engines. His last assignment was as Chief Engineer for USAF TF39 engines, where his team implemented a number of reliability improvements, cost-based workscope tools, and engine modifications. David joined StandardAero in June 2000, and until Aug 2006 served as the Director of Product Engineering. In this role, he oversaw engineering support to StandardAero's T56 Engine facility, also located in San Antonio. David now leads a team to research, design, and field advancements in cost-effective aircraft propulsion systems maintenance management.

Author Index

Author Guidelines

The International Journal of Prognostics and Health Management (IJPHM) publishes scientific papers dealing with all aspects of prognostics, diagnostics, and system health management of complex engineered systems. High quality articles focused on assessing the current status and predicting the future condition of an engineered component and/or system of components. Such articles may come from a variety of disciplines, including electrical, electronics, mechanical, civil, and chemical engineering, computer and materials science, reliability, test and measurement, artificial intelligence, physics, and economics.

Copyright

The Prognostic and Health Management Society advocates open-access to scientific data and uses a Creative Commons license for publishing and distributing any papers. A Creative Commons license does not relinquish the author's copyright; rather it allows them to share some of their rights with any member of the public under certain conditions whilst enjoying full legal protection. By submitting an article to the International Conference of the Prognostics and Health Management Society, the authors agree to be bound by the associated terms and conditions including the following: As the author, you retain the copyright to your Work. By submitting your Work, you are granting anybody the right to copy, distribute and transmit your Work and to adapt your Work with proper attribution under the terms of the Creative Commons Attribution 3.0 United States license. You assign rights to the Prognostics and Health Management Society to publish and disseminate your Work through electronic and print media if it is accepted for publication. A license note citing the Creative Commons Attribution 3.0 United States License, as shown below, needs to be placed in the footnote on the first page of the article.

> *First Author et al. This is an open-access article distributed under the terms of the Creative Commons Attribution 3.0 United States License, which permits unrestricted use, distribution, and reproduction in any medium, provided the original author and source are credited.*

Ethics

Contributions to IJPHM must report original research and will be subjected to review by referees at the discretion of the Editor. IJPHM considers only manuscripts that have not been published elsewhere (including at conferences), and that are not under consideration for publication or in press elsewhere. Moreover, it is the responsibility of the author to ensure that any data or information submitted complies with the export-control regulations of the author's home country (e.g., International Traffic in Arms Regulations (ITAR) in the United States). IJPHM honors code of conduct provided by the Committee of Publication Ethics (COPE). More details on IJPHM policies and publication ethics can be found online.

Submission Types

IJPHM publishes full-length regular papers, technical briefs, communications, and survey papers.

Full-Length Regular Papers should describe new and carefully confirmed findings, and experimental procedures and results should be given in detail sufficient for others to replicate the work. A full paper should be long enough to describe and interpret the work clearly, placing it in the context of other research.

Technical Briefs usually describe a single result, experiment, or technique of general interest for which a short treatment is appropriate. A short paper should be long enough to describe experimental procedures and clearly, and interpret the results in the context of other research.

Communications are a separate class of short manuscripts that are subject to an expedited review process. Appropriate items include (but are not limited to) rebuttals and/or counterexamples of previously published papers. A short communication is suitable for recording the results of complete small investigations or giving details of new models or hypotheses, innovative methods, techniques or apparatus. The style of main sections need not conform to that of full-length papers. Short communications are 2 to 4 printed pages in length. The Editors will review these submissions internally, and request outside review when appropriate.

Survey Papers covering emerging research topics in PHM are also published, and unsolicited manuscripts of a tutorial or review nature are welcome. However, prospective authors of survey papers should contact in advance the Editor-in-Chief in order to assess the possible interest of the topic to IJPHM. Papers describing specific current applications are encouraged, provided that the designs represent the best current practice, detailed characteristics and performance are included, and they are of general interest.

Prospective authors should note that for any type of IJPHM content, poorly documented papers using "proprietary" techniques will be rejected. Moreover, excessive "branding" within a paper also cause for rejection; e.g., "The team used the magical CompanyBrand™ preprocessing to prepare the data to extract the amazing CompanyBrand™-proprietary features (which we can't tell you about)." Papers should present techniques and results clearly and objectively.
Although bound editions will be available for purchase, IJPHM is fundamentally an online journal. As such, we are able to have a very fast turnaround time. We will acknowledge receipt of submissions within three business days, and we intend to rigorously review and return a decision to the authors in approximately 8-12 weeks. Thus, papers may be published in a very short time, allowing your research to be available to the scientific community when it is most relevant.

Option to Present Your Work at a Conference

PHM Society publications have maintained high quality standards for both its Conferences and the Journal. Highest quality conference papers are also invited to be published in the Journal. However, since 2012 IJPHM provides an option to the journal authors to present their journal paper at one of the upcoming PHM conferences.

Authors are reminded that the paper must be journal quality and adhering to the journal template. The paper will be reviewed as per journal review standards and if accepted a presentation slot will be reserved at the target conference. The paper will be published in the journal archives and linked through conference proceedings.

Benefits

- A journal publication of your high quality research work
- A peer review of your work by experts in the field
- A chance to present your work to the targeted audience
- No reworking required to publish in the Journal
- A shortened review cycle to journal publishing

Risks

- Rejected papers will not automatically be considered for the conference and may additionally miss the submission deadline.
- If re-submitted for the conference, they will be reviewed subject to conference review criteria

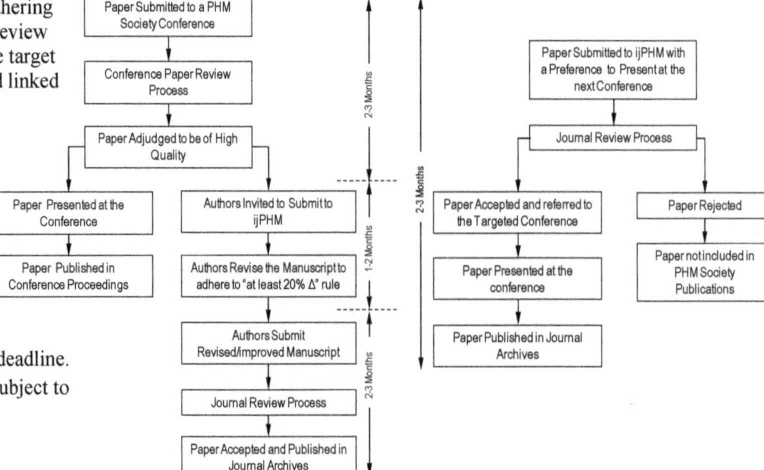

www.ingramcontent.com/pod-product-compliance
Lightning Source LLC
Chambersburg PA
CBHW080553220326
41599CB00032B/6466

9781936263073